Irrigation Management in Developing Countries

Studies in Water Policy and Management

Charles W. Howe, General Editor

Water and Western Energy: Impacts, Issues, and Choices, Steven C. Ballard, Michael D. Devine, and associates

Water and Agriculture in the Western U.S.: Conservation, Reallocation, and Markets, Gary D. Weatherford, in association with Lee Brown, Helen Ingram, and Dean Mann

Economic Benefits of Improved Water Quality: Public Perceptions of Option and Preservation Values, Douglas A. Greenley, Richard G. Walsh, and Robert A. Young

Municipal Water Demand: Statistical and Management Issues, C. Vaughan Jones, John J. Boland, James E. Crews, C. Frederick DeKay, and John R. Norris

Flood Plain Land Use Management: A National Assessment, Raymond J. Burby and Steven P. French with Beverly A. Cigler, Edward J. Kaiser, David H. Moreau, and Bruce Stiftel

Natural Radioactivity in Water Supplies, Jack K. Horner

Energy and Water Management in Western Irrigated Agriculture, edited by Norman K. Whittelsey

Irrigation Management in Developing Countries: Current Issues and Approaches, edited by K. C. Nobe and R. K. Sampath

About the Book and Editors

This interdisciplinary volume brings together current issues in and approaches to the development, utilization, and management of water resources in developing countries. The contributors, who have had extensive research and work experience in such countries, analyze these irrigation issues in the context of past experience and offer possible future strategies to help bridge the gap between potential and reality in Third World agriculture. The book will be of interest to agricultural economists, hydro-engineers, agronomists, sociologists, and extension practitioners concerned with agricultural and irrigation development.

K. C. Nobe is chairman and professor in the Department of Agricultural and Natural Resource Economics at Colorado State University. Prior to assuming his present position, he was chief of the Agricultural and Economics Branch of Harza Engineering Company International and general consultant to the Water and Power Development Authority of West Pakistan. R. K. Sampath is professor in the Department of Agricultural and Natural Resource Economics and managing director of the International School for Agricultural and Resource Development at Colorado State University. Drs. Nobe and Sampath are co-editors of *Issues in Third World Development* (Westview, 1983).

Published in cooperation with
the International School for
Agricultural and Resource Development

The International School for Agricultural and Resource Development (ISARD) program at Colorado State University engages in educational, research, and technical assistance activities relating to agricultural and natural resource development in Third World nations.

ISARD was established in 1980 as a component of CSU's program for International Education, Research, and Technical Activities in response to the International Development and Food Assistance Congressional Act of 1975, which encouraged U.S. land grant institutions to become more involved in famine prevention and food production in Third World countries.

Initial financial support for ISARD's activities came from the U.S. Agency for International Development. ISARD is now totally self-supporting.

Irrigation Management in Developing Countries: Current Issues and Approaches

Proceedings of an Invited Seminar Series Sponsored by the International School for Agricultural and Resource Development

edited by K. C. Nobe
and R. K. Sampath

Routledge
Taylor & Francis Group
LONDON AND NEW YORK

First published 1986 by Westview Press

Published 2018 by Routledge
52 Vanderbilt Avenue, New York, NY 10017
2 Park Square, Milton Park, Abingdon, Oxon OX14 4RN

Routledge is an imprint of the Taylor & Francis Group, an informa business

Library of Congress Cataloging-in-Publication Data
Main entry under title:
Irrigation management in developing countries.
(Studies in water policy and management)
Includes index.
1. Irrigation--Developing countries--Management.
I. Nobe, Kenneth C. II. Sampath, Rajan K. (Rajan Kasturi), 1948- . III. Series.
TC812.I76 1986 333.91'3'091724 85-26988

ISBN 13: 978-0-367-00859-8 (hbk)

Contents

List of Tables and Figures x
Foreword, *Charles W. Howe* xvii
Acknowledgments xix

Introduction
Kenneth C. Nobe and Rajan K. Sampath 1

PART 1
METHODOLOGIES AND CONCEPTS

1 Irrigated Agriculture: A Comparative Analysis of Development Concepts, Wayne Clyma 19

2 On the Development and Use of Improved Methodologies for Irrigation Management, Roberto Lenton 47

3 From Diagnostic Analysis to Designing and Conducting On-Farm Research, Willard R. Schmehl 67

4 Management of Gravity Flow Irrigation Systems, J. Mohan Reddy 95

5 Some Tools and Concepts for Better Irrigation Water Use, John A. Replogle 117

PART 2
ECONOMIC ASPECTS

6 On the Allocation, Pricing, and Valuation of Irrigation Water, Robert A. Young 151

7 Social and Economic Impacts of Investments in Groundwater: Lessons from Pakistan and Bangladesh, Sam H. Johnson III 179

8 Economic Aspects of Irrigation with Saline Water, Dan Yaron 217

9 Irrigation, Drainage, and Food Supplies, Ian Carruthers 265

10 Developing Farm-Level Information for Improved Irrigation Water Management in Developing Countries, Melvin D. Skold and Donald W. Lybecker 285

PART 3
MANAGEMENT AND INSTITUTIONAL ASPECTS

11 Irrigation System Management, Jack Keller 329

12 Improving Management of Irrigation Projects in Developing Countries: Translating Theory into Practice, Warren Fairchild and Kenneth C. Nobe 353

13 The Management of Paddy Irrigation Systems: A Laissez-Faire, Supply-Side Theory, David W. Seckler 413

14 Improved Irrigation Management: Why Involve Farmers? Max K. Lowdermilk 427

15 Legal and Institutional Aspects of Irrigation Water Management, George Radosevich 457

16 State and Locality in Asian Irrigation Development: The Property Factor, E. Walter Coward, Jr. 491

17 Water Management: Problems and Potential for Communications in Technology Transfer, Dan Lattimore 509

List of Contributors 535
Index 537

Tables and Figures

TABLES

I.1 Irrigated areas in selected countries, up to 1974 3

3.1 Setting the background for design 76

3.2 Types of farming systems research activities 78

3.3 Classification of on-farm research 81

3.4 Cropping systems on-farm research methodologies 82

4.1 Factors influencing irrigation system performance in distribution and application of water 110

5.1 Scheduling policies for three projects . . . 134

7.1 Implementation of public tubewell projects 186

7.2 Cropping intensities in SCARP projects . . . 187

7.3 Actual delivery compared to design delivery: SCARP-II 189

7.4 Number of private tubewells in Pakistan . . 193

7.5 Thakurgaon Tubewell Project, Bangladesh 1981, 1982, and 1983 dry seasons 198

7.6 Present and projected status of mechanical irrigation lift devices in Bangladesh 202

7.7 Participation in DTICP schemes: averages for 1982-83 data 204

7.8 Effects of increasing command area size on per acre net present economic values (NPEV) associated with selected irrigation technologies 207

7.9 Operational status of public and private tubewells 209

8.1 Selected results from simulations of salt accumulation and leaching, sprinkler irrigation, fruit crops, southern Israel 220

8.2 Estimated losses to net returns for alternative water salinity levels, Mexicali and San Luis valleys, North Mexico, 1975 225

8.3 Estimated income losses (%) for alternative salinity levels of irrigation water, South and Negev regions in Israel 226

8.4 Estimated relative income losses (%) accrued to fruit crops on moshav farms for alternative salinity levels of irrigation water, South and Negev regions in Israel 226

8.5 Estimated relative return to water and land on small, medium, and large farms in response to increasing salinity, Imperial Valley, California 238

8.6 Priorities in allocation of high-quality water on a farm with three sources and varying water salinity 238

8.7 Estimated returns to salt leaching under selected situations in Israel 242

8.8 Marginal productivity and marginal rates of substitution between low and high quality water, kibbutz farms, South and Negev regions in Israel 248

8.9 Frequency distribution and means of the estimated MRS values under selected situations 249

10.1 Crop enterprise budget for cotton grown at the Abu Raya Project site; 1980-81 crop year 304

10.2 Estimates of amount of labor per feddan as derived from sample survey and farm record data, Kafr el Sheikh Governorate, 1980-81 crop year 309

11.1 Areas of irrigated land in the world, in millions of acres 331

12.1 Subproject area by unit 363

12.2 Irrigation water supplies and requirements at watercourse head 364

12.3 Average crop yields 369

12.4 Planned lining and estimated water savings 373

12.5 Watercourse renovation planned under the CWM Project 377

13.1 Water use and grain yield of IR20 under four water treatments, IRRI, 1972 dry season 419

13.2 Average paddy parameters in research sites served by similar water delivery methods; selected turnout areas in three national irrigation systems, Philippines, 1972-73 423

17.1 Kearl's summary of communication considerations involved in successive models of agricultural development 514

FIGURES

1.1 Paradigm of the research-development process 24

1.2 The development model 25

1.3 Comparison of the various planning frameworks as presented by Skogerboe et al. (1979, 1980b, 1980c, 1982) and the development model as presented by Skogerboe et al. (1982) 28

1.4 The learning process approach (Korten, 1982) or bureaucratic reorientation approach (Uphoff, 1984) to irrigation water management 31

1.5 Comparison of the concepts of the learning process approach (Korten, 1982) and the development model (Clyma et al., 1977) 32

1.6 The phases of action research as suggested by Bottrall (1981) for large irrigation schemes 34

1.7 Comparison of the phases of action research (Bottrall, 1981) with the development model 35

1.8 The farming systems research and development strategy 36

1.9 Comparison of the farming systems research and development strategy with the development model 37

1.10 A farmer-oriented model for generating appropriate technology 39

1.11	Stages and activities in generating appropriate technology for small farmers	40
1.12	The scientific method	42
3.1	Stages in the farming systems process (Shaner et al., 1982)	73
3.2	Activities in the diagnostic stage	74
3.3	Structure of the managerial system (Fairchild and Nobe, 1986)	87
3.4	Intersection of the management (MS), irrigation (IS), and farming systems (FS)	89
4.1	Layout of a typical irrigation system . . .	97
4.2	Relationships between the agricultural, economic, technical, and institutional and organizational aspects	100
4.3	The irrigation water requirement for lowland rice as affected by the level of control inputs (Levine, 1977)	102
4.4	Cost-effectiveness of monitoring and evaluation information	109
4.5	Flow chart of management plan	113
6.1	Concepts employed in analyzing water rates	157
6.2	Demand (marginal value) and marginal cost curves for water supply	164
7.1	Map of Asian subcontinent showing location of studies	180
7.2	Water table profiles, Chaj, Rechna, and Bari doabs (D. W. Greenman, V. M. Swarzenski, and G. D. Bennett, 1967)	182

8.1 Hypothetical salinity-induced farm losses under three situations 224

8.2 Net irrigation water requirement as a function of time during the irrigation season for five levels of initial soil salinity and four levels of salinity of irrigation water 233

8.3 Optimal cropping and irrigation pattern for high, medium, and low initial soil salt conditions where corn roots are shallow and where flood irrigation only is allowed, a Utah farm (Hanks and Andersen, 1981) 236

8.4 A schematic presentation of irrigation with saline water over a sequence of years 240

9.1 Cross section of an irrigated river valley showing sources of seepage 269

9.2 Increase in surface soil salinity (MacDonald, 1983) 271

9.3 Rising water table effects on wheat and cotton (MacDonald, 1983) 271

9.4 Effect of changes in irrigation intensity in part of the lower Indus in Pakistan (adapted from MacDonald, 1983) 273

10.1a Total product curve 314

10.1b Marginal and average product curves 314

11.1 Typical surface irrigation components . . . 335

11.2 Adequate irrigation system with effective management "stretching" the water like a uniform membrane over the entire command area 337

11.3 Relationship between relative water supply and management intensity required for equitable distribution 338

11.4 Command area with various types of water distribution problems 339

11.5 Demand and relative water supply hydrographs for 100 percent and for 67 percent of area planted, assuming a supply hydrograph (SH) of 50 percent of peak demand 344

11.6 Existing and potential curriculum content for training physical scientists dealing with agricultural water management 349

12.1 The transformer concept of cybernetics . . . 384

12.2 Structure of an administrative organization 388

12.3 Structure of the managerial organization . . 389

12.4 Organization chart, Command Water Management Project at the province level (e.g., Sind) 392

12.5 Pakistan Command Water Management Project organizational chart 395

13.1 A hypothetical production function of other crops 416

13.2 Yield response of IR8 to treatments of varied water application intensity, IRRI, 1969 dry season 418

15.1 A descriptive map of major legal systems and their variations or paths of influence 465

15.2 "Path of solutions" for determining the most desirable "water law system" for any particular case 466

15.3 Formulation of water law 469

15.4 Natural system/legal system interface . . . 470

15.5 Improving water management for increased food production in Egypt 481

17.1 Problem-solver model 522

Foreword

Improving the management of existing irrigation projects is the subject of this volume, the eighth in the Westview series "Studies in Water Policy and Management." The volume focuses on a critical problem: Investments in Third World irrigation projects of all sizes in the post-war period have not been as productive as expected. As the costs of new projects escalate and as the debts from past projects drain resources from national treasuries, aid recipients and donors have come to realize the importance of obtaining greater payoffs from existing infrastructure.

The results of extensive, widely varied field research by irrigation specialists from the major disciplinary fields are presented in this volume and can be used in the training of irrigation planners, managers, and extension personnel. The chapters emphasize the need for total management of agricultural inputs, the need for appropriate incentives, the utility of farmer involvement, the possibility of changing institutions, and so on. The chapters are not highly technical and they constitute excellent training materials for short courses, as well as supplemental reading in standard courses.

Charles W. Howe
General Editor

Charles W. Howe is professor of natural resource economics at the University of Colorado and a specialist in water resources.

Acknowledgments

We would like to thank the U.S. Agency for International Development, New Delhi Mission, for financial support in conducting this seminar series as part of the India Irrigation Management Training Program at Colorado State University.

We are extremely grateful to Cathy Busch, the staff coordinator for the seminar series, for her outstanding contributions in making this program successful. Without her hard work and dedication, the series could not have gone as smoothly as it did.

Turning a set of seminar papers into a book is far from a quick or simple job. Our task would have been extremely arduous but for the professional skills of Tracy Holmes in helping us in editing. We are very grateful to her for all the hard work, care, and dedication she put into getting the camera-ready manuscript out within the limited time given to her.

We also want to thank our colleagues, especially those who participated in the India Irrigation Management Training Program, for all the enthusiastic support they gave to this effort.

K. C. Nobe
R. K. Sampath

Introduction

K. C. Nobe and R. K. Sampath

> Irrigation management and its technologies have no inherent value in themselves, nor value to farmers or to society until they are applied for the purpose for which they were developed. The central question then is what kinds of institutions and their delivery systems and policies are needed to bring these new agriculture production possibilities to farmer clients so they can benefit by higher production, higher income, more equity, and better levels of living. This means also finding ways to reduce the very negative impacts or externalities on nonintended audiences, the environment, and society at large (Leagans, 1979).

Irrigation remains a high priority consideration in the development strategy of a large number of countries, particularly in South and Southeast Asia, but also in the arid regions of Africa, the Middle East, and Latin America as well. While these efforts continue to attract large investments in civil works, increasing attention is now being given by donor agency personnel and public agency staff in the developing countries to achieving more effective management of these large, complex systems. There is ample evidence that low irrigation efficiency levels and overall poor performance levels of major projects developed since World War II are primarily the result of poor project and program management, rather than being attributable only to poor farming practices and/or to faulty engineering design of the associated civil works. What all this means is that in the arid areas of developing countries, provision of an irrigation water supply is a necessary but insufficient condition for achieving sufficient increases in crop production so that increased benefits will outweigh increased investments for construction, operation, and maintenance. Management entails the provision of an operational framework which makes it possible for farmers to combine the water input with other required inputs, such as seeds, fertilizer, labor, and capital, under conditions where the potential returns outweigh the perceived high risks associated with the costs of these inputs to be provided by the farmers.

The technical requirements of irrigation development, including its engineering, agronomic, economic, and sociological parameters, are reasonably well understood, even though until recently they were being studied and implemented under conditions of discipline isolation. The management of a completed project and an extension information delivery program for farmers managing small farm subunits included in a command area, however, requires an overall understanding of the linkages of these discipline components in the highly complex system that a large irrigation project represents. Because so many of the variables that contribute to the success or failure of a project are beyond the farmer's direct control, the arguments for some form of government involvement in project operation after the construction phase are highly convincing. Whether such government intervention will be successful depends critically on the quality of the project management that it provides. Therefore, it is the purpose of this volume of collected writings by experienced irrigation specialists to provide an organized body of knowledge that we feel should be a key input into the training of irrigation project planners and managers in those developing countries that are highly dependent on irrigated crop production.

THE IMPORTANCE OF IRRIGATION: AREA COMMANDED VS. EFFICIENCY LEVELS

The largest beneficial use of water, worldwide, is for irrigated crop production. While irrigation is practiced in a large number of countries, the area commanded under irrigation is highly concentrated in a relatively few countries, most of which are classified as underdeveloped and most of which are located in South and Southeast Asia. Data collected by Rao (1975), for example, show that 12 countries, including the U.S. and USSR as developed nations, are irrigating in excess of 135 million ha, which account for 80 percent or more of the total irrigated area worldwide (Table I.1).

While Rao's data are relatively old (1954 to 1974), there are reasons to believe that current figures, if available, would not differ significantly. First, most of the readily developable, lower-cost project areas had been brought into production by 1970; and second, in some countries, such as India, where major expansion of irrigation projects is still under way, loss of cropland to waterlogging and soil salinization is almost offsetting the acreage being added by new projects. Given these two conditions, there is considerable merit in the argument that public agencies in these countries should now give greater weight to providing effective management for existing projects than to adding expensive new projects. These would

Table I.1 Irrigated areas in selected countries, up to 1974

Sl No.	Country	Area Cultivated (in million ha)	Area Irrigated (in million ha)	Remarks
1	2	3	4	5
1	India	161.0	43.0	The figures are up to 1974
2	China	134.0	29.0	The figures are up to 1954 and are exclusive of blocks less than 630 ha which were done by the people
3	U.S.	176.0	16.9	
4	Pakistan	19.7	11.4	
5	USSR	225.5	9.9	
6	Iraq	7.5	4.0	
7	Indonesia	14.0	3.8	The figures are given up to 1968
8	Japan	6.0	3.4	
9	Mexico	15.0	3.3	
10	Italy	27.5	9.2	
11	Iran	6.8	3.1	
12	U.A.R.	2.9	2.9	

surely fail if operated under the management levels provided for past projects.

Another major reason for concentrating on improving project management is to increase the irrigation efficiency levels of available water supplies. While modern technical approaches--such as level basin flooding and drip irrigation--are helping to achieve extremely high efficiency levels in developed countries, the efficiency levels in most developing countries are still extremely low. In India, for example, irrigation using groundwater supplies typically requires a 0.65 hectare meter when surface water is used, due to high conveyance losses. As both cropping intensities and the levels of nonwater inputs increase, the incentives for increasing irrigation efficiencies will also intensify; however, project response capability will require improved project management capability.

Finally, it should be recognized that well-managed irrigation projects can serve as a primary mechanism for modernizing the agricultural sector in many developing countries. But to do this, mobilizing the availability of a total package of necessary agricultural inputs and facilitating their adoption by farmers being supplied with irrigation water will be necessary. The recent adoption of the command water management project approach by major donor agencies, such as the World Bank and the U.S. Agency for International Development (USAID), is based on this comprehensive systems approach. For such projects to be successful, however, total agricultural input managers, not merely water delivery managers, would be required. And, in turn, such managers would have to be trained in modern concepts of project management and given the opportunity to operate within an institutional framework that encompasses managerial control of all forms of input, ranging from the water supply to credit to the extension delivery system.

Transformation of the rigid administrative systems now in place in most developing countries, with split responsibilities between irrigation and agricultural agencies, will not be easy, but irrigation management cannot be significantly improved unless such changes occur, however slowly. Without improved management, improved irrigation efficiencies, increased agricultural output, and higher returns to farmers will not occur. There is reason to believe that, because of the high stakes involved, government administrators in the key developing countries heavily dependent on irrigation are beginning to respond to efforts to improve irrigation management. The material provided in this text will, hopefully, aid in their understanding of the processes involved in effectively managing complex irrigation farming systems.

A Systems Approach to Irrigation Management

An irrigation delivery system is defined as a created entity with complex interdependent social, economic, legal, biochemical and physical factors, processes, and procedures designed to transport water from a known source to the root zones of plants and remove excess water through horizontal or vertical drainage. At the farm level, the water input is combined with other farm inputs and managed to produce crops of economic value. Thus, a systems approach to irrigation management encompasses the total set of process interactions involved in irrigated agriculture--not just the water input. Failure to grasp the vital principle of interaction of systems components is the greatest present technical (and institutional) handicap to agricultural development in the newly developing countries. This is particularly true in those countries heavily dependent upon irrigated agriculture.

Irrigation management, as a recognized systematic methodology, is only about 15 years old, although many of the technologies and concepts employed have been around for 75 years or more. This approach is now coming of age because decision makers in developing countries and donor agencies alike now realize that concentrated irrigation development and operation of existing irrigation systems everywhere requires urgent attention. Escalating energy costs have further aggravated the pressures of population growth, and food shortages are again becoming major problems in some developing countries, particularly in Africa. As the costs of new projects continue to escalate and developable water supplies become scarcer, irrigation decision makers are turning increasing attention to improving and maintaining existing irrigation systems. The interdisciplinary systems approach to irrigation management will be a useful methodological tool in such efforts.

Chambers (1983) has stressed the need for rapid and useful appraisal techniques for helping countries identify priority strategies for further public intervention into existing irrigation systems. But few attempts have been made to develop workable appraisal approaches, nor have many studies been conducted on system performance at the field level. One notable exception is the interdisciplinary approach to diagnostic analysis developed by the Colorado State University On-Farm Water Management team during the 1970s, working with an integrated surface and groundwater irrigation water supply in the Punjab area of Pakistan (Clyma, Lowdermilk, and Corey, 1977). While the project objective was to improve the performance of one locale, sufficient field research was conducted to identify causes and effects so that the results could be operationalized for widespread application of the resulting diagnostic approach to other irrigated areas of the world.

Irrigation management is the orchestration of scarce physical and biological resources, using the skills of several disciplines to bring water to the root zones of plants for increased food and fiber production for all classes of farmers. Clyma, Corey, and Lowdermilk (1977) discussed this new interdisciplinary approach to irrigation management (IM) as follows:

> IM is not water resources, dams, canals, command areas, soils, engineering, agronomy, economics, watercourses, social science, farmers, or plants. Instead, it is how these resources are manipulated and orchestrated by all these disciplines to bring water to the root zones of crops with other inputs at the proper time, the proper rate, and the proper place and cost to produce food and fiber.

A more recent definition states that irrigation management is:

> a process that has at its core an efficiency objective of improving a production system's performance by adjusting inputs to produce a more desired level or mix of outputs . . . an interdisciplinary system process with built-in learning mechanisms to improve system performance by adjusting physical, technological, and institutional inputs to achieve the desired levels of output (Seckler and Nobe, 1983).

Diagnostic analysis (DA) is a proven first phase for improving existing irrigation systems, but it must be followed with the development of solutions, assessment of appropriate technological packages, and implementation of national programs of water management improvement. The DA approach evolved from projects that focused on irrigation problems faced by farmers at the field level, but the methodology is adaptable to a total irrigation delivery system. The command water management programs now under way in India and Pakistan are attempting total systems application of an interdisciplinary DA approach. Seckler and Nobe (1983) have developed a system-wide monitoring and evaluation approach to such projects, which they term "Management by Results" (MBR).

The MBR approach requires continued monitoring and evaluation at the project level and sufficient power by the project manager to force delivery changes as various bottlenecks appear in the delivery system for all irrigation crop production inputs. The new Pakistan Command Water Management Project, jointly funded by the World Bank and USAID, was designed to operate under the MBR approach. The project was slow in getting off the ground, largely due to administrative delays by USAID and a severe shortage of trained manpower for project operation. Nonetheless, the project is now under way and will be recognized as one of the first large- scale efforts to use a systems approach to irrigation management.

Training Requirements for Modern Irrigation Management

The most economic resource for irrigation development today may not be financial capital for new schemes, but rather require the abilities of people to do the job--their knowledge, skills and professional commitment. It is professionals and technicians who build and operate good and/or bad irrigation systems. Until recently, training requirements for developing country personnel to "manage" modern irrigation systems has been the least understood and least supported component of donor agency-funded agricultural development efforts. In this regard, Aaron Wiener (1976), formerly chairman of the famous Tahal National Water Program in Israel, has stated:

> Irrigation not only requires much larger capital inputs than other agricultural methods, but also the heaviest inputs in trained and experienced human resources. With a few exceptions the more critical scarce resources in the low income nations are management talent of professional and sub-professional manpower.

Training of professionals and farmers for improved irrigation management received little attention during the 1960s and 1970s. A recent survey of 63 World Bank and USAID irrigation projects found that only about 1 percent of the total project costs was devoted to improving human capital. But the situation in the 1980s is definitely improving; for example, in the recently initiated World Bank/USAID-funded Pakistan Command Water Management Project, a higher percentage of total project costs was set aside for training project management personnel, extension agents, and farmers (World Bank, 1984).

Aaron Wiener (1976) has also noted the importance of training farmers for operating successful irrigation systems:

> Engineering is not the fundamental problem underlying irrigation development in LDCs. Engineering principles are known and can be adapted, but the major problem, however, is to discover ways to utilize farmer clients more effectively in operations and maintenance and development programs which will create rural transformation. Rural transformation essentially requires changes in farmers' behavior, motivations, and expectations, which is hardly possible until institutions exist to provide them with the improved production possibilities and incentives . . .

Irrigation and other agricultural professionals working with farmers in existing irrigated areas urgently require retooling and training in modern irrigation management. In addition to project managers, who traditionally are personnel trained as civil engineers in developing countries, such interdisciplinary training is also required of project monitoring and evaluation personnel, extension workers, and farmers in the project areas. Such training is distinct from discipline-oriented training, which is the primary function of universities. We do not believe that universities should be advised to develop new interdisciplinary educational programs leading to degrees in irrigation management. But we do think that traditional training for irrigation engineering and irrigation extension degrees should include more courses from other disciplines. Further, university professionals can assist in continuing on-the-job training efforts in interdisciplinary, modern irrigation management via offering nondegree, specialized training programs. We believe that the material presented in this volume can be used effectively in such efforts, based on our recent special training program on India held at Colorado State University.

Max Lowdermilk (1983), presently a training officer with the USAID Mission in India, has identified the following necessary ingredients for an effective professional training program in irrigation management:

1. <u>Training and educational programs for irrigation management should be based on careful action research and need analysis.</u>
   Before initiating any training activity, one should be very clear about why training is needed. Is it to provide more productive work, better service to farmers, better job performance to save time and money, or what? Some other questions in this context are: What are the priority management deficiencies of irrigation systems for which training is needed? What are the criteria for improvements? What is the quantity and quality of available manpower required at all levels? What new rules and skills are needed? Where, by whom, for what skills, and how long is training needed? What basic incentives are required for trainers and trainees?

2. <u>A training program in irrigation should fit into and support the strategy or framework for improving irrigation systems.</u>
   There are four interlinked aspects:
   a. training and research are organically linked
   b. training is real-world oriented
   c. life systems are a laboratory for training
   d. training is interdisciplinary.

3. <u>Maintain an organic linkage between training and field research and projects: keep training relevant; it must be oriented toward field exercises and training material should evolve from local situations.</u>
   One useful approach to training would be through diagnostic analysis, as is being done by personnel in the USAID-financed Water Management Synthesis II Project. The diagnostic analysis mode of training is designed to train professionals in field methods useful for monitoring irrigation systems, project appraisals, and management of systems. The training is interdisciplinary and takes place in the field on live systems.

4. Successful training and educational programs require strong institutional commitment.
   Without strong commitment for training by the agency using the trained staff, the training is a waste. We need to focus on finding better ways and means to build up the demand and commitment for training. In order to sensitize senior officials and policymakers, they should be exposed to irrigation management concepts before training operational staff.

Institutional building is a management concept which means the planning, structuring, guidance, and reshaping of new or existing organizations in terms of functions and physical and/or social technologies. Such changes cannot occur without training in the new approaches desired for personnel at all levels in the organization, not the least of which are the top agency personnel who will be responsible for the change decisions. As an example of needed institutional change in irrigation management, traditional irrigation departments need to reorganize so as to create a new cadre and career path for interdisciplinary professionals trained in system-oriented irrigation management. Further, a means must be found to provide continued professional development training in this new function for in-service personnel. When these innovations become established and become valued parts of normative relationships within the agency, they have, in fact, become institutionalized! Without such changes, the massive donor agency and LDC top echelon commitment to in-service training in modern irrigation management will fall far short of its potential contribution to improved agricultural production in these countries.

Organization and Description of Chapters

The chapters included in this volume were originally prepared for and presented in the 1984 International School for Agricultural and Resource Development (ISARD) Invited Seminar Series, "Current Issues in and Approaches to Irrigation Water Management in Developing Countries." This effort was an integral part of a special training program for 19 Indian irrigation water management personnel responsible for establishing a number of state-level training institutes. We organized this seminar series, however, to focus on a number of leading issues and concerns that are currently the major preoccupation of irrigation development experts and practitioners on a worldwide basis. Leading issues addressed were centered around strategies to improve irrigation input efficiency in terms of both disciplinary and interdisciplinary aspects involving the engineering-agronomic-economic-social-institutional-legal dimensions of irrigation management.

Part I presents an overview of methodological and conceptual aspects of irrigation management in developing countries. In Chapter 1, Wayne Clyma discusses "Irrigated Agriculture: A Comparative Analysis of Development Concepts." He reviews several existing approaches to improving irrigated agriculture by classifying the conceptual phases according to the phases in the research-development process. He notes the strengths and weaknesses of each. Based on these results, he suggests an improved strategy involving specific planning objectives.

In Chapter 2, "On the Development and Use of Improved Methodologies for Irrigation Management," Roberto Lenton makes a case for the development and use of general approaches to irrigation management that are not specific to any given irrigation system, so that they can be applied in a broad range of existing situations. In this context, he clarifies the concept of modern irrigation management, examines to what extent methodologies for implementation are available and used in managing irrigation systems around the world, explores research needs, and draws lessons that may be applied in the development of improved methodologies. The emphasis in this chapter is on large-scale, publicly administered irrigation systems, such as those prevalent in much of South and Southeast Asia.

Willard Schmehl's paper in Chapter 3 focuses on the necessary linkage "From Diagnostic Analysis to Designing and Conducting On-Farm Research." The major objective of his paper is to emphasize that diagnostic analysis is only the first step in the irrigation development process. He presents some general methodologies that are used in the research phase and then reviews the current status of on-farm research in irrigated farming systems.

Chapter 4 is J. Mohan Reddy's presentation of "Management of Gravity Flow Irrigation Systems." This paper deals with water control and management aspects of gravity flow irrigation systems in general.

In Chapter 5, John Replogle deals with "Some Tools and Concepts for Better Irrigation Water Use." Dr. Replogle discusses different water delivery systems and scheduling policies, based on intensive research under way by the U.S. Aricultural Research Service in Arizona and California, that he considers applicable to managing irrigation systems elsewhere in the world.

Part II deals with economic aspects of irrigation systems. In Chapter 6, "On the Allocation, Pricing, and Valuation of Irrigation Water," Robert Young discusses different water pricing concepts and systems to distribute water to achieve efficiency, equity, and cost-recovery goals.

Sam Johnson's paper in Chapter 7 deals with "Social and Economic Impacts of Investments in Groundwater: Lessons from Pakistan and Bangladesh." The purpose of his paper is to examine social, technical, and economic aspects of investments in groundwater in these countries. The paper focuses on government policies concerning groundwater development and documents the economic and social impacts of these policies.

Dan Yaron discusses "Economic Aspects of Irrigation with Saline Water" in Chapter 8. This paper reviews economic dimensions of irrigation with water of varying salinity levels, with emphasis on on-farm irrigation problems in Israel.

Ian Carruthers discusses economic and technical issues inherent in "Irrigation, Drainage, and Food Supplies" in Chapter 9. Specifically, this article concentrates upon the growing problem of waterlogging and salinity, which portends to destroy the food-producing capacity of much of the irrigated lands of the Nile, Euphrates, Indus, Ganges, and many other arid zone river basins.

In Chapter 10, Melvin Skold and Donald Lybecker discuss "Developing Farm-Level Information for Improved Irrigation Water Management in Developing Countries." This paper, based on extensive field research in Egypt, focuses on data useful for evaluating and understanding the farm economic situation and the data base necessary to perform financial and economic evaluations of alternatives pertaining to irrigation system development. Its primary focus, however, is on the necessary data base for farms and farmers.

Part III introduces the management and institutional aspects of modern irrigation management. In Chapter 11, Jack Keller discusses "Irrigation System Management." This paper presents an overview of concepts related to irrigation system management, based on conclusions reached by the author as a result of extensive interdisciplinary field study and consulting activities involving irrigated agricultural management in several developed and developing countries.

In Chapter 12, Warren Fairchild and Kenneth Nobe focus on "Improving Management of Irrigation Projects in Developing Countries: Translating Theory Into Practice." In this paper, the authors discuss the jointly funded World Bank/USAID Command Water Management Project in Pakistan, in which the new "Management by Results" approach is being tested on a pilot basis. Institutional constraints and trained managerial manpower shortages are specifically highlighted.

Chapter 13 is a paper by David Seckler on "The Management of Paddy Irrigation Systems: A Laissez-Faire, Supply-Side Theory," based on his recent experiences in India, Thailand, and Indonesia. Seckler argues that paddy irrigation is different from other irrigated crop systems in at least two basic physical parameters that substantially effect the design and efficient operation of these irrigation management systems: (1) water is primarily stored on and drained from the surface of fields and (2) paddy irrigation systems have a self-regulating property that leads to a reasonably optimal allocation of water supply between farms. Thus, he notes that, in contrast to other irrigated crop systems, it is doubtful if management improvements in the form of rationing and rotation of water supplied to paddy farmers would result in cost-effective improvements over the allocation now being achieved by existing, naturally functioning laissez-faire systems.

Chapter 14, by Max Lowdermilk, deals with "Improved Irrigation Management: Why Involve Farmers?" In this paper, the author deals with three basic questions that are currently widely discussed:

- o Why involve farmers in irrigation development and improvement?
- o Why don't farmers cooperate with irrigation authorities more effectively?
- o What are some useful lessons about farmer involvement which may have relevance for developing countries in general, and India in particular?

In Chapter 15, George Radosevich presents an overview of the "Legal and Institutional Aspects of Irrigation Water Management" in developing countries. Specifically, he deals with three areas that affect development and utilization of water-related resources: (1) laws, (2) organizations, and (3) the system of planning and management adopted or available to the decision-making parties involved.

Chapter 16 is Walt Coward's paper on the "State and Locality in Asian Irrigation Development: The Property Factor." In this paper, the author argues that irrigation development is the result of activity by both the state and the locality and that improving irrigation development outcomes is dependent upon discovering and using better means for joining state and locality actions. He argues that particular attention must be paid to how much actions affect the property rights of individuals.

Dan Lattimore's paper in Chapter 17 discusses "Water Management: Problems and Potential for Communications in Technology Transfer." This concluding paper synthesizes what we presently know about the communication aspects of technology transfer as it relates to water management around the world.

In conclusion, we can only express the hope that the material presented in this volume will be of use to those interested in promoting modern irrigation management techniques in the developing countries. Based on our experience in using these papers in a special training program for Indian irrigation personnel, we are confident that these materials can be utilized in future training programs for personnel from many other developing countries.

REFERENCES

Chambers, Robert. "Rapid Appraisal for Improving Existing Canal Irrigation Systems," Discussion Paper Series, D.P. No. 8, Ford Foundation, New Delhi, India, August 1983.

Clyma, Wayne, Max Lowdermilk, and Gilbert Corey. "A Research Development Process for Improvement of On-Farm Water Management," Water Management Technical Report No. 47, Colorado State University, Fort Collins, Colorado, June 1977.

Clyma, Wayne, Max Lowdermilk, and Dan Lattimore. "On-Farm Water Management for Rural Development," Agricultural Engineering, February 1981.

Leagans, J. P. "Adoption of Modern Agricultural Technology by Small Farm Operators: An Interdisciplinary Model for Researchers and Strategy Builders," Program in International Agriculture, Cornell University, Ithaca, New York, 1979.

Lowdermilk, Max K. "Professional Development Needs for Improved Irrigation Management," United States Agency for International Development, New Delhi, India, December 1983.

Rao, K. L. India's Water Wealth: Its Assessment, Uses and Projections. New Delhi: Orient Longman, 1975.

Seckler, David, and Kenneth C. Nobe. "The Management Factor in Developing Economies," Chapter 12, Issues in Third World Development, K. C. Nobe and R. K. Sampath, eds. Boulder, Colorado: Westview Press, 1983.

Wiener, Aaron. "The World Food Situation and Irrigation Programmes," ICIDD Bulletin, 1976.

World Bank. Staff Appraisal Report: Pakistan Command Water Management Project, No. 4971-PAK, South Asia Department, Irrigation I Division, World Bank, Washington, D.C., April 30, 1984.

Part 1

Methodologies and Concepts

1
Irrigated Agriculture: A Comparative Analysis of Development Concepts

Wayne Clyma

Irrigated agriculture has a significant impact on many developed and less developed countries. A review of yield differences between irrigated and dryland production suggests the difference is usually three to four times in favor of irrigation as a country or regional average. Potential yields are even greater. This average does not reflect the increased variability of yield that usually accompanies dryland agriculture. Thus, the mean yield is increased, its potential increased by several magnitudes, and its variability greatly reduced under irrigated agriculture.

Irrigated agricultural development, however, greatly lags behind the potential. As a result, major efforts and significant programs have been developed that attempt to improve the performance of irrigated agriculture as a development strategy.

The technologies available to irrigated agriculture continue to increase at an accelerated rate. Current concepts such as computerized scheduling of irrigation, mechanized and automated irrigation, the Green Revolution, farmer participation through farmer organizations, and policies to improve the economic benefits of irrigation to farmers have been developed, articulated, and attempted. The use of these improvements by farmers or appropriate organizations to benefit farmers and improve the productivity of irrigated agriculture is limited. Irrigated agriculture has yet to benefit near its potential from these developments in most countries of the world, even in the developed countries.

Several writers have suggested that the gap between the state of the art and the state of the science in irrigation continues to widen (Hagan and Stewart, 1972). Further, over the last 10 years, a number of development strategies have been suggested for irrigated agriculture or agriculture in general that attempt to suggest a systematic structure for improving their level of performance. The fundamental emphasis of each strategy is to suggest a systematic approach for improving irrigated agriculture or agriculture in general.

This paper will review these approaches by classifying their conceptual phases according to the phases in the research-development process (Clyma et al., 1977) as currently defined (Clyma et al., 1981), and discuss the strengths and weaknesses of each. Based on these results, the conceptual approach to water management improvement in irrigated agriculture will be presented and the important emphases needed will be suggested. An improved strategy involving specific planning will also be suggested.

HISTORICAL PERSPECTIVE

Systematic approaches to agricultural or irrigated agricultural development in the past have revolved around the concepts of specific project emphases or integrated rural development. The results have been less than satisfactory, with performance of irrigated agriculture still limited.

The author was first introduced to the concept of a need for a systematic approach to improving irrigated agriculture in 1974, as the Pakistan On-Farm Water Management Research Project began to have a significant impact on the development programs of the government of Pakistan, USAID, and the World Bank. The fundamental question was this: If there were concepts, principles, and procedures that came from the experiences of the team in Pakistan, would they provide a basis for improving the development efforts in irrigated agriculture? The answer for many, over the next four years, was an emphatic, "No!" The efforts of Dr. Max Lowdermilk; Dr. Gil Corey; many Colorado State University (CSU) personnel, such as Dr. W. Schmehl, Dr. Dan Lattimore, and Dr. David Freeman; and many other reviewer comments suggest that the answer from a 10-year perspective is an emphatic, "Yes!" The initial definition of the development model was provided from the Pakistan experience (Clyma et al., 1977).

Since that time, numerous models for development in irrigated agriculture or agriculture have been articulated. These development approaches fall into three general categories as follows: (1) those specifically for irrigated agriculture, such as those suggested by Skogerboe, Walker, and Evans (Skogerboe et al., 1979, 1980b, 1980c; Skogerboe, 1982), the Korten (1982) and Uphoff (1984) models emphasizing farmer participation and organizational development, and the Overseas Development Institute (ODI) (Bottrall, 1981) action research emphasis; (2) those for agricultural development in general, including the farming systems approach, such as given by Shaner, Phillip, and Schmehl (Shaner et al., 1982) or the International Research Center models (Rhoades and Booth, 1981, 1983); and (3) those giving a general environmental approach, which attempt to deal with a broad area in need of development. The latter category will not be reviewed here.

The above strategies for development have been published since 1977, with many of them only in the last two to three years. An interesting observation is that none of the strategies references any other as a source for any concepts or principles. Sometimes, as in Skogerboe, Walker, and Evans (Skogerboe et al., 1979, 1980b, 1980c; Skogerboe, 1982), there is some indication that one strategy draws on another but not in any specific sense. Shaner, Phillip, and Schmehl (1982) do reference broadly the area of farming systems but not other sources for their concepts. Neither Korten (1982) nor ODI (Bottrall, 1981) reference the Pakistan material, although these were widely distributed in journals and other types of publications.

The author would suggest a conclusion from personal experience. In development, much effort is directed at competition for responsibility and credit. Thus, there is a significant resistance to acknowledging other organizations for initiating a significant program, and there is too frequently an unwillingness to acknowledge other authors for their concepts or principles as a part of their program. The result is a competition for ideas and credit, too frequently on a personal basis, that limits direct progress in development. Some conclusions will be suggested in other areas after further analysis.

REVIEW OF DEVELOPMENT STRATEGIES

The concept that a specific approach or strategy can be developed and applied to improve the performance of irrigated agriculture does not have widespread acceptance. Evidence for this thesis is the repeated efforts to assist farmers to improve irrigation through specific disciplinary practices, often recommended after research only, or project designs that basically recommend practices from prior experiences be applied on a project-wide basis. Neither approach addresses the specific needs of the system nor the complex interaction of system changes. A simple concept of addressing specific system needs and making sure solutions solve the problems does not seem to be considered.

Some experiences from the earlier work of the Pakistan program suggest that many people believe development to be too complex to lend itself to simple strategies. Other reactions were that a simplified phase or step strategy is what "everyone" uses or that such a simple strategy is nothing more than the "scientific method." Such depreciating comments fail to recognize the preconceived biases and disciplinary perspectives of personnel involved in development. Old ideas are tried again on a grand scale to the detriment of the farmers and the project, or new ideas are tried at great expense without helping farmers solve their own problems in particular circumstances.

This section will review the various published strategies for development presented over the past 10 years. The evolution of the development model (Clyma et al., 1977, 1981) will be reviewed, the phases summarized, and the strengths and weaknesses described. The basic steps and important concepts of the other processes will be reviewed and compared to the development model. Significant strengths and weaknesses of each strategy will also be discussed.

The models or strategies for development to be reviewed are as follows:

For irrigated agriculture:

1. The development model (Clyma et al., 1977)
2. The planning frameworks for salinity control, development, dryland agriculture, and waterlogging and salinity of Skogerboe, Walker, and

Evans (Skogerboe et al., 1979, 1980b, 1980c; Skogerboe, 1982)

3. The Korten (1982) and Uphoff (1984) learning process approach
4. The Overseas Development Institute action research approach (Bottrall, 1981)

For agriculture in general:

5. The farming systems approach:
   a. Generalized by Shaner, Phillip, and Schmehl (1982)
   b. The International Research Center's farming systems (Rhoades and Booth, 1981, 1983).

The general environmental approach will not be discussed in this paper. The scientific method will be reviewed briefly to demonstrate its concepts in the above approaches. A review of each model will now be presented.

THE DEVELOPMENT MODEL

The original concepts and principles of this strategy were developed by Clyma, Lowdermilk, and Corey (Clyma et al., 1977). The basis for the model were the experiences in a water management improvement program in Pakistan. A similar program based on these concepts was developed for Egypt and has been implemented over the past seven years (Clyma et al., 1981; Egypt Water Use and Management Project, 1984). The phases of the model initially developed were: (1) problem identification, (2) development of solutions, (3) assessment of solutions, and (4) program implementation (Figure 1.1). Subsequent efforts to define an in-country training program for host country professionals revised the process to: (1) diagnostic analysis, (2) development and assessment of solutions, and (3) program implementation (Figure 1.2).

Many host country professionals objected to the term "problem identification" because of its negative connotation. The result was that many individuals became defensive about an emphasis on problems, with reluctance to discuss the program plans and their outcomes.

Strengths of the development model include the following: (1) field studies with farmer participation of operating irrigation systems by interdisciplinary teams to

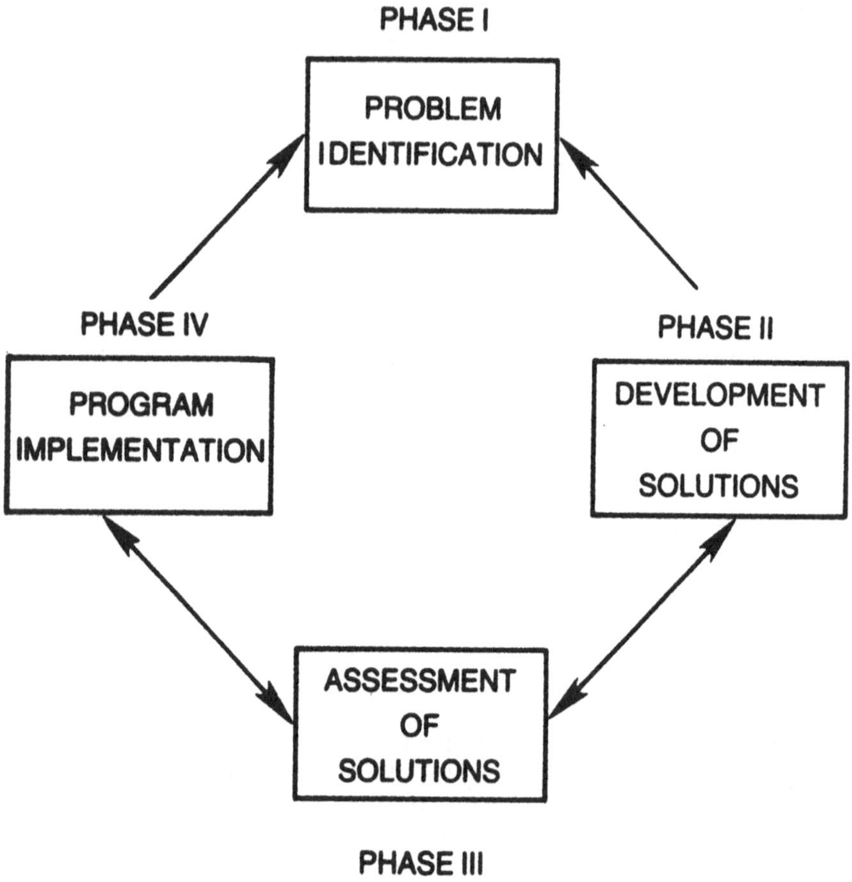

Figure 1.1 Paradigm of the research-development process

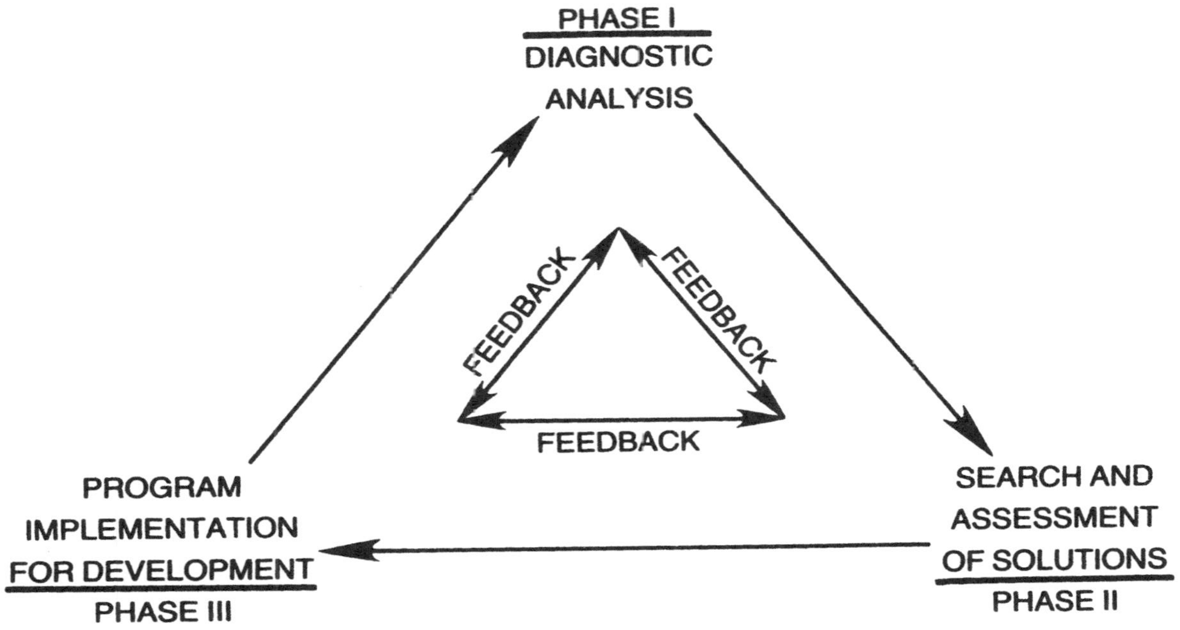

Figure 1.2 The development model

understand how the system operates and develop solutions which solve key problems; (2) developing solutions to problems through a field action research program such that the solutions do work; and, (3) trying to develop and implement a program with trained, capable personnel in the organization. Each of these strengths will now be discussed briefly.

The emphasis on developing an understanding of irrigated agriculture as practiced in the field--including the distribution of water, the farmer decision-making, organizational services, and the physical and biological conditions of the system--is an important aspect. Several countries have developed new emphases and programs in irrigation because of the understanding developed from diagnostic analysis studies. Experience in a number of countries suggests that most personnel understand the general problems, but they do not understand the magnitude of the problems nor the causes of them. The result is that frequently priority problems are not addressed, and the causes of the problems are not changed to provide effective solutions.

Solutions to problems need careful testing under live system operations to insure that they are successful. Otherwise, solutions may create more problems than they solve. Developing the organizational capability to implement programs is essential, but this strength is also a weakness. Organizational change is difficult, and resistance to change has restricted the use of the development model in Pakistan, Egypt, and Sri Lanka. Assessing the improvements for implementation needs further definition in an action research program. This important phase needs improvement in many respects.

Considerable effort has been invested in developing scientific principles and procedures for diagnostic analysis (Lowdermilk et al., 1983; Podmore and Eynon, 1983). There is a need to develop more specific principles and procedures for the development and assessment of solutions. Concepts have been defined, but the specifics are still vague. Perhaps the use of some of the developments in farming systems (Shaner et al., 1982) will improve the research of developing and assessing solutions in the development model.

Another limitation of the application of the development model is the implied organizational change that is necessary. The interdisciplinary emphasis requires the restructuring of irrigation and agriculture departments in most countries. The result is that the organizational changes delay the implementation of improvement programs using the development model. Both developed and less-developed countries have trouble restructuring their organizations to implement the interdisciplinary activities.

An early weakness of the development model was its on-farm focus. The concepts were developed in Pakistan, where priority constraints and bureaucratic restraints restricted research to the on-farm system. Subsequent applications in a number of countries have evolved an integrated methodology for relating farm and main systems in an interdisciplinary manner.

SKOGERBOE'S MODELS

Development processes described as planning frameworks for salinity control (Skogerboe et al., 1980b), waterlogging, and salinity (Skogerboe et al., 1979), best management practices in agriculture (Skogerboe et al., 1980c), and water quality from rainfed lands (Skogerboe, 1982) have been suggested. A manual for salinity management also advocates the same concepts. Figure 1.3 compares these planning frameworks with the development model (Clyma et al., 1977) as initially defined by Clyma, Lowdermilk, and Corey and subsequently expanded into manuals by a number of authors (Lowdermilk et al., 1980; Sparling et al., 1980; Hautaluoma et al., 1980; Skogerboe et al., 1980a). A summary of the model was subsequently published (Skogerboe et al., 1982) without reference to the original source.

The conceptual content of each of the models designated as planning frameworks is essentially the same. The terms used have much overlap, although the detailed diagrams describing each framework are changed to reflect the change in appropriate terms. The articles, in general, do not reference each other except in a narrow sense. The salinity control paper (Skogerboe et al., 1980b), for example, does reference the EPA manual on salinity management but does not reference the related best management practices (Skogerboe et al., 1980c) (nor vice versa), nor does the subsequent article on waterlogging and salinity (Skogerboe et al., 1979) reference the earlier articles. None of the articles cite the conceptual model developed from experiences in Pakistan (Clyma et al., 1977). Even the journal paper on the development model (Skogerboe et al., 1982) cites only those reports developed by the

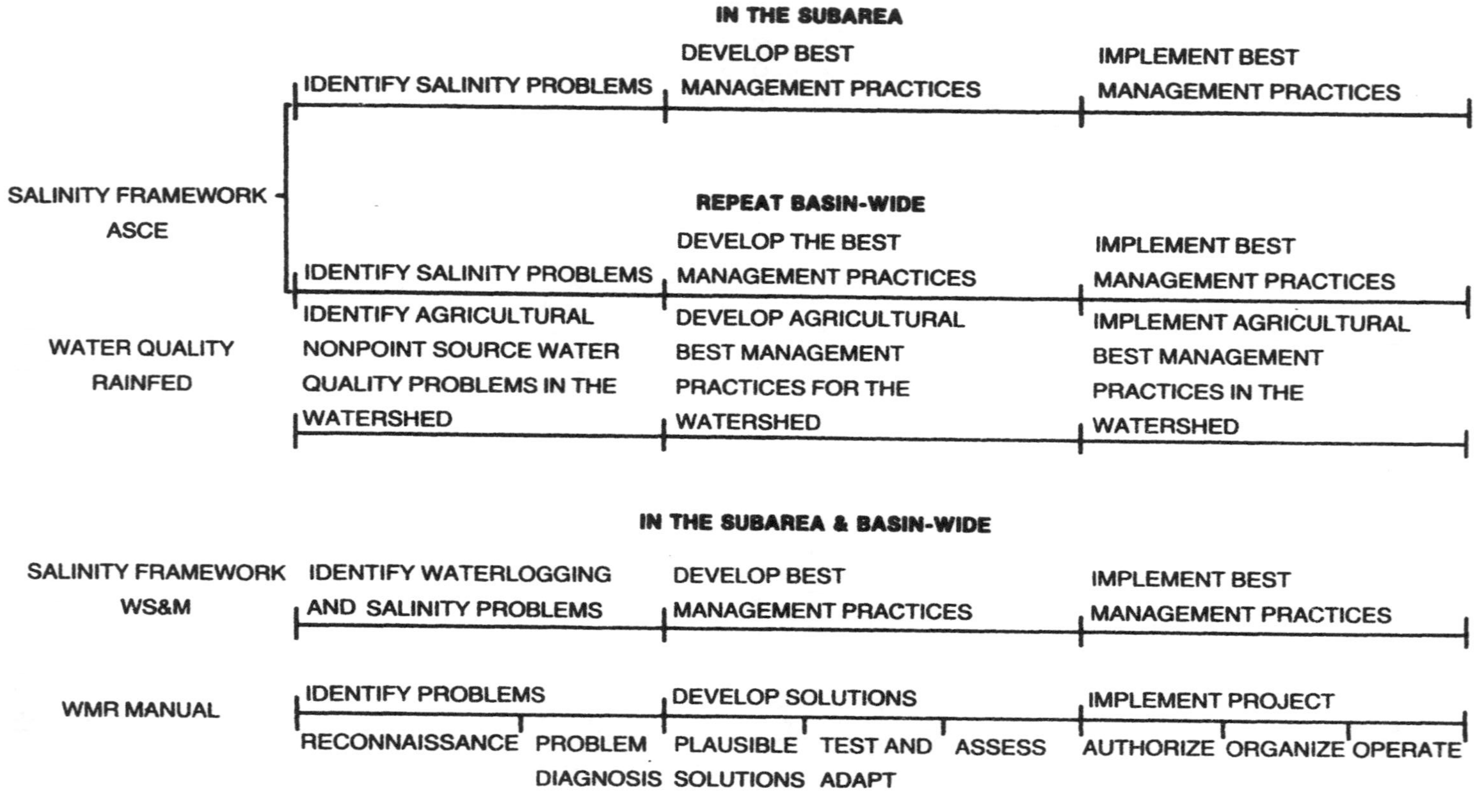

Figure 1.3 Comparison of the various planning frameworks as presented by Skogerboe et al. (1979, 1980b, 1980c, 1982) and the development model as presented by Skogerboe et al. (1982)

authors and not the original source. The article on salinity control (Skogerboe et al., 1980b) in the ASCE and the article on waterlogging and salinity (Skogerboe et al., 1979) are very similar in content, but the earlier ASCE article is not referenced. Engineers were the authors of the above reports, even though each model is described as an interdisciplinary process.

The water management technical reports (Lowdermilk et al., 1980; Sparling et al., 1980; Hautaluoma et al., 1980; Skogerboe et al., 1980a) were authored primarily by individuals without long-term experience in Pakistan, as only one report had such a senior author. Only one other author, out of a total of nine, had long-term experience. The manuals present general procedures covering a broad base of activities in a general way. The processes described in each instance are complex, with many alternate steps, feedback cycles, and decision points. The authors seem unsure of whether they are describing the concepts or the detailed procedures, with the result that both are less than clear. While they contribute ideas and approaches to development, there is still a need for a clear description of concepts and the definition of appropriate procedures based on field experience.

These differing descriptions of concepts as developed from the experiences in Pakistan were used to design a program for Egypt (Clyma et al., 1981) and were further refined for an in-country training program under the Water Management Synthesis Project (Lowdermilk et al., 1983; Podmore and Eynon, 1983). The development model concepts have also been applied to irrigation water management programs in a number of countries and also were the conceptual base for the design of the Water Management Synthesis II Project (Contract No. DAN-4127-C-00-2086-00). The experience of teaching the strategy and specific practices of diagnostic analysis in field irrigation systems seems to provide a more effective basis for improving irrigation water management on both a conceptual and operational basis.

KORTEN MODEL

The learning process approach of Korten (1982), as adopted in Cornell University's water management program in Sri Lanka and described by Korten and Uphoff (1981), and Uphoff (1984) as bureaucratic reorientation, follows the classical steps of the other models for development

(Figure 1.4). The strengths of this approach are in the emphasis on organizational reform and farmer participation. These are concepts important to development and have been sources of constraints to application of the development model. More explicit incorporation of concepts and procedures for organizational reform are needed in future applications of the development model. Farmer participation in management will be essential to improving management in irrigated agriculture.

The learning process approach (Korten, 1982) depends largely on organizational reform and farmer knowledge to provide the learning for improving irrigation systems. As a result, applications have frequently not dealt with major system problems, and significant misunderstandings about the system needs appear to have developed. For example, specialists have looked at main systems and concluded that that is where technical knowledge needs to be developed (Uphoff, 1984). Understanding of productivity has been limited to farmer knowledge, such that productivity potentials have not been recognized (Uphoff, 1984).

The concepts of the learning process approach (Korten, 1982) and of bureaucratic reorientation (Uphoff, 1984) are illustrated in Figure 1.4. The examination of field experiences is largely a social science inventory of farmer perceptions of problems that need solutions. This is an important aspect of system improvement, but does not provide the technical knowledge needed to collect additional information to understand problems not obvious to farmers. Farmers may also identify symptoms of problems or not understand the specific causes of a particular problem or the magnitude.

Comparison of the learning process approach to the development model is provided in Figure 1.5. Similarities between these sequences are obvious, and closer examination of the concepts suggests a close relationship. Each of the processes would benefit from incorporation of the respective strengths.

ACTION RESEARCH

Bottrall (1981) has suggested that large irrigation schemes should be improved through an action research process. He suggests the process is a social science research process proposed nearly 40 years ago, as outlined by Susman and Evered (1978), who also suggested affinities with the planning process. Bottrall (1981) compares the

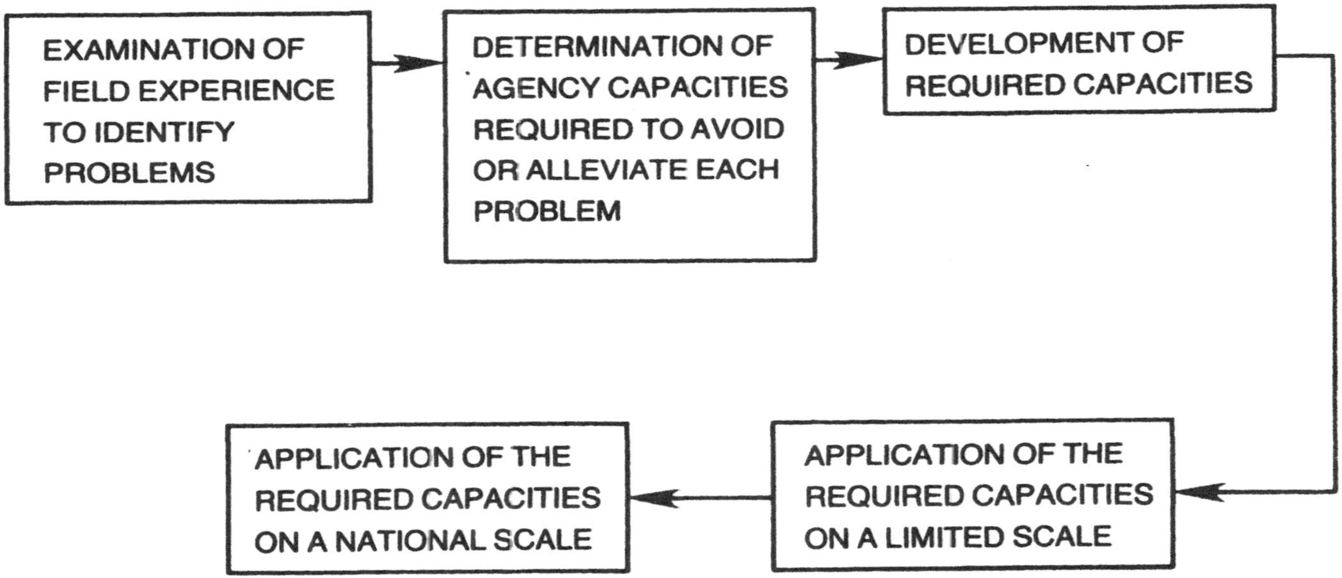

Figure 1.4 The learning process approach (Korten, 1982) or bureaucratic reorientation approach (Uphoff, 1984) to irrigation water management

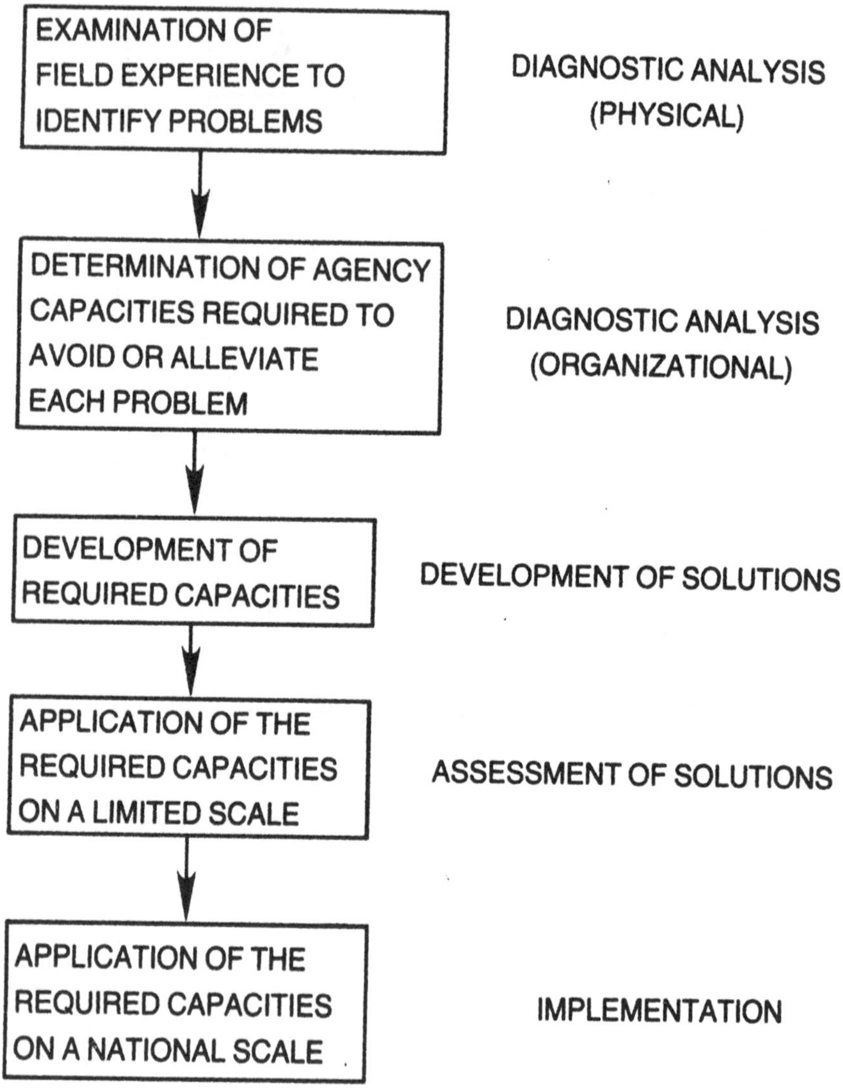

Figure 1.5 Comparison of the concepts of the learning process approach (Korten, 1982) and the development model (Clyma et al.,1977)

action research process outlined to the process suggested by Korten (1982), as well as to work reported by Early (1980, 1981). The phases of the process are given in Figure 1.6.

"Action research" is a synonymous term used to describe the development model (Clyma et al., 1977) and commonly referring to the steps in developing solutions or assessing solutions. Bottrall (1981) lists a number of important emphases that have been problem areas in irrigation action research. Phase I of the development model (Clyma et al., 1977) was changed from problem identification to diagnostic analysis. Action research is defined by Bottrall (1981) as including diagnosis.

Comparison of the phases of action research with the development model are given in Figure 1.7. Again, much overlap exists. The action research emphasis is not clear about interdisciplinary systems perspectives and, by implication, is mostly a social science emphasis. An improved emphasis on defining how solutions will be developed and tested in the "live" system is outlined. Specifying what is learned is the end of the process. In the discussion, implementation is not an explicit phase but is a condition reached when the action research is not ongoing. Thus, the implementation program may not be assessed for implementation; instead, only what becomes a part of the standard program would be, by default, that which is implemented. The similarities, improved emphases, and suggested applications are important contributions to the suggested process for improving irrigation water management.

FARMING SYSTEMS RESEARCH

Farming systems research and development is a relatively recent strategy with most relevant literature cited by Shaner, Philipp, and Schmehl (1982) published in 1979 or later. The phases are illustrated in Figure 1.8. The similarity of farming systems to the development model is great. The structure of the on-farm research in farming systems is much more substantive and definitive. The design and implementation of the research is thorough. These phases, problem identification and the phases before, are essentially the same as the development model, including an overlap in terminology and concepts (Figure 1.9). "Problem identification" was the same term used for this important phase during the earlier versions of the

- DIAGNOSING (IDENTIFYING AND DEFINING A PROBLEM)
- ACTION PLANNING (CONSIDERING ALTERNATIVE COURSES OF ACTION FOR SOLVING A PROBLEM)
- ACTION TAKING (SELECTING A COURSE OF ACTION)
- EVALUATING (STUDYING THE CONSEQUENCES OF AN ACTION)
- SPECIFYING LEARNING (IDENTIFYING GENERAL FINDINGS)

Figure 1.6 The phases of action research as suggested by Bottrall (1981) for large irrigation schemes

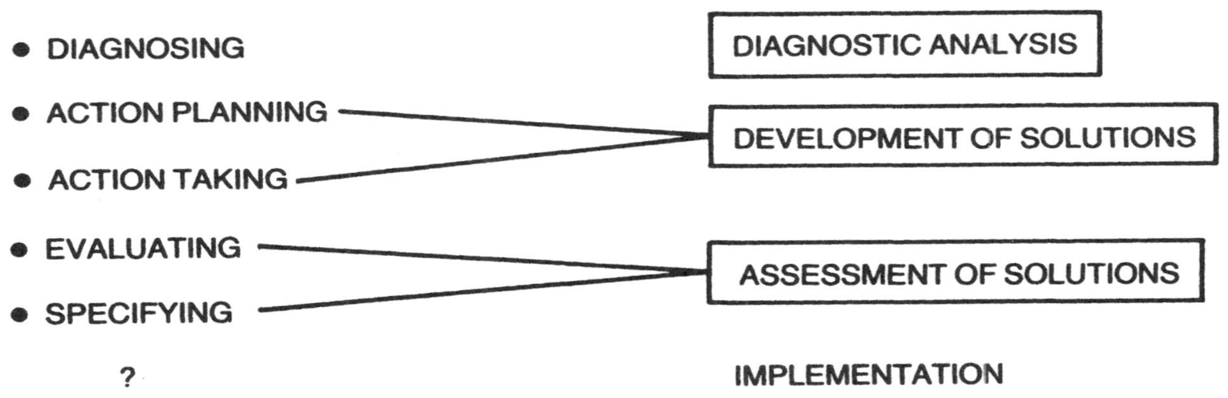

Figure 1.7 Comparison of the phases of action research (Bottrall, 1981) with the development model

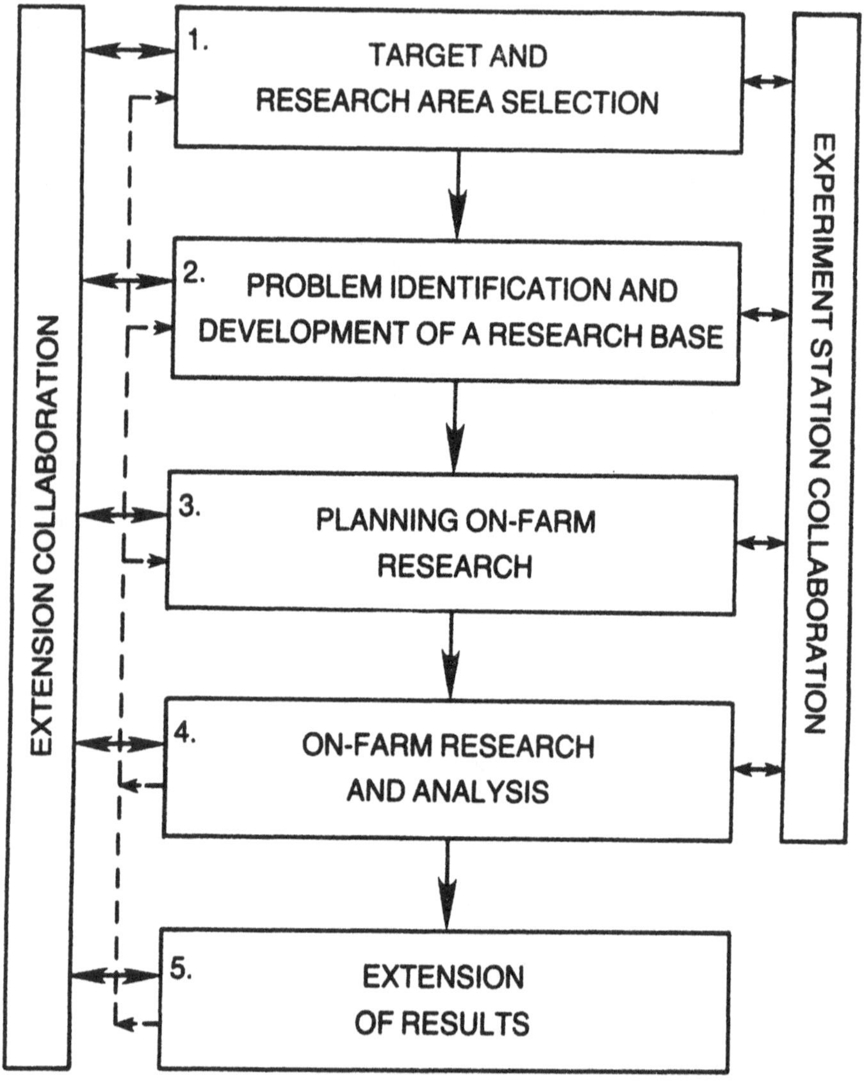

Figure 1.8 The farming systems research and development strategy

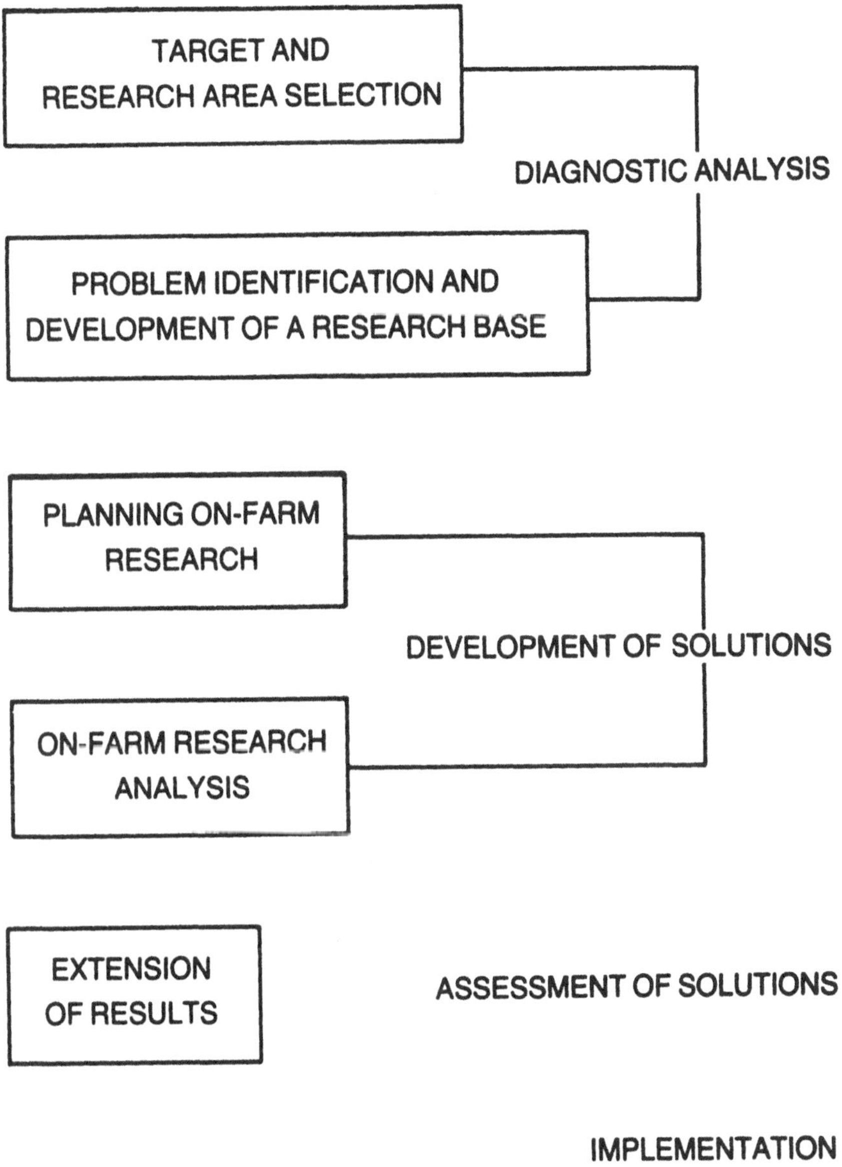

Figure 1.9 Comparison of the farming systems research and development strategy with the development model

development model. Reconnaissance again is used in both approaches, and the concept is the same.

A major difference between water management systems concepts in the development model and farming systems concepts is that farming systems attempts to deal with extension of results as a process of extending and intensifying the research activities. This is similar in concept to the assessment process of the development model (Figure 1.9). Implementation is the process of organizing and applying farming systems theory, from target and research area selection through the phase of extension of results. Thus, development, or implementation, is not a phase in farming systems research and development (Figure 1.9).

Social science concepts related to sociology and anthropology are mostly for farmer interviews and household data in farming systems. The importance of farmer knowledge, information, and custom on farmer decision-making is not clearly stated. The role of organizations and their performance also is not emphasized.

Water management systems and farming systems are related, like two sides of the same coin. Each can gain from the other. That they are described as two different paths is the most regrettable aspect of both.

The International Research Center's concepts (Rhoades and Booth, 1981, 1983) for farming systems research and development are illustrated in Figure 1.10. They follow closely the concepts of Shaner, Philipp, and Schmehl (1982), or vice versa. Somewhat casual observations of their operational procedures suggest that a balanced set of disciplines is not used in the problem identification phase. Instead, the focus is agronomic, some economic, and usually with a social emphasis. Engineers are frequently not members of the farming systems team for either rainfed or irrigation water management. Further, the data on farm problems is based more on the farmer's perceptions of his problems than on a complementary use of farmer knowledge and scientific measurements of the farming system. Again, extension of the results is an expansion of the research system. Implementation is a process of the application of the phases (Figure 1.11).

THE SCIENTIFIC METHOD

Numerous individuals have reacted to a presentation of the concepts and procedures of the development model by responding that the formalized process is used by most

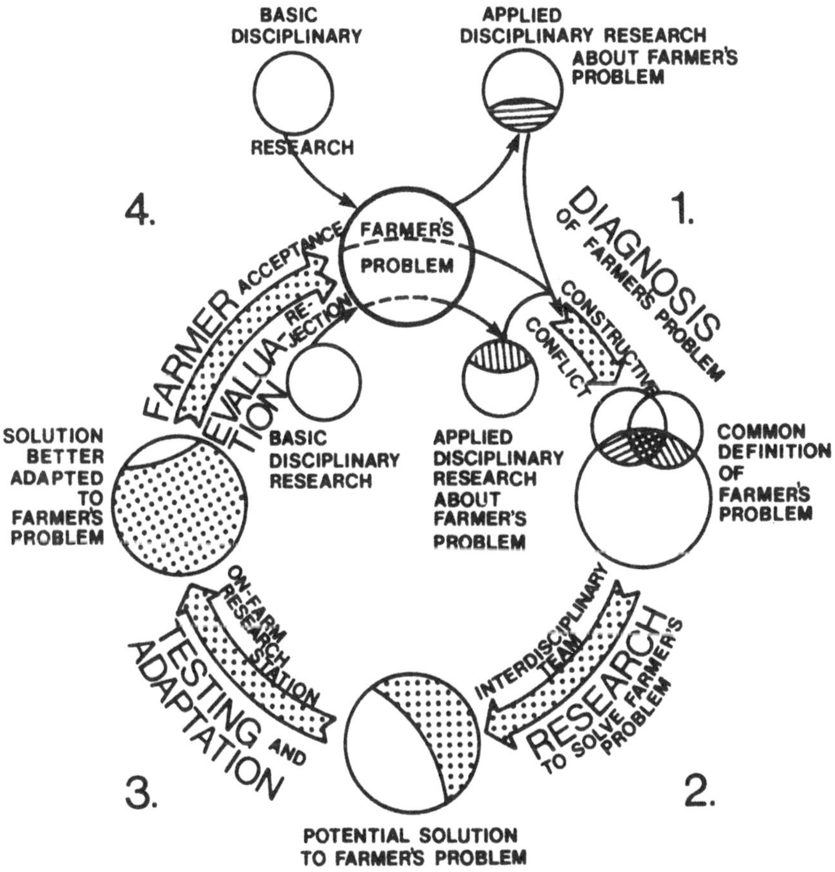

Figure 1.10 A farmer-oriented model for generating appropriate technology

1. DIAGNOSIS	DIAGNOSTIC ANALYSIS
2. RESEARCH	DEVELOPMENT OF SOLUTIONS
3. TESTING AND ADAPTATION 4. FARMER EVALUATION	ASSESSMENT OF SOLUTION
BACK TO (1)	
?	IMPLEMENTATION

Figure 1.11 Stages and activities in generating appropriate technology for small farmers

scientists as it represents the "scientific method." A formalization of the scientific method is given in Figure 1.12, as described by Cosen and Nagel (1970). A review of the process suggests that the scientific method is the basic concept of diagnostic analysis (Lowdermilk et al., 1983), with the reconnaissance used to collect data to form a hypothesis, and the detailed study the emphasis to collect data that confirm or reject the hypothesis. A review of the steps outlined for development of solutions also shows that again the scientific method is used in formulating the solution to a problem as a hypothesis, and then tests are conducted to verify or reject the hypothesis. Implementation is again an organizational design that can be tested as a hypothesis. The conclusion is that, while the development model is not the scientific method, the scientific method is an essential concept used in application of the development model.

SYNTHESIS OF IMPROVEMENTS NEEDED

An analysis of a number of approaches to improving irrigated agriculture suggests that several areas of the development model (Clyma et al., 1977) can be improved from study of experiences with other approaches. These include the structured approaches to research, as outlined in the farming systems research (Shaner et al., 1982; Rhoades and Booth, 1981, 1983). This is especially appropriate in the development and assessment of solutions. The learning process approach (Korten, 1982) emphasizes the organizational improvements needed if a particular program is to succeed. The experiences (Korten, 1982) in the Philippines provide important lessons in this regard. The experiences of Cornell University (Uphoff, 1984) suggest that farmer participation is important and can be systematically developed, such as was accomplished in Sri Lanka and the Philippines. The creation of farmer organizations and their acceptance within the structure of government has not been accomplished. The action research emphases of Bottrall (1981) are changes needed in assessing improvements for project-wide implementation. Systematic implementation of the development model in improving irrigation projects is still urgently needed. A more comprehensive methodology for planning new or improving old irrigation projects will now be suggested.

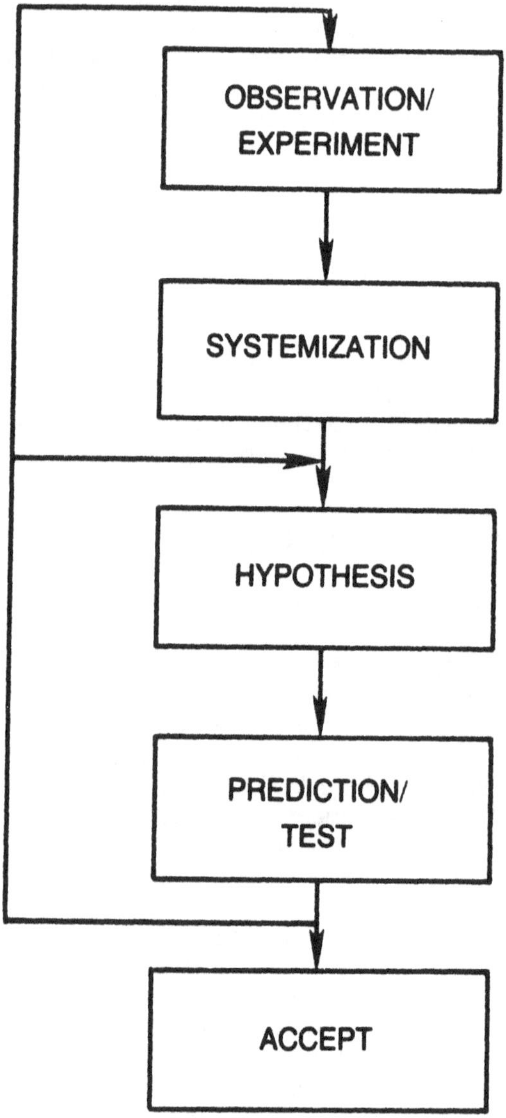

Figure 1.12 The scientific method

IMPROVING IRRIGATED AGRICULTURE

The development model (Clyma et al., 1977) is a conceptual approach for systematically improving irrigated agriculture. Improvements can and should be made, both in the concepts and the procedures for implementation. These improvements need definition and articulation. The development model concepts (Clyma et al., 1977) have been used to systematically improve irrigated agriculture, and similar concepts are believed to be valuable by others in irrigation and farming systems. A key constraint is the need for systematic, general, but usable procedures for improving irrigated agriculture.

Recent research by Sritharan (1984) suggests that irrigation projects can be more effectively managed through a systematic process involving more effective planning, design for accomplishment of objectives, development of operational plans for management, and management to achieve design objectives with monitoring. The concepts of this approach have been outlined by Clyma and Sritharan (1984). The value of the results by Sritharan is that systematic procedures have been developed using input from an interdisciplinary team and providing computer models to evaluate alternate planning objectives, design procedures, and management alternatives. Such a procedure is implemented through the use of the development model concepts.

The improvement of an irrigation project would be initiated with a diagnostic analysis study. Additional data would be collected to allow a computer model of the irrigation project to be developed. The understanding of how the system operates would be used to define the constraints and conditions of the operation of the existing project. This would allow the definition of priority constraints as well as a comparison with how the project was planned. Solutions would be developed through the development and assessment of solutions concepts, but the modeling process would be used to evaluate both alternate proposed solutions and the effects of the actual solutions as defined from field studies. The model of the project is also used to define implementation strategies, as well as to test and refine a management plan. The modeling procedures can be continuously improved through their application in the management of existing irrigation projects.

Systematic approaches to improvement of irrigation water management can be implemented by currently available concepts, including a field-based action research program with computerized planning. New concepts can be incorporated as they evolve, and increased productivity and improved well-being of farmers should be the result.

REFERENCES

Bottrall, A. "Action Research Towards Improved Water Distribution," Overseas Development Institute, 1:81:2, 1981, pp. 1-20.

Clyma, W., M. K. Lowdermilk, and G. L. Corey. "Research Development Process for Improving Farm Water Management," Water Management Technical Report No. 47, 4:439-451, Water Management Research Project, Colorado State University, Fort Collins, Colorado, 1977.

Clyma, Wayne, M. K. Lowdermilk, E. V. Richardson, W. R. Schmehl, M. D. Skold, and D. K. Sunada. "An Agricultural Development Model for Egypt," AESE Paper No. 81-5025, American Society of Agricultural Engineers 1981 Summer Meeting, St. Joseph, Missouri, 1981.

Clyma, W., and S. Sritharan. "Planning, Design, and Operational Plan for Management of Irrigation Projects," Water Management Synthesis Project, Colorado State University, Fort Collins, Colorado, April 1984.

Cosen, M. A., and Ernest Nagel. "The Scientific Method," Methods of Science, 1970, pp. 13-27.

Early, A. C. "An Approach to Solving Irrigation System Management Problems," International Rice Research Institute, Los Banos, Philippines, 1980.

Early, A. C. "Lessons from System Management Research in Central Luzon," ADC/Kasetsort University, Bangkok, Thailand, 1981.

Egypt Water Use and Management Project (EWUP). "Improving Egypt's Irrigation System in the Old Lands: Findings of the Egypt Water Use and Management Project," Halim and Hilleman, eds., Final Report, March 1984.

Hagan, R. M., and J. I. Stewart. "Water Deficits-Irrigation Design and Programming," Journal of the Irrigation and Drainage Division, American Society of Civil Engineers, 98(2):215-237, 1972.

Hautaluoma, J. E., D. M. Freeman, W. D. Kemper, J. J. Layton, M. K. Lowdermilk, G. E. Radosevich, G. V. Skogerboe, E. W. Sparling, and W. G. Stewart. "Development Process for Improving Irrigation Water

Management on Farms: Project Implementation Manual," Water Management Technical Report No. 65D, Water Management Research Project, Colorado State University, Fort Collins, Colorado, 1980.
Korten, Frances F. "Building National Capacity to Develop Water Users' Associations: Experience from the Philippines," World Bank Staff Working Paper No. 528, 1982.
Korten, David C., and Norman T. Uphoff. "Bureaucratic Reorientation for Participatory Rural Development," NASPAA Working Paper No. 1, USAID and NASPAA, November 1981.
Lowdermilk, M. K., W. T. Franklin, J. J. Layton, G. E. Radosevich, G. V. Skogerboe, E. W. Sparling, and W. G. Stewart. "Development Process for Improving Irrigation Water Management on Farms: Problem Identification Manual," Water Management Technical Report No. 65B, Water Management Research Project, Colorado State University, Fort Collins, Colorado, 1980.
Lowdermilk, M. K., W. Clyma, L. E. Dunn, M. I. Haider, W. R. Laitos, L. J. Nelson, D. K. Sunada, C. A. Podmore, and T. H. Podmore. "Diagnostic Analysis of Irrigation Systems," Vol. 1, Concepts and Methodology, Water Management Synthesis Project, Colorado State University, Fort Collins, Colorado, December 1983.
Podmore, C. A., and D. G. Eynon, eds. "Diagnostic Analysis of Irrigation Systems," Vol. 2, Evaluation Techniques," Water Management Synthesis Project, Colorado State University, Fort Collins, Colorado, October 1983.
Rhoades, Robert E., and Robert H. Booth. "Farmer-Back-To-Farmer: A Model for Generating Acceptable Agricultural Technology," Agricultural Administration 11:127-137, 1981.
Rhoades, Robert E., and Robert H. Booth. "Interdisciplinary Teams in Agricultural Research and Development," Culture and Agriculture, 1983, pp. 1-7.
Shaner, W. W., P. F. Philipp, and W. R. Schmehl. Farming Systems Research and Development Guidelines for Developing Countries. Boulder, Colorado: Westview Press, 1982.
Skogerboe, G. V. "Planning Framework and Modeling of Water Quality from Rainfed Agricultural Lands," Water Supply and Management, 6(5):359-374, Colorado State University, Fort Collins, Colorado, 1982.
Skogerboe, G. V., M. K. Lowdermilk, E. W. Sparling, and J. E. Hautaluoma. "Development Process for Improving

Irrigation Water Management on Farms: Executive Summary," Water Management Technical Report No. 65A, Water Management Research Project, Colorado State University, Fort Collins, Colorado, 1980a.

Skogerboe, G. V., W. R. Walker, and R. G. Evans. "Planning Framework for Alleviating Waterlogging and Salinity Problems," Water Supply and Management, Technical Report No. 3, Colorado State University, Fort Collins, Colorado, 1979.

Skogerboe, G. V., W. R. Walker, and R. G. Evans. "Salinity Control Planning Framework," Journal of the Water Resources Planning Management Division, American Society of Civil Engineers, Vol. 105, No. WR2, 1980b.

Skogerboe, G. V., W. R. Walker, and R. G. Evans. "Developing Agricultural Best Management Practices," Proceedings of Irrigation and Drainage Specialty Conference, July 23-25, American Society of Civil Engineers, 1980c.

Skogerboe, G. V., M. K. Lowdermilk, E. W. Sparling, and J. E. Hautaluoma. "Development Process for Improving Irrigation Water Management," Water Supply and Management, 6(4):329-342, Colorado State University, Fort Collins, Colorado, 1982.

Sparling, E. W., W. D. Kemper, J. E. Hautaluoma, M. K. Lowdermilk, G. V. Skogerboe, and W. G. Stewart. "Development Process for Improving Irrigation Water Management on Farms: Development of Solutions Manual," Water Management Technical Report No. 65C, Water Management Research Project, Colorado State University, Fort Collins, Colorado, 1980.

Sritharan, Subramaniaiyer. "Synthesis of Design Operation and Management of Surface Irrigation Conveyance Systems," Ph.D. dissertation, Colorado State University, Fort Collins, Colorado, 1984.

Susman, G. I., and R. D. Evered. "An Assessment of the Scientific Merits of Action Research," Administrative Science Quarterly, December 1978.

Uphoff, N. "Experience with Peoples' Participation in Water Management: Gal Oya, Sri Lanka," Participation and Development Planning and Management: Cases from Asia and Africa, Jean Claude Garcia-Zamor, ed. West Hartford, Connecticut: Kumarian Press, 1984.

2
On the Development and Use of Improved Methodologies for Irrigation Management

Roberto Lenton

This paper is concerned with the development and use of general approaches for irrigation management that are not specific to any given irrigation system and can be applied in a broad range of situations.[1] Its three principal objectives are (1) to clarify the concept of an irrigation management methodology; (2) to examine to what extent methodologies are available and used in practice in the management of irrigation systems around the world; and (3) to explore research needs and draw lessons which may be employed in the development of improved methodologies.

The focus of the paper is on large-scale, publicly administered irrigation systems such as those prevalent in much of South and Southeast Asia. Following the terminology increasingly used with respect to these systems, I will use the term "irrigation management" to mean the management of irrigation systems as a whole, including not only water, but also management of information and controls, of people, and of other inputs (Consultative Group on International Agricultural Research, 1982). I will also use the term "irrigation performance" to mean the extent to which an irrigation system achieves established objectives, often defined in terms of meeting equitable water delivery schedules in time and space, increasing agricultural productivity, and minimizing adverse effects. Much of the literature referred to in the paper, and many of the examples employed, are drawn from South and Southeast Asia.

The paper is divided into three parts. In the first, I explore the concept of a "methodology"--a term not yet widely used by irrigation practitioners or well defined in the literature. In the second, I discuss five key irrigation management activities and the methodologies needed and available to undertake them. In the third and final section, I draw lessons which may be employed in the analysis and development of new and improved irrigation management methodologies and explore further research needs.

CONCEPTS AND ISSUES

Countless examples of methodologies exist in everyday life: medical diagnosis, the process and art of identifying a disease in a particular person; or benefit/cost analysis, the process (by now almost routinized) of evaluating public investment projects in order to make rational choices among alternatives. Under a number of different names, planning and design methodologies are employed routinely by irrigation agencies and engineering firms around the world. In the field of irrigation management, however, the concept which the term "methodology" embodies is less recognized and understood. Let me propose the following definition:

> An irrigation management methodology comprises the generalizable practices, processes, techniques or approaches--not specific to any given irrigation system--employed by irrigation and other agencies and/or farmers to undertake a given irrigation management activity.

Three features of irrigation management methodologies may be identified. First, methodologies are required only in relation to specific activities. The design of improved management methodologies therefore must begin with a proper understanding of the corresponding activities.

Second, users of irrigation management methodologies include farmers, lending agencies, consulting firms, and primarily, irrigation agencies. Although undoubtedly irrigation agencies have widely varying staff and resources, they retain some common characteristics, and the design of improved irrigation management methodologies must start with an understanding of irrigation agencies and their limitations, both financial and in terms of human resources.

Third, the term "methodology" covers a range of activities, and a given methodology is unlikely to be applicable in exactly the same way by different agencies in different irrigation systems. It resembles a body of knowledge more than a specific procedure or method used by a specific agency in a specific irrigation system.

Irrigation researchers have paid relatively little attention to irrigation management methodologies as an appropriate area for research and development. Research on irrigation management has generally focused on providing specific answers to specific questions under specific ecological conditions, or, at a greater level of abstraction, a more generic understanding of the behavior of irrigation systems. But there has been comparatively less specific research on methodologies in an attempt to develop more effective and less costly and time-consuming methodologies for irrigation management.

This situation is now changing. The Study Team on Water Management Research and Training commissioned by the Consultative Group on International Agricultural Research (CGIAR), for example, identified research on practices for management as a key priority area, noting that "research is needed to identify, analyze, and improve practices for key activities associated with irrigation management" (CGIAR, 1982). A key outcome of the study was the establishment of the International Irrigation Management Institute, an organization with an international mandate to develop generalizable methodologies which transcend the boundaries of specific irrigation systems and can be applied in different situations (Ford Foundation, 1983).

KEY IRRIGATION MANAGEMENT ACTIVITIES

The following sections will focus on five key activities which have particular significance in irrigation management. For each of these I will briefly review the activities and the methodologies available and required for these activities, touching on the following aspects: (1) definition of the activity; (2) extent to which these activities are recognized as legitimate; (3) types of methodologies available in use, both in terms of process (how agencies can implement) and output (to what extent methodologies accomplish desired activity); and (4) further research and development work needed. From the detailed consideration of methodological developments in these and other areas, I will try to draw lessons which may be employed in the analysis and development of new and improved irrigation management methodologies.

The five activities chosen for review are performance monitoring, diagnostic appraisal, water scheduling and delivery, action research, and farmer participation. All are critical for effective irrigation management; the first three, in particular, were singled out for attention in the CGIAR Study Team Report. Other irrigation management activities of special significance, but not reviewed here, include the development and use of communication systems. For the purpose of this review, I will distinguish between analytical or diagnostic methodologies (performance monitoring, diagnostic appraisal, and action research) and action methodologies (water scheduling and farmer participation).

ANALYTICAL METHODOLOGIES

Performance Monitoring

Performance monitoring may be defined as an activity to estimate the performance of an irrigation system in order to enable irrigation managers to determine whether the system is performing at a satisfactory level or not. Monitoring of performance requires the establishment of performance criteria, which may be defined in terms of water delivery, crop yields, equity, or other variables. Three features of performance monitoring should be noted:

Widespread Ignorance of Actual Performance. One of the extraordinary characteristics of irrigation systems management is that, despite the fact that large irrigation projects generate revenues far in excess of the largest business corporations, there is virtually no information on the extent to which these irrigation systems are achieving performance objectives--a practice which would shock most production management specialists (Seckler, 1981). For example, the managers of the Second Bhakra Main Circle (SBMC) in India--a project with a command area of 1.9 million acres and some 400,000 farmers, generating about twice the pre-tax profit of the largest multinational operating in India--until recently had no way of obtaining specific information on how much water farmers were actually receiving, and at what times.

Consensus on Appropriate Performance Indicators. Until recently, "irrigation efficiency" was the most used indicator of system performance, but of late there has been increasing consensus that broader concepts focusing on the use of water are more accurate performance indicators. It is now increasingly recognized (see, for example, Lenton, 1983, and Abernathy, 1984) that the two principal measures of irrigation system performance are productivity (measured by water delivery, yields or potential yields) and equity (measured by productivity variations within an irrigation system).

Development of Methodologies. In recent years, there has been steady progress in some countries in the development of methodologies to enable irrigation managers to monitor the performance of large irrigation systems. An excellent example (and one which also provides a good illustration of how a methodology for any irrigation management activity may be developed) is that of Malhotra, Raheja, and Seckler (1984) on the SBMC project in India. The objective of this study was to develop a procedure to enable the SBMC management to determine performance over the entire SBMC project at a reasonable cost and with existing staff. Ultimately, it was hoped to develop a procedure to estimate SBMC's performance as a continuous and ongoing component of a management information system. The collaborators in the study included a former manager of the SBMC, a statistician from the Indian Agricultural Research Statistics Institute, and an economist from the Ford Foundation.

The study initially focused on one subsystem of the SBMC, the Phabra Distributary, which is 55 km long and has a command area of about 52,000 acres. A stratified sample of 10 of the distributary's 50 water courses, each of which served about 50 farms, were selected at the head, middle, and tail of the canal. In each of these, a sample of farms was chosen for detailed performance estimation. In order to enable performance to be estimated in each of the samples with limited staff and a small budget, a very simple performance criterion--the relationship between the sum of the areas of the farm wetted in each irrigation during the crop season, and the command area of the farm[2]--was chosen as a proxy for water delivered to the sample farm. By analyzing this relationship across the sample of farms, a quick and reliable estimate of performance in terms of the amount of water delivered to farms and the variability in water delivery among farms was obtained. The results showed that the performance of the distributary was probably as high as would be economically feasible for a system of its size, complexity, and water control characteristics.

Although the specific results of the study were of great interest to the SBMC management, from a methodological point of view, the most important contribution of the study was the development of a procedure which allowed the estimation of irrigation performance over a fairly large area with ease and objectivity, using only moderately trained people who could be deployed to take measurements even after an irrigation event had taken place. Four factors can be said to have contributed to the success of the study:

- The methodology was developed through action research on an irrigation system of considerable size and complexity, rather than in an experimental station or computer laboratory.
- The study was conducted in direct collaboration with the State Irrigation Department, and one of the three principal investigators was the former engineer-in-chief of the department. Thus, the study started with an understanding of the resources and staff of the client agency.
- Right from the start, the study had a clear goal of developing a methodology for monitoring the performance of the entire project, rather than of a limited area. Had this not been the case, there would have been a temptation to develop more accurate performance indicators which might have yielded more reliable results over a small area but which ultimately would not have been feasible over hundreds or thousands of farms.
- Likewise, there was a similar focus on reducing costs and staff requirements, while maintaining a reasonable level of accuracy consistent with the nature of the problem. In studies such as these, keeping close tabs on the real costs of implementing the methodology is important (see Bottrall, 1981).

Diagnostic Appraisal

Diagnostic or performance appraisal, in the sense used by Chambers (1983), may be defined as the activities involved in finding out about an irrigation system in order to identify interventions (and their sequences) which maximize performance.[3] As Chambers notes, approaches should be "opportunity" rather than "problem" oriented; thus, diagnostic appraisal is not simply the identification of the problems and constraints of a particular irrigation system, but rather the identification of opportunities to improve that system's performance. In many ways, diagnostic appraisals are to improved management programs what project appraisals are to new projects.

Diagnostic appraisals, like project appraisals, need to be systematically organized in order to ensure that the appropriate interventions to improve performance are identified. Recognizing this, there has been fairly substantial research on the development of systematic interdisciplinary appraisal methodologies, including those of Lowdermilk et al. (1980) and Chambers (1981), and the work of Hildebrand (1981) and Collinson (1981) on the systematization and organization of field visits. However, little follow-through at the operational level has yet taken place, other than attempts involving one, or at most two, disciplines (see, for example, Central Water Commission, n.d.). At least three factors may be said to contribute to this:

- Diagnostic appraisal is not yet viewed as a legitimate, explicit activity which falls within the responsibility of irrigation departments. Few, if any, job descriptions for irrigation staff refer, implicitly or explicitly, to diagnostic appraisal. Within external lending agencies, too, the concept of a performance appraisal has not yet achieved the status and credibility of project appraisals, primarily because lending for management improvement is still small relative to project lending.
- Widespread lack of understanding of, and perhaps confidence in, available methodologies for diagnostic appraisal prevails.
- Available methodologies for diagnostic appraisal require staff with educational and disciplinary backgrounds beyond the range of most developing countries' irrigation agencies.

On the basis of experience to date, both in irrigation and in other fields, research on the design of better methodologies for diagnostic appraisal should give priority to:

- Making staffing requirements more nearly fit those of irrigation agencies, or, alternatively, finding operational ways for irrigation agencies to work with outside organizations on diagnostic appraisal.

- o Methodological analysis starting with an understanding of irrigation agencies and their limitations, both physical and in terms of human capital.
- o Helping to make diagnostic appraisal a legitimate and professionally recognized activity, essential to the determination of interventions to improve irrigation systems performance (including not only improved irrigation management programs, but also irrigation rehabilitation and modernization projects).
- o Increasing confidence in the results of diagnostic appraisal through a focus on validation and reliability. (This would require, in part, a more thorough conceptual understanding of the behavior of irrigation systems.)
- o Developing of systematic procedures for analysis of information and evaluation of alternatives, along the lines of the techniques used in project appraisals.

It is interesting to note that these points are supported by the history of medical diagnosis. Perhaps the greatest advances in medical diagnosis were derived from an increased understanding of anatomy and physiology as the conceptual basis for diagnosis. Furthermore, though the techniques of medical diagnosis evolved considerably over the last several centuries, for most of this time diagnosis was considered a professionally legitimate activity and a prerequisite to prescription, and, in addition, was within reach of the individual physician. All this suggests that developing good diagnostic methodologies will require a far better analytical underpinning for irrigation systems--an irrigation system "anatomy," if you will--than is now available. Furthermore, effective use of these methodologies will require that they be simple and readily available to irrigation managers.

Action Research

Action research may be defined as an activity designed to evaluate the range of alternative interventions to improve irrigation system performance, in which interventions based on diagnostic appraisal are made in a representative area of a project on an experimental basis and then carefully monitored and evaluated. Action research typically involves six activities (CGIAR, 1982): (1) diagnostic appraisal and choice of interventions, (2) benchmark surveys, (3) action taking, (4) monitoring and evaluation, (5) identification of issues, and (6) extension of learning to others and elsewhere.

Several features characterize action research efforts to date:

- Action research programs--ranging from "demonstration" or "pilot projects" carried out by irrigation agencies with little systematic monitoring and evaluation, through field research projects conducted by research institutions with little field implementation of interventions, to comprehensive programs involving both action and research--are increasingly used by both irrigation researchers and practitioners.[4] An excellent example is the IRRI/NIA program referred to later in the text.
- There is often little recognition of the methodological difficulties inherent in action research--including the selection of representation areas, the interpretation of results given year-to-year variations of inputs, and the problems of unmonitored special inputs and multiple causation (see Chambers and Lenton, 1981; CGIAR, 1982; and Tamil Nadu Agricultural University, 1981).
- Though existing action research methodologies are clumsy, expensive, time-consuming, and often non-operational, there is a conspicuous absence of "research on action research" to develop better action research methodologies. Clearly, action research is a difficult process and one which needs to be better systematized if it is to be effective as a management tool.

On the latter point, much can be learned from the development of the physical and analog models extensively used in hydraulic engineering, which engineers have nearly perfected over several decades (Lenton, 1983). Here engineers have established procedures to insure that proposed interventions are accurately represented in their models and that results are accurately interpreted; found ways to reduce model start-up and operating costs and duration and to increase the accuracy of the results; and developed models into respected and legitimate forms of analysis, both by validating their use in practice through experience, and by developing formal rules for analysis so that modeling studies could be carried out routinely by technical staff.

A similar approach with action research is needed. In particular, ways to generalize rapidly from action research through simulation modeling need to be systematically explored.

ACTION METHODOLOGIES

Water Scheduling and Delivery

Water scheduling and delivery may be defined as the way in which water is both scheduled and actually delivered in time and space to farms located in the command area of an irrigation system. Water delivery in most large irrigation projects is carried out by an irrigation agency on the basis of an implicitly or explicitly established policy or schedule, which may be said to have three essential components (Replogle, 1986): the delivery flow rate to the field; the frequency of times of delivery to the field; and the duration of the delivery.

Water scheduling and delivery are activities which are recognized by most irrigation agencies as necessary and legitimate. In many cases, schedules have been in place for several decades; furthermore, water scheduling and delivery are likely to be present in the job descriptions of most agency personnel responsible for irrigation management. However, in many irrigation schemes, existing practices result in inequitable water distribution, with substantial differences between head and tail enders (see, for example, CGIAR, 1982, or Lenton, 1983)--either because existing practices do not follow established policies (and are subject to "local pressures") or because the policies themselves are deficient (having been established by convention or rule of thumb) and inappropriate for the irrigation systems to which they are applied. Moreover, in many countries, water scheduling and delivery (much less advances in scheduling and delivery practices) are not even considered subjects worthy of inclusion in basic professional irrigation training.

Substantial attention has been given to providing a framework for considering water scheduling systems, particularly for large-holder irrigation projects such as those in the United States (see, for example, Replogle, 1986, and Replogle and Merriam, 1980). These frameworks, though useful in designing schedules for new irrigation systems, provide limited guidance to irrigation agencies on how they may change current practices. Closing the gap between the "actual" and the "design" schedule in an irrigation system or changing practices from one scheduling and delivery system to another may result in losses to some, as well as gains to others, and therefore require political commitment, in addition to changes in professional training, improvement in control infrastructure, and the like.

Several recent programs designed to develop and introduce changes in existing scheduling and delivery practices in developing countries deserve mention. In the Philippines, an innovative research project has been conducted by the International Rice Research Institute and the Philippines National Irrigation Administration on the 2,500 ha Lower Talavera irrigation system (Early, 1981). In this project, the scheduling and control of water along the distributary and minor canal systems have been improved by reallocating excess water in areas of surplus to areas of deficit, thus achieving a more equitable distribution of water. In India, the government of the state of Andhra Pradesh issued an order in 1981 for "integrated water management" above and below the outlet to equitably distribute water shortages in the state's canal systems and to make water available to the tail ends of the systems (see Hashim Ali, 1983). Two recent workshops in India on water delivery and scheduling--one organized by the Indian Water Resources Society at the University of Roorkee (IWRS, 1982) and the other by the Water and Land Management Institute in Aurangabad (WALMI, 1983)--demonstrate the lively ongoing enquiry and experimentation in that country to identify alternatives to current practices. For example, in the Mula Project in Maharashtra, a proposal has been made to evolve better water allocation practices to ensure a more equitable distribution of irrigation benefits, recognizing that the present system "is not equitable from head-reach to tail-reach of canal" (Dhamdhere and Padhye, 1983).

Of significance, too, are recent efforts to document those scheduling and delivery systems which are generally recognized as leading to high systems performance levels. Perhaps the best example is S. P. Malhotra's thorough documentation of the approach used in the warabandi system of Northwest India (1982). Similar detailed documentation of scheduling and delivery systems employed in other successful projects is much needed, since procedures for scheduling and water delivery in irrigation, even when rigorously applied, are often not well documented. Thus, to a large extent, methodologies for scheduling and water delivery remain unknown to those outside the concerned agencies.

In sum, there appears to be increased information and research on alternative water scheduling and delivery systems for small-holder irrigation systems, along the lines of that available for large-holder systems. However, few quantitative evaluations and comparisons of alternative approaches (i.e., which methods are better, under what circumstances, and why) exist. Furthermore, there is little information and research on the political feasibility of changing water distribution practices, on the ways in which irrigation agencies might design new irrigation scheduling and delivery practices which would allow all farmers to gain (Chambers, 1983), or on the ways in which changes implying losses to head enders might be enforced.

Farmer Participation

For the purpose of this paper, farmer participation may be defined as an approach (by irrigation agencies) to increase irrigation performance by providing effective incentives and conditions that enable farmers, both individually and collectively, to accept and fulfill irrigation management responsibilities where and when appropriate (Chambers, 1985). Although irrigation agencies often recognize the importance of enhancing the participation of farmers in irrigation management, in many cases these same agencies are unable to efficiently translate these ideas into practice because of lack of know-how. Little information on the performance consequences of more effective farmer participation is available to irrigation agencies, and, as a result, it is often not recognized as a legitimate, efficient approach to improving system performance. Furthermore, many agencies are concerned about the financial and staff requirements of working in a more participatory way and may also believe the approach will delay project implementation.

Experience to date with farmer participation in several countries suggests the following:

- o In some regions of Southeast and South Asia, there appears to be developing a remarkable body of experience and know-how on ways in which irrigation agencies might work more effectively with farmers (see Food and Agriculture Organization, 1982). In the Philippines, this experience now goes back eight years, when the Philippines National Irrigation Administration began fielding community organizers in communal irrigation systems who helped develop irrigator associations to work with agency staff in planning, designing, and constructing rehabilitation programs (Bagadion, 1984; Korten, 1981). Similar experience has been developed in Sri Lanka and Pakistan, among other countries, as a result of successful experimental programs conducted through the MONA project in Pakistan (Government of Pakistan, 1978) and the Minipe project in Sri Lanka (de Silva, n.d.). In most of these programs, a principal focus of the agency's efforts was the establishment of formal water user associations to perform such tasks as water distribution, maintenance of facilities, and conflict resolution.

- In these programs (particularly in the Philippines), there has been a strong emphasis on documenting the process by which farmer participation can be enhanced. In at least five of the Filipino systems, there has been intensive process documentation, with the principal objective of bringing operational problems to the attention of program managers so that needed design changes and/or problem solving measures could be introduced (Korten, 1984). In a broad sense, this can be described as work to develop and document a methodology of effective farmer participation, which may be applied to other systems elsewhere. Indeed, the Philippines' "methodology" is now beginning to be applied in Indonesia and Thailand (Korten, 1984).
- There have been few attempts to evaluate the extent to which participatory approaches improve system performance in productivity and equity terms. An exception is a study now under way at the Institute of Philippine Culture of the Ateneo de Manila University to assess the performance of "participatory" and "nonparticipatory" irrigation systems in terms of cropping intensity, yields, water distribution, and "function-

ality" of rehabilitated facilities (Korten, 1984).

- o Although much of the literature on farmer participation is dominated by current efforts in south and southeast Asia to change from nonparticipatory to participatory methods, in some countries, such as Argentina and Chile, there are long-established practices whereby farmers participate actively and effectively in the organization and management of large-scale irrigation projects (Grassi, 1977). In addition, there are numerous examples all over the world of small-scale indigenous systems which have developed effective forms of local self-management.

In sum, there appears to be a clear need to further develop and document improved methodologies for more effective participation of farmers, taking into account agency staff and financial limitations, their objectives of improving performance, and their need to meet project deadlines. In particular, methodological developments must give attention to the performance consequences of more effective participation, in order to gain legitimacy and credibility.

NEEDED RESEARCH ON DEVELOPMENT OF METHODOLOGIES

The foregoing sections have demonstrated that, although some pioneering methodological work has been undertaken in relation to several key irrigation management activities, there is need for further work on developing more effective, and less costly and time-consuming, methodologies for irrigation management. The problem, however, is that there is not yet a clear "methodology for developing methodologies." Furthermore, the lack of an analytical underpinning for irrigation systems hampers work in this area.

One characteristic of much of the methodological research projects described in the previous section was that they started out as field research to obtain answers to specific questions and did not necessarily, at the outset, have a clearly established methodological goal. In some cases, only once the project was completed was it recognized that an important output of the research was an approach that could be used elsewhere to address similar issues. This seems to suggest that good methodological research must start (and then build on) field or action research on a specific irrigation system. Methodologies cannot be developed in the abstract, through conceptual analysis or model building. They must be developed in the context of a live problem; otherwise, the research and the research output will lack credibility.

It can thus be argued that good methodological research must be initially structured along the lines of field or action research, and must, at a minimum, possess the features (interdisciplinarity, effective collaboration between researchers and agency staff, rigorous process documentation, and performance monitoring) that such research requires in order to be effective. Methodological research demands something more of the researchers, however: the ability to step back from the research to focus on the methodology, on those generic aspects of the approach which are not specific to the irrigation system under study. And, if there is a clear methodological goal for the research at the outset, then some aspects of the research might well be structured differently. In the SBMC performance monitoring study described earlier, for example, the approach was deliberately kept simple and inexpensive, in order to ensure its viability when applied on a much larger scale.

Another problem which may arise in considering research on methodology is the lack of a clear sense of the goals of such research. What, for example, is meant by an "improved methodology" for irrigation management? Although this question needs further careful thought, reasonable goals for such research include the following:

> Increasing Effectiveness: Researchers can help in the further development of methodologies by finding ways to reduce costs, staff requirements, and implementation time, and thereby increase the reliability of the results.
>
> Increasing Professional Status and Recognition: Researchers need to work with national irrigation agencies to demonstrate that activities such as diagnostic appraisal and farmer participation are key to improving performance, and that methodologies to assist agencies to undertake these activities more effectively can and should be developed and applied.

<u>Increasing Client Confidence</u>: Researchers can help increase confidence in the use of analytical or diagnostic methodologies through a focus on validation and reliability of results, and in action methodologies by demonstrating their performance impact through research and documentation.

<u>Increasing Compatibility with Agency Constraints</u>: Researchers can help develop methodologies based on an understanding of irrigation agencies and their financial and human resource constraints, and whose staffing and other requirements fit those of irrigation agencies.

In closing, and despite the arguments in favor of methodological research, it may be useful to reflect on some of the dangers associated with such research. One is that a focus on methodologies might bring with it an excessive preoccupation with process instead of output; with the way of carrying out a specific activity rather than with what that activity achieves. Thus, goals and outputs must not get lost in such research. A second is that an excessive reliance on methodologies might inhibit our capacity to think creatively--much as the revolution in the development of mathematical modeling techniques led to a dependency on procedures that sometimes have come in the way of logical thinking. A third is that a focus on development of methodologies might make researchers lose sight of the need for work to ensure effective dissemination and use by client agencies. There may well be lessons to be learned in this regard from the pioneering methodologies for the planning and design of water resource systems developed over 25 years ago by the Harvard Water Program (Maass et al., 1962); their use by water resource agencies has been disappointing.

NOTES

1. In this paper, the word "methodology" is used to convey this sense of a general approach, following Webster's definition of a methodology as "the processes, techniques, or approaches employed in the solution of a problem or in doing something." An alternative term is simply "practices."

2. This criterion assumes fairly homogeneous water application rates, an assumption that may not always be true.

3. Some authors (see, for example, Lowdermilk et al., 1980) use the term "diagnostic analysis" in roughly the same sense.

4. See CGIAR, for an extensive listing.

5. I would like to acknowledge the many conversations held with Robert Chambers on the subject of methodologies, which provided the basis for much of this paper. The discussions on the subject at the IIMI/WMS-II workshop, "Research Priorities for Irrigation Management in Asia" held in Sri Lanka in January, 1985, together with Robert Chambers' comments on an earlier draft of this paper and those of an anonymous reviewer, were also very helpful in structuring the paper and in clarifying several of the concepts and issues discussed therein. All errors and omissions are entirely my responsibility, and the views expressed do not necessarily reflect those of the Ford Foundation.

REFERENCES

Abernathy, Charles. "Methodology for Studies of Irrigation Water Management," Proceedings of the African Regional Symposium on Smallholder Irrigation, University of Zimbabwe, Harare, Zimbabwe, published by the Overseas Development Unit of Hydraulics Research Limited, Wallingford, U.K., September 1984.

Ali, Syed Hashim. "One Season of Integrated Water Management in Andhra Pradesh," Irrigation Water Management Network Paper 7B, Agricultural Administration Unit, Overseas Development Institute, London, England, April 1983.

Bagadion, B. "Developing Farmers' Participation in Managing Irrigation Systems under the National Irrigation Administration, Philippines," Transactions, Twelfth International Congress on Irrigation and Drainage, International Commission on Irrigation and Drainage, Colorado State University, Fort Collins, Colorado, June 1984.

Bottrall, Anthony. "Improving Canal Management: The Role of Evaluation and Action Research," Water Supply and Management, Vol. 5, No. 1, 1981, pp. 67-79.

Central Water Commission (CWC). "Recommendation of Central Team on Water Utilization on 24 Existing Projects," Central Water Commission, New Delhi, India, 1980.

Chambers, Robert. "In Search of a Water Revolution: Questions for Managing Canal Irrigation in the 1980s," Report of the Planning Workshop on Irrigation Water Management, the International Rice Research Institute, Los Banos, the Philippines, 1980.

Chambers, Robert. "Rapid Appraisal for Improving Existing Canal Irrigation Systems," Discussion Paper Series No. 8, Ford Foundation, Delhi, India, August 1983.

Chambers, Robert. "Farmers Above the Outlet: Irrigators and Canal Management in South Asia," typescript, Institute of Development Studies, University of Brighton, Sussex, United Kingdom, January 1985.

Chambers, Robert, and Roberto Lenton. "Action Research on Canal Irrigation: Traps, Tactics, and a Code," Tamil Nadu Agricultural University, 1981.

Collinson, Michael "A Low-Cost Approach to Understanding Small Farmers," Agricultural Administration, June 1981.

Consultative Group on International Agricultural Research (CGIAR). "Report of the Study Team on Water Management Research and Training," Technical Advisory Committee, CGIAR, Rome, Italy, February/March 1982.

deSilva, R.G.R. "Farmer Participation in Water Management: The Minipe Project in Sri Lanka," Kandy Range, Department of Irrigation, Sri Lanka, n.d.

Dhamdhere, H. V., and V. S. Padhye. "Scheduling of Irrigation," Water and Land Mangement Institute, India, November 1983.

Early, Alan. "Irrigation Systems Management Diagnosis and Improvement Methodologies Developed in Central Luzon of the Philippines," Tamil Nadu Agricultural University, India, 1981.

Food and Agriculture Organization (FAO). "Farmers' Participation and Organization for Irrigation Water Management," International Support Program for Farm Water Management, Land and Water Development Division, FAO, Rome, Italy, April 1982.

Ford Foundation. "International Irrigation Management Institute," Draft Proposal, India, February 1983.

Government of Pakistan. "Proceedings of the Seminar on Water Users Associations for Improving Irrigated Agriculture," Ministry of Food and Agriculture, Government of Pakistan and USAID, Islamabad, Pakistan, June 1978.

Grassi, Carlos J. "Operacion y Conservacion de Sistemas de Riego," (Operation and Maintenance of Irrigation Systems), Publicacion RD-14, CIDIAT, Merida, Venezuela, 1977.

Hashim Ali, Syed. "One Season of Integrated Water Management in Andhra Pradesh," Irrigation Water Management Network Paper 7B, Agricultural Administration Unit, Overseas Development Institute, London, England, April 1983.

Hildebrand, Peter E. "Combining Disciplines in Rapid Appraisal: The Sondeo Approach," Agricultural Administration, June 1981.

Indian Water Resources Society (IWRS). Proceedings of the Workshop on Water Distribution Practices, Water Resources Development Training Center, University of Roorkee, Roorkee, India, July 1982.

Korten, Francis F. "Building National Capacity to Develop Water Users Associations: Experience from the Philippines," World Bank Staff Working Paper No. 528, Agricultural and Rural Development Department, World Bank, Washington D. C., 1982.

Korten, Francis F. Personal Communication, 1984.

Lenton, Roberto L. "Management Tools for Improving Irrigation Performance," Discussion Paper Series No. 5, Ford Foundation, Delhi, India, June 1983.

Lowdermilk, Max, William Franklin, James Layton, George Radosevich, Gaylord Skogerboe, Edward Sparling, and William Stewart. "Problem Identification Manual: Development Process for Improving Irrigation Water Management on Farms," Water Management Technical Report No. 65B, Water Management Research Project, Colorado State University, Fort Collins, Colorado, March 1980.

Maass, Arthur, Maynard M. Hufschmidt, Robert Dorfman, Harold A. Thomas, Jr., Stephen A. Marglin, and Gordon Maskew Fair. Design of Water Resource Systems. Cambridge, Massachusetts: Harvard University Press, 1962.

Malhotra, S. P. "The Warabandi System and its Infrastructure," Central Board of Irrigation and Power, New Delhi, India, 1982.

Malhotra, S. P., S. K. Raheja and David Seckler, "A Methodology for Monitoring the Performance of Large-Scale Irrigation Systems: A Case Study of the Warabandi System of Northwest India," Agricultural Administration, 17:231-259, 1984.

Replogle, John A. "Some Tools and Concepts for Better Irrigation Water Use," Chapter 5, Irrigation Management in Developing Countries, K. C. Nobe and R. K. Sampath, eds. Boulder, Colorado: Westview Press, 1986.

Replogle, John A., and J. L. Merriam. "Scheduling and Management of Water Delivery Systems," Proceedings of the American Society of Agricultural Engineers, Second National Irrigation Symposium, American Society of Agricultural Engineers, St. Joseph, Michigan, 1980, pp. 112-126.

Seckler, David. "The New Era of Irrigation Management in India," Draft Paper, Ford Foundation, Delhi, India, January 1981.

Tamil Nadu Agricultural University. Proceedings of the International Seminar on Field Research Methodologies for Improved Irrigation Systems Management. College of Agricultural Engineering, Tamil Nadu Agricultural University, India, September 15-18, 1981.

Water and Land Management Institute. Proceedings of the National Workshop: Scheduling of Irrigation. Water and Land Management Institute, Aurangabad, India, November 12-13, 1983.

3
From Diagnostic Analysis to Designing and Conducting On-Farm Research

Willard R. Schmehl

INTRODUCTION

In an earlier seminar, Clyma (1986) presented a comparative analysis of development concepts, with special reference to their use in improving irrigated agriculture. Later,* Lattimore (1986) followed with a general presentation of several current technology transfer methods, with advantages and disadvantages of each. He noted that technology transfer models are often examples of the top-down approach. Lattimore then proposed the client-oriented "problem-solver" model. Clyma and Lattimore both emphasized the need for researcher interaction with the farmer to understand the production environment to attain success in developing improved technologies that fit into the client's management system. Similar themes were presented by Lenton (1986) and Coward (1986) in their seminars. In each of these seminars, however, considerable emphasis was placed on the diagnosis of the farming system, with appropriate farmer inputs to identify constraints or problems and opportunities for improvement. This type of activity is commonly called a "diagnostic analysis" (DA). The term "descriptive stage" has also been used. Historically, similar types of farm analyses were used in the U.S. as much as 50 years ago, under the term "farm management," although probably lacking the intensive interdisciplinary technical input of current procedures.

*Editors' note: The sequence of the papers presented at the ISARD Invited Seminar Series was different than the sequence in which the papers are arranged in this book.

The diagnostic analysis of farming systems has been conducted in many developing countries, but relatively few have led to successful research programs that have resulted in developing farmer-accepted technologies. For example, in a recent survey of projects in east Africa, Collinson (1982) reported that although many diagnostic analyses of farming systems have been conducted, follow-up research programs have been slow to develop. The competition among institutes and lack of institutional support are primary reasons why on-farm research has progressed so slowly. Also, many station or institute researchers believe on-farm research is not necessary, that demonstrations of station results are quite adequate. Another difficulty is that on-farm research methodologies are not well developed, as are the traditional station research methods. There may also be the assumption that the diagnostic analysis provides solutions to the farmers' problems rather than giving direction to the research. The purpose of this seminar is to emphasize that the diagnostic analysis is only the first step in the development process, to present some general methodologies that are used in the research phase, and then to review the status of on-farm research in irrigated farming systems.

Before proceeding, we need definitions of a few terms that will be used frequently.

System: Any "set of elements or components that are interrelated and interact among themselves. Specification of a system implies a boundary delimiting the system from its environment. Two systems may share a common component or environment, and one system may be a subsystem of another" (Technical Advisory Committee, 1978).

Farming system: "A unique and reasonably stable arrangement of farming enterprises that a household manages according to well-defined practices, in response to the physical, biological, and socioeconomic environments and in accordance with the household's goals, preferences, and resources. These factors combine to influence output and production methods" (Shaner et al., 1982).

Irrigated farming system: A farming system in which the primary source of water for plant growth is surface water or groundwater. A higher level of technology is generally required to manage an irrigated system than a rainfed system.

Technology transfer: A process in which an innovation originating in one institution is adopted elsewhere. It is a planned and rational movement of information and techniques on how to perform some tasks, and includes the act of transmitting ideas and information from one person to another (Lattimore, 1986).

Before proceeding further, I would like to review the essential attributes of a successful "technology transfer model," "development model," and "farming systems model." These models all involve the client-oriented approach to development. I would like to emphasize commonality among models.

Briefly, the client-oriented approach is directed to the development of improved technologies that have a high probability of acceptance by farmers with limited resources. Emphasis is placed on the use of an interdisciplinary team to describe the production environment from the farmer's point of view. This is followed by an analysis of the production system and the identification and ranking of constraints. Research (on-farm trials and studies) is then conducted to evaluate alternative solutions proposed to resolve the identified constraints. The solutions should be feasible within the farmer's available resources, support services and government policy. The research results are evaluated in terms of the biological, physical resource, economic, and sociocultural feasibility for the farming system under investigation. Methods of extending the new or improved technologies are then outlined for the specific farmer groups. The client-oriented approach is proposed as being economically more effective than the top-down approach often used.

Why is the client-oriented approach needed? Generally, the top-down approach, often used to present improved technologies to small farmers, has not been successful. In the top-down approach, technologies are developed in a research institute or on a research station and then, through some diffusion mechanism, are given to the farmers. Many of the proposed changes are rejected by the farmer because the proposed improvements are not profitable, are too risky, or the farmer lacks the resources required to use the improvements. In effect, many proposed technologies have not been suitable because the researchers did not know or fully understand the farmer's production environment.

A number of development specialists have analyzed the various causes for the inability to develop technologies that are acceptable to the farmer. For example, Lowdermilk (1980) gives the following reasons for lack of success:

- o inability of the research team to see the farm as a system
- o the researcher's image of how the system works is inconsistent with how it actually does work
- o lack of interaction and communication among disciplines
- o inability to appreciate the contribution of other disciplines
- o lack of appreciation of the farmer's culture and lack of acceptance of the farmer's input when planning on-farm tests
- o assumption that development problems can be solved by technology without socioeconomic input
- o lack of understanding and sensitivity to cross-cultural differences that exist between researchers and farmers.

In order to resolve some of these difficulties, a number of specialists working in various small farm situations throughout the world have proposed a client-oriented development that has the following concepts in common:

- o a system approach (holistic)
- o an interdisciplinary team is used
- o client oriented (farmer involvement)
- o on-farm research to solve site-specific problems
- o task oriented (as opposed to discipline oriented)
- o iterative and dynamic

When these concepts are followed in developing improved technologies, the technology is more likely to fit the production environment of the farmer, i.e., the climatic and ecological conditions, input and output markets, the structure of the farming community, and farm-household factors.

An Overview of the Client-Oriented Approach to Development

The target area for a project is usually designated by key national or regional decision makers. The target area is a geographic area selected for a project based on the needs of the people living there or to research a specific regional production problem. After the target area is designated, the approach consists of conducting a sequence of activities, beginning with the diagnostic stage, then on to the design, testing, and extension stages as described by Shaner et al. (1982) and shown in Figure 3.1. There is continuous collaboration during each stage with the research station and with extension. Feedback, as new information is obtained, improves the description of the system and facilitates the design of new research. Although the figure shows discrete stages, they overlap considerably, and a field team may have activities in several stages at any one time, depending upon the technologies under study.

Output of the client-oriented research process is the development of technologies that will fit into the farmer's production environment in the research area and then into the target area. Several adjustments in the proposed technology may be required before it is ready for the extension stage. Multilocational testing then is conducted in the target area to determine what adjustments in the proposed technologies may be needed before broad diffusion. This is followed by a pilot program to test the effect of infrastructural and agricultural policy factors on the adoption of the new technology when introduced into an area on a large scale, for example, 100 to 500 ha.

The Diagnostic Stage

After the research area has been selected as a unit representing the conditions typical of the target area, the interdisciplinary team makes a concentrated study of the human and technical aspects of the farming systems in the area. The purpose in this stage is:

- to gain an understanding of the farmers' production environment
- to gain an understanding of those conditions of the system that influence farmers' decisions
- to identify major physical, biological, and socioeconomic constraints in the system that limit changes in the farmers' management practices
- to set priorities for research that appear to offer the greatest potential for developing appropriate interventions (both off-farm and on-farm).

The first activity in the diagnostic stage (Figure 3.2) is to collect available secondary data pertaining to the research area. This may include published materials, local and regional reports, and unpublished local information. After a preliminary analysis of the secondary data, the team obtains primary data by conducting farmer interviews and making both technical and socioeconomic observations of the farming system. The team then analyzes both primary and secondary data, describes the farming systems in the research area in terms of the biophysical and socioeconomic settings, and drafts a conceptual model of the system in the research area. The system is analyzed for problems or constraints and opportunities. Those problems or constraints that appear to be well defined and are believed to have potential solutions can be taken up at the research stage. Poorly defined problems require detailed study in the research stage before deciding how to proceed. The diagnostic team then categorizes the research needs and sets preliminary priorities based on the potential for adoption, as well as on societal and national interests.

The diagnostic stage will generally take four to eight weeks, depending upon factors such as staffing, complexity of the system, the research area, and time of year. Typical methodological procedures to carry out the diagnostic stage have been outlined by Lowdermilk et al. (1981) for irrigated farming systems and by Hildebrand (1981), Collinson (1982), and others (Shaner et al., 1982) for rainfed systems.

The Design Stage

The purpose of the design stage (Figure 3.1) is to use information from the diagnostic stage to design alternative solutions to identified constraints, then plan the research. Output from the design stage is primarily a plan of action for the on-site field team, except where little component technology is available. Then greater emphasis may be placed on experiment station research. It will include a plan of on-farm research, how the research should be conducted and analyzed, supply and equipment needs, logistics, and team management. The plan should be detailed for the immediate cropping season, but general for subsequent seasons. When appropriate, upstream or off-farm research may also be recommended. Upstream research refers to research designed to generate general

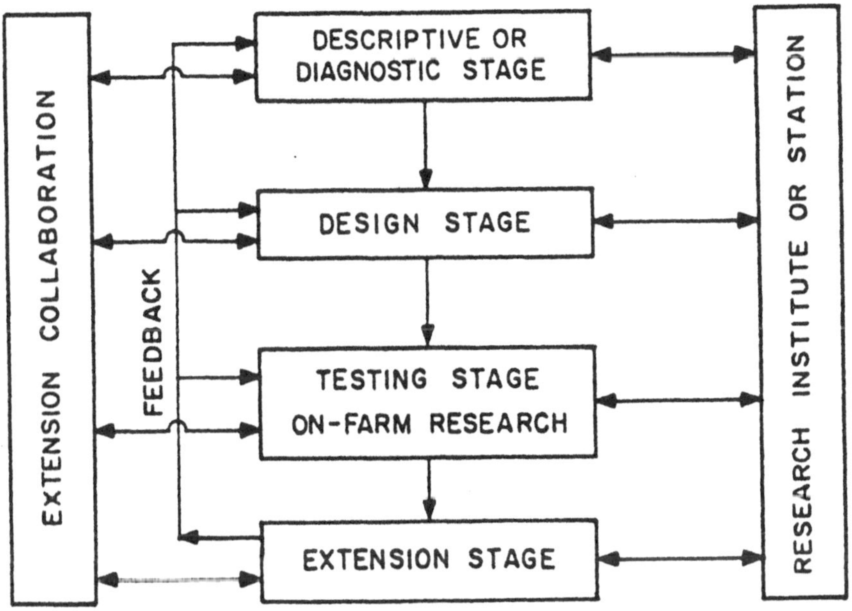

Figure 3.1 Stages in the farming systems process (Shaner et al., 1982)

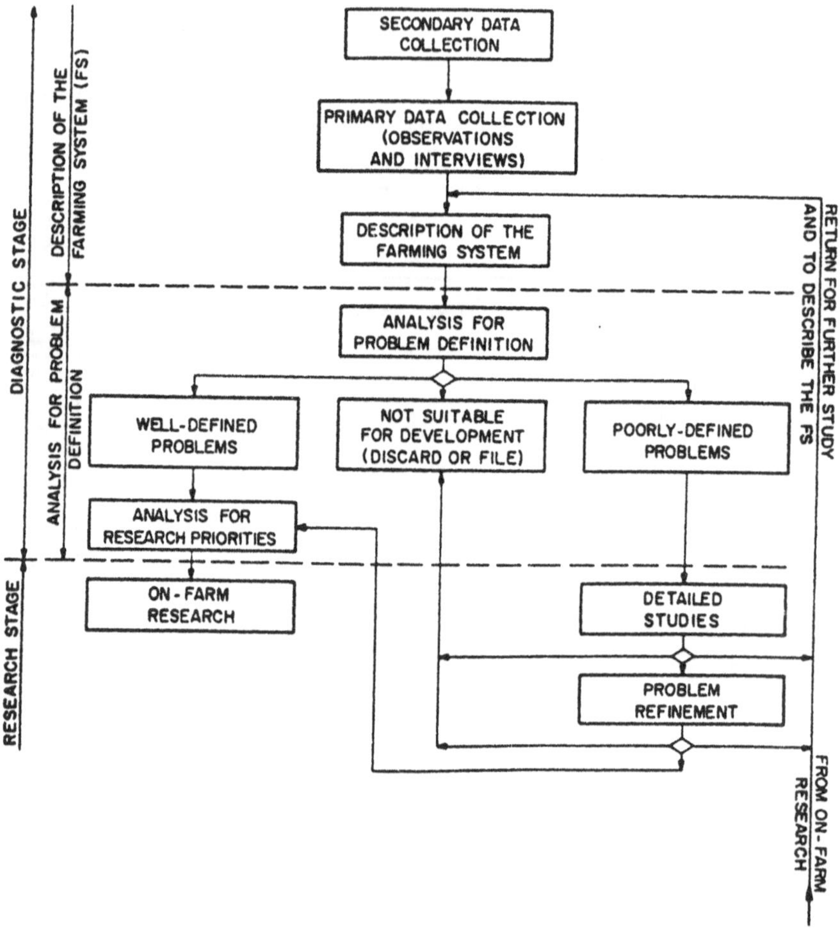

Figure 3.2 Activities in the diagnostic stage

solutions to problems identified in the target area and may be directed to the solution of either technical or socioeconomic concerns. It is conducted at regional, national, or international agricultural research centers, or by regional or national planning groups.

The design stage activity is usually conducted by the design team during a workshop that may take from two weeks to a month. To arrive at a plan for research, the team conducts the following activities:

- o sets the background for design
- o designs potential solutions
- o plans the on-farm studies
- o recommends upstream research.

The first activity in the design stage is to set the background for design. Categories of background information are summarized in Table 3.1. The team then uses this information to design the research potential solutions.

In designing research, the team attempts to develop solutions that will be attractive to the farmer and are potentially feasible within the farmer's constraints. Four design criteria that provide a useful guide in making ex ante evaluations of potentially viable solutions are:

- o biological potential
- o resource availability
- o economic and financial feasibility
- o sociocultural acceptability.

Potential solutions to constraints that, upon analysis, meet these four criteria have a higher probability of acceptance than those, for example, that do not fit into the community norms, or for which required inputs are not available. These criteria also can be used to distinguish technologies that can probably be integrated into the system early, using farmer adaptation testing, from a technology that may require several years' research to develop (technology development).

Potentially acceptable technologies that are proposed for testing will depend upon:

- o the production environment
- o resource requirements of the technology
- o management requirements of the technology

Table 3.1 Setting the background for design

---

a. Information from the diagnostic analysis:

- Description of the farming system
- Preliminary analysis for constraints and opportunities
- Preliminary ranking of research needs

b. Give specific cognizance to (and possible expansion from) the diagnostic analysis report:

- The farm system
- The dominant cropping and livestock patterns
- Current technology level of the farmer
- Available farm resources and farmers' capabilities
- Farmers' perceptions of their constraints
- Constraints in the natural resource environment

c. New information inputs, e.g.:

- Expected socioeconomic-cultural environment
- Available component technologies (from national and regional research stations, etc.)
- Capabilities and resources of the research organizations
- Potential research collaboration

d. Categorize the constraints:

- Those that appear to be fixed, at least over the short term (outside the farmers' management boundaries)
- Those that have the greatest potential for changing within the short term (less than 5 years)
- Those that have the potential for changing only over the long term

---

- o production criteria (yield, labor, monetary resource conservation, effectiveness in use of water)
- o magnitude of effect required for acceptance
- o perceived farmer acceptance
- o effects on family goals
- o perceived societal acceptance.

For example, using the design criteria would predict that a technology that maximizes production but requires extensive or costly inputs would not fit into a farming system that has very limited resources. The introduction of a technology that depends upon new markets will not be successful until such markets are evaluated and assured. Or a technology that is labor intensive at a time when there are more attractive opportunities for transient labor, e.g., for harvesting coffee, will have little possibility of acceptance.

After potentially acceptable solutions to the high priority problems have been proposed for testing, the team then develops the on-farm research program. In selecting hypotheses for testing, the team will probably want to review, again, types of constraints which should be considered fixed (Table 3.1), the several possible types or levels of research activities (Table 3.2), and a research program that has the potential for producing early-on benefits for the farmer, as well as benefits to follow later in the project's life. The early-on benefits are essential for maintaining farmer cooperation during the life of the project as, step by step, improved farm management practices are developed. Generally, this involves placing emphasis, early in the project, on simple technologies that are effective but require fewer resources and develop, over the longer term, progressively more complex packages of practices that probably will require more inputs and increased management skills.

After the hypotheses for testing are selected, the design team plans each research activity for the field (on-site) team during the upcoming season. The plan should outline clearly how the field team will carry out each research activity, for example:

- o the objective of each task or activity
- o research to be conducted under each task
- o how the research is to be conducted
- o type of data to collect and how to collect it

Table 3.2 Types of farming systems research activities

---

a. Farmer adaptation testing (on-farm trials/tests)

- o The technology is judged ready to go; maybe all that is needed is to get it into farmers' hands; monitoring of the socioeconomic as well as biophysical information is required to make a productive evaluation; desired output to extension within one to two years.
- o A recommended technology has been unsuccessful; the team believes that it can be easily adjusted to the farmers' environment in the traget area; output to extension may be 3 years as a minimum.
- o Farmer-oriented, site-specific rate, variety, cultural practices, irrigation and other studies of the technologies with a high probability of acceptance; output may be either short or long term.

b. Technology adaptation:

Research designed to adapt currently known technology to the research area; for example, component technologies that have been developed and tested only on research stations need adjustment to fit the farmers' conditions. This research may require several years of adaptation trials, and tends, at first, to be more discipline oriented than farmer adaptation testing.

c. Technology development:

These are studies to search for types of technology not currently deliverable; the technology could be unique or innovative for the biophysical or socio-economic environment of the research site; it may also be used for exploratory technologies using *ex post* analysis; this research generally will be farther from development than technology adaptation, and some of the output may not be suitable for development in the current target area.

Table 3.2 Types of farming systems research activities (continued)

---

d. A better description of the farming system:

This is an extension of the "diagnostic analysis" into the "on-farm research" stage; it may also complement the on-farm research; may be short term, e.g., one or two months, or may be continuing throughout one or more years.

e. Recommendations for upstream research:

Recommendations are submitted to national planning organizations and to national or regional research organizations for research on pricing policies, infrastructural changes, irrigation project management, new component technologies, etc.

- how to summarize and evaluate accomplishments in preparation for the next design/planning workshop
- resource requirements of each activity, including an estimated budget.

Several resources that provide guidelines for designing and planning the detailed on-farm activities are Podmore and Eynon (1983), Lowdermilk et al. (1981), Sparling et al. (1980), Shaner et al. (1982), Zandstra et al. (1981), and Byerlee et al. (1980).

The Testing Stage

On-farm research is conducted to test the proposed technologies, to develop site-specific technologies, and to continue to develop an understanding of how the farmer's system operates (Table 3.2). This is called the testing stage, and plans for the on-site research were developed in the design workshop.

Research methodologies used in on-farm research are based on the traditional discipline-oriented procedures but modified for use in the more heterogeneous farming systems environment. The on-farm research is designed to solve problems of the production environment and requires more extensive interaction among disciplines and farmer, both in conducting the research and in the analysis.

The on-farm research conducted in an irrigated farming system is generally classified as biological testing, economic and sociocultural research, irrigation management research, and analysis of the physical resource base (Table 3.3). Although it may first appear that the research shown in the table is quite discipline oriented, one should recall that the proposed research is the output of an interdisciplinary activity by the design team. Some research may be quite discipline oriented, but the decision to conduct such research was made by the design team to assist in the support of a project task and was not an independent decision of a given discipline. Also, even though biological research by designation may appear discipline oriented, for example, cropping systems farmer-managed tests, it is quite interdisciplinary. As noted in Table 3.4, farmer-managed tests require input from all disciplines, from planning to data-taking through analysis. Interdisciplinarity is required for a successful

Table 3.3 Classification of on-farm research

a. Biological testing

- o Researcher-managed trials
- o Superimposed trials
- o Farmer-managed tests

b. Economic on-farm research

- o Farm records
- o Farm surveys
- o Analysis of biophysical on-farm tests
- o Monitoring farmer activities
- o Enterprise analyses
- o Markets and marketing studies

c. Irrigation management research

- o Monitoring performance of the system
- o Monitoring farmer irrigation practices
- o Testing proposed modifications, e.g., the delivery system, scheduling, method of irrigation, leveling
- o Evaluation of prior changes in the irrigation system
- o Conjunctive use of water

d. Socialcultural on-farm research

- o Surveys of the farmers' sociocultural environment
- o Participant observation studies
- o Monitoring community and farmer relationships
- o Farmer decision-making processes

e. Analysis of the physical resource base

- o Climatic monitoring
- o Soil surveys
- o Land class surveys
- o Soil erosion evaluations

Table 3.4 Cropping systems on-farm research methodologies (Shaner et al., 1982)

Characteristics of trials and tests*	Researcher-managed trial
Plot size	Generally small--on the order of 75 square meters
Number of treatments	5-20
Number of replications per field	1-5**
Total replications across farms, per land type	4-5**
Field design	Completely randomized, randomized complete block, randomized incomplete block, split block
Sensitivity to treatment differences	Medium to high
Types of data collected	Physical and biological

*These characteristics will vary with experimental objectives, type of treatment, farm size, and cooperating farmers.

**Usually all replications will be placed on one farm field to give the complete experiment. However, if the field is small and only one or two replications on a field are possible, additional replicates will be placed on other fields of the same land type to give a total of four or five replications for the experiment.

Table 3.4 Cropping systems on-farm research methodologies (continued)

Superimposed trial	Farmer-managed test
Both large and small	Generally large--on the order of 1,000 square meters
4-6	2-4
1-2	1-2
4-10	4-25
Completely randomized, randomized complete block, randomized incomplete block	Competely randomized, randomized incomplete block, paired treatments
Medium to high	Low to medium
Predominantly physical and biological, but some socioeconomic	Physical, biological, socioeconomic

evaluation of how the proposed technology can be integrated into the farmer's management system. Farmer-managed tests are classified as biological research because the biological discipline takes the lead, yet the effort is interdisciplinary. There are similar examples for other types of research.

On-farm research methodologies are still being developed. Those for biological research (Table 3.3, Part a.) with field crops are best developed. The methodologies used for cropping systems research are described in Table 3.4 (Shaner et al., 1982). Economic methods for on-farm research methods are probably the next best developed. Much of the difficulty in conducting on-farm research has risen because researchers in the academic climate are not inclined to define as "research" the iterative-integrative process required to solve site-specific agricultural problems. Consequently, this type of researcher finds it difficult to conduct research in the highly variable production environment of the farmer, and at suboptimum levels of nonexperimental variables. The same philosophy probably is also a reason for the relatively greater emphasis on the diagnostic stage, where methodologies are better defined, and why output from a diagnostic analysis is often considered a solution. This philosophy has also delayed the development of on-site research methodologies that are appropriate to solving problems at the farmer level. The livestock task group of the Farming Systems Support Project (FSSP) has recognized the lack of appropriate methodologies to conduct on-farm research in mixed systems involving livestock. An FSSP task group is planning a series of workshops, with input from those actively engaged in livestock research in small farmer situations, to review current methodologies and to recommend how to develop improved on-farm research methods involving livestock in the small farm environment. The workshops will be interdisciplinary, with input from social, as well as biological, scientists. I believe a similar approach is needed to develop the on-farm irrigation research methods required to identify the appropriate technologies needed to improve the management of farm irrigation water. Too often the top-down approach is used by scientists who propose interventions they believe will improve the irrigation system, but who are not aware that the proposed interventions may not coincide with other higher-priority goals of the farmer.

The research listed in Table 3.3 is not intended to be all inclusive, nor will the team conduct, on a given project, all types of research given in the table. Research on a given project will depend upon the priorities for the system under study, as well as upon resources available to the team.

When conducting on-farm research, the field team will want to consider (Shaner et al., 1982, Chapter 7, Part 1):

- o farmer, field, hydrologic unit selection
- o incentives and agreements
- o farmer/researcher relationships
- o field test design
- o monitoring seasonal progress
- o methods to measure and evaluate results.

The team should locate and select farmers and conditions that represent project goals and objectives. In addition, the farmers should be willing to cooperate in the tests, and also have the capability to cooperate in the type of test proposed. As a general rule, incentives or some form of encouragement should not be provided for farmer-managed tests, since these tests are conducted to determine how the farmers will react to the new technologies. There may be situations, however, where the project does compensate the farmer; for example, if the technology inadvertently reduces yields below those normally obtained by the farmer, or where the team removes large biological samples for analysis, or possibly in the exploratory stages of technology development.

Farmer-researcher relationships are very important in field testing where the researcher wants to observe the farmer's response to a technology. The researcher may be so convinced the proposed technology is appropriate that he may try to induce the farmer to carry out a certain practice needed to improve production, or the researcher may do it himself, if the farmer is not so inclined. In either case, the proposed technology is not properly evaluated and would not represent the farmer's true situation.

Progress of on-farm experiments, such as the socio-cultural conditions, labor requirements, and costs of inputs of the proposed technology, should be monitored for comparison with the farmer's usual practices. There are various other considerations discussed in detail by Shaner et al. (1982).

The results of the on-farm studies are analyzed with respect to the objectives of each of the activities planned in the design stage. As noted previously (Table 3.2), objectives of some of the on-farm studies include activities such as technology adaptation, technology development, and further description of the farming system. Analysis of the farmer adaptation studies will be of central interest, because these results are the best guide to identifying whether a proposed technology will be successful or if more adjustment for the technology is required. In evaluating the results, the team again considers the same four criteria used for the *ex ante* evaluation--biological feasibility, resource accessibility, economic/ financial feasibility, and sociocultural acceptability. The technologies that appear to meet these criteria are carried to the extension stage (Figure 3.1) for multilocational testing. Proposed technologies that do not meet these criteria are returned to the design stage for reevaluation.

Irrigation Management Research

Early (1983) defined an irrigation system as the entire set of interacting social, economic, biological, and physical factors, objectives, and entities, from the source of water through the conveyances to the farm and the land that is irrigated, including the drainage network that removes water excesses from the boundary of the irrigation service area. One of the subsets of the irrigation system is the farm irrigation system, which, in turn, is a subset of the farming system. The fundamental objective of an irrigation system is to increase agricultural output through improved management of water. As Fairchild and Nobe (1986) emphasize, success in management should be judged by output from the system, rather than by inputs to the system. They refer to the concept as "management by results" (MBR). In the MBR approach, the farmer is the "transformer" of inputs to outputs. Outputs from the clients involve crop yields, cropping intensity, net farm income, social well-being, etc. The authors propose the structure of an improved managerial organization in which feedback is used to modify the implementation of inputs by inserting a performance monitoring/ evaluation link in the management system (Figure 3.3). The management system refers to the organization that (1) transforms general policy goals into specific objectives defined as desired project inputs and (2) designs, prioritizes, and schedules program elements to produce output objectives (Fairchild and Nobe, 1986).

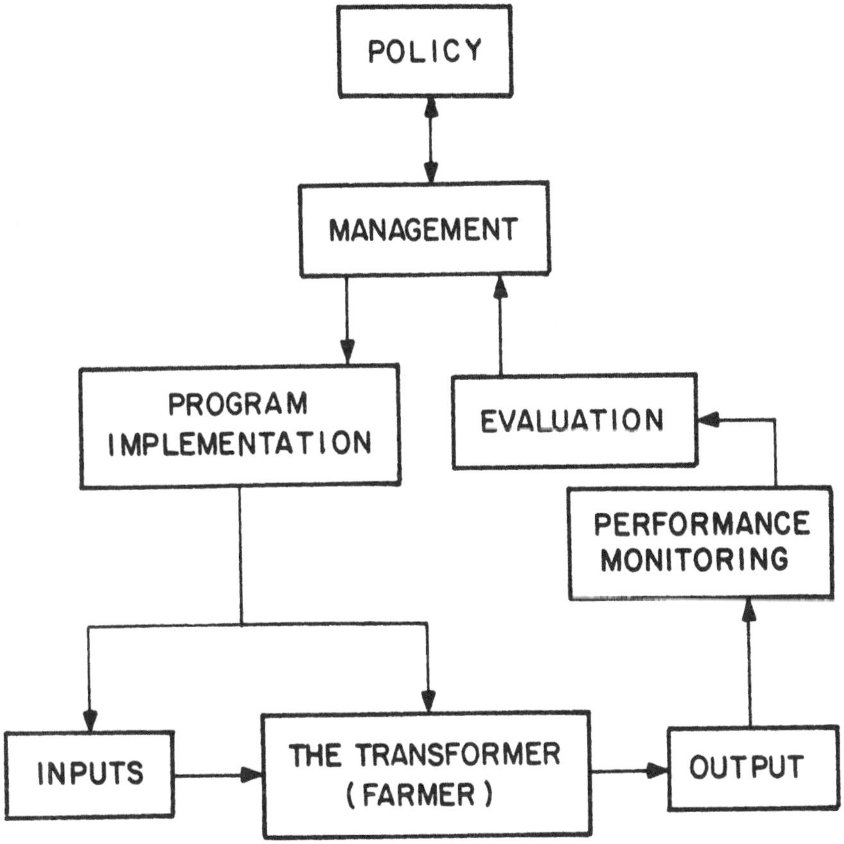

Figure 3.3 Structure of the managerial system (Fairchild and Nobe, 1984)

The intersection of the irrigation system, the farming system, and the management system can be illustrated with a Venn diagram (Figure 3.4). The dark area, common to all three systems, represents the irrigation subset in each of the systems. The goal of the on-farm research in an irrigated farming system is to develop technologies that will increase agricultural output through improved irrigation management. But as the diagram indicates, proposed changes in the irrigation or management systems must be appropriate for the farming system as a whole. For an irrigation project to be successful, the three systems must be integrated at the subset level to attain project objectives. As Fairchild and Nobe point out, the "transformer" of inputs to outputs is the farmer. Even though the irrigation system, and also the management system, may be in place and ready to function, unless the transformer is functioning properly, the project will not function satisfactorily. The purpose of the on-farm research is to develop improved technologies that will help the "transformer" perform more effectively. Performance monitoring was proposed as an essential link in MBR. Fairchild and Nobe note that performance monitoring of an irrigation project includes monitoring of all enterprises in the farming system, and is not limited to water concerns. Many irrigation projects have, by the MBR concept, been unsuccessful because emphasis was placed on improving inputs, with the supposition that improved outputs would result, that is, improving some aspect of the irrigation system input, a common procedure, does not necessarily improve the transformer's operational system, i.e., the farming system.

What are the goals of irrigation management research (IMR)? It is apparent that irrigation management is complex and involves an interaction among three systems--the irrigation, management, and farming systems. Projects to improve irrigation management are frequently mandated to improve some physical input by the irrigation system, with the assumption that production output will also be improved. Often, physical limitations observed in the irrigated farm or the delivery system are assumed to have "obvious" solutions. But when the system is improved by inserting the "obvious" solution, the intervention may have little effect on output of the system. The "obvious" improvement in the farm irrigation system may have been suggested by experiment station research or by a diagnostic analysis. Either would be an example of the top-down

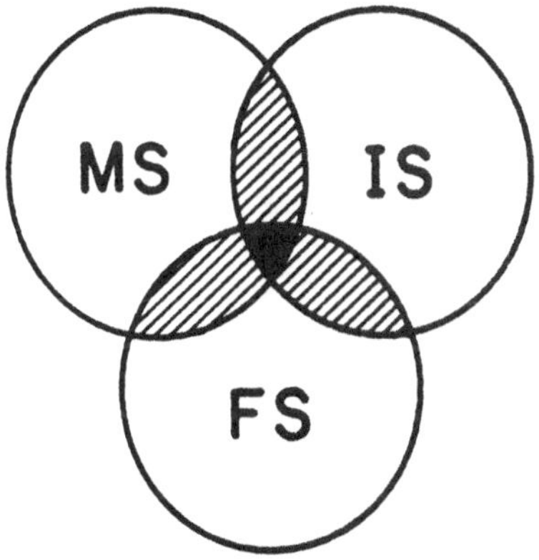

Figure 3.4 Intersection of the management (MS), irrigation (IS), and farming systems (FS)

approach which has been shown many times to be unsuccessful. The lack of an understanding of farmers' problems is generally the reason for poor success when proposing interventions. For example, the farmer may have other problems that so dominate management practices that improving an irrigation structure or some part of the water delivery system or land leveling has low priority. If the farmer's major problem is attacked and solved first, the farmer may then be interested in improvements in the irrigation system. By this concept, the goal of IMR is to develop, by using appropriate procedures, improved irrigation management practices that will fit the farmer's environment.

Another difficulty arising from a mandate to improve management of the irrigation system is that so much emphasis may be placed on identifying limitations and problems associated with water that other limiting factors in the production environment are not properly evaluated. The usual view is that, when conducting research to improve an irrigation system, other farmer management practices are assumed to be optimized. This concept requires the package approach to development. The difficulty when proposing a set of practices for the package approach is that the small farmer has limited resources, and those resources which are available are allocated to practices that give the most return, as perceived by the farmer, with the least risk. The incremental or single-factor approach is proposed as the more effective, even for irrigation improvement research. Using this procedure, the more significantly limiting factors, as viewed by the farmer, are researched first. Their solution is used as the base for building, incrementally, to improvements in the irrigation system. This is not to imply that irrigation improvements may not come early on, but in many situations, other less complex improvements may be instituted first. The single-factor approach fits better into improving the small farmer's management system and generally is a more effective procedure for introducing interventions than the package approach often proposed.

Performance monitoring, as proposed by Fairchild and Nobe, is an excellent research tool for irrigation management research. A team should monitor not only agricultural production itself, but also those factors in the farming system that affect the level of production. Performance monitoring gives a better description of the system, and with feedback to research design, the technologies being tested are revised and tuned to the farmer's production environment.

One of the limitations in developing improvements in an irrigation system generally is its high cost, a cost usually more than a small farmer can afford. This implies that off-farm resources will be required, and unless the design team is confident of such resources, research should emphasize low-input irrigation technologies. Nor is it appropriate to provide costly incentives to the farmers, other than for early-on exploratory research, if the resources required for implementation are beyond those available to the small farmer.

As Clyma (1985) states in reviewing the farming system approach to development, "The structure of the on-farm research in farming systems is much more substantiative and definite," and implies the need to improve on-farm irrigation research methodologies. In reviewing the literature for this seminar, this statement appears to be quite correct for IMR models. Possibly because of the complexity of the irrigated farming system, most of the effort in IMR models has been to develop procedures for conducting the diagnostic analysis, with little effort on developing methodologies for the research phase. Current methodologies used in the farming systems approach as described by Shaner et al. (1982) were developed primarily for rainfed systems. The general procedures can, however, be adapted, after some adjustment to on-farm research in irrigated systems. New innovative methods for conducting irrigation on-farm research will probably be required to test proposed irrigation-oriented technologies coming from the design stage. The conclusion of this paper is that, in the future, greater emphasis should be placed on developing the on-farm irrigation research methodologies that are needed to progress more effectively into the research phase. This conclusion is not unique. It has been one of the objectives of the Colorado State University Water Management Synthesis Project, but progress has been slow because of limited funding.

REFERENCES

Byerlee, D., J. P. Collinson, R. K. Perrin, D. L. Winkelman, S. Biggs, E. R. Moscardi, J. C. Martinez, L. Harrington, and A. Benjamin. "Planning Technologies Appropriate to Farmers: Concepts and Procedures," CIMMYT, El Batan, Mexico, 1980.

Clyma, W. "Integrated Agriculture: A Comparative Analysis of Development Concepts," Chapter 1, Irrigation

Management in Developing Countries, K. C. Nobe and R. K. Sampath, eds. Boulder, Colorado: Westview Press, 1986.

Collinson, M. P. "Farming Systems Research in Eastern Africa: The Experience of CIMMYT and Some National Agricultural Research Services, 1976-81," Michigan State University International Development Paper No. 3, Department of Agricultural Economics, Michigan State University, East Lansing, Michigan, 1982.

Coward, E. W. "State and Locality in Asian Irrigation Development," Chapter 16, Irrigation Management in Developing Countries, K. C. Nobe and R. K. Sampath, eds. Boulder, Colorado: Westview Press, 1986.

Early, A. "Conceptual Framework for Irrigation System Management," unpublished manuscript, Agricultural and Chemical Engineering, Colorado State University, Fort Collins, Colorado, 1983.

Fairchild, W., and K. C. Nobe. "Improving Management of Irrigation Projects in Developing Countries: Translating Theory Into Practice," Chapter 12, Irrigation Management in Developing Countries, K. C. Nobe and R. K. Sampath, eds. Boulder, Colorado: Westview Press, 1986.

Hildebrand, P. "Combining Disciplines in Rapid Appraisal: The Sondeo Approach," Agricultural Administration, 8:423-432, 1981.

Lattimore, D. "Water Management: Problems and Potential for Communications in Technology Transfer," Chapter 17, Irrigation Management in Developing Countries, K. C. Nobe and R. K. Sampath, eds. Boulder, Colorado: Westview Press, 1986.

Lenton, R. "On the Development and Use of Improved Methodologies for Irrigation Management," Chapter 2, Irrigation Management in Developing Countries, K. C. Nobe and R. K. Sampath, eds. Boulder, Colorado: Westview Press, 1986.

Lowdermilk, M. K., W. Clyma, M. I. Haider, D. L. Lattimore, J. J. Layton, D. W. Lybecker, A. G. Madsen, L. J. Nelson, D. J. Redgrave, F. A. Santopolo, and D. K. Sunada. "Diagnostic Analysis of Farm Irrigation Systems," Vol. 1, Water Management Synthesis Project, Colorado State University, Fort Collins, Colorado, 1981.

Lowdermilk, M. K., W. T. Franklin, J. J. Layton, G. E. Radosevich, G. V. Skogerboe, E. W. Sparling, and W. G. Stewart. "Problem Identification Manual," Water

Management Technical Report 65B, Water Management Resource Project, Engineering Resource Center, Colorado State University, Fort Collins, Colorado, 1980.

Podmore, C. A., and D. G. Eynon, eds. "Diagnosis of Irrigation Systems," Vol. 2, Evaluation Techniques, Water Management Synthesis Project, Colorado State University, Fort Collins, Colorado, 1983.

Shaner, W., P. F. Philipp, and W. R. Schmehl. Farming Systems Research and Development: Guidelines for Developing Countries. Boulder, Colorado: Westview Press, 1982.

Sparling, E. W., W. D. Kemper, J. E. Hautaluoma, M. K. Lowdermilk, G. V. Skogerboe, and W. G. Stewart. "Development of Solutions Manual," Water Management Technical Report 65C. Water Management Research Project, Colorado State University, Fort Collins, Colorado, 1980.

Technical Advisory Committee (TAC) (Review Team of the Consultative Group on International Agricultural Research). "Farming Systems Research at the International Agricultural Research Center," The World Bank, Washington, D.C., 1978.

Zandstra, H. G. E. C. Price, J. A. Litsinger, and R. A. Morris. "A Methodology for On-Farm Cropping Systems Research," International Rice Research Institute, Los Banos, Philippines, 1981.

4
Management of Gravity Flow Irrigation Systems

J. Mohan Reddy

INTRODUCTION

The demand for water is increasing because of population increase, improvement of living standards, and industrial development. The water supplies are scarce. The rainfall is erratic. To cope with the increasing demand for food, conservation and optimum use of the available water for crop production is of paramount importance. To this end, several irrigation projects have been, and are being, constructed to increase irrigated area, with the potential of providing an assured water supply so farmers can obtain higher crop yields.

Unfortunately, on several irrigation projects around the world, the potential for increased areas and yields has remained just potential, and the projects are plagued with large gaps between the potential and actual area irrigated and between the potential and actual yield levels obtained. Two explanations are possible. First, the potential levels (area and yield) are so ambitious that, in reality, they cannot be achieved. Second, the vital elements that affect system performance have been inadequately provided. Quite often, it is the latter that is responsible for the low level of performance achieved on the irrigation projects.

Gravity flow is the most widely used irrigation method in the world. Data from several projects around the world suggest that the common problems facing gravity flow irrigation projects are excessive water losses, low crop yields, differing amounts of water received and crop yields at the head and tail ends of the system, unreliable water supply, waterlogging, and salinity. However, the causes of the above problems may differ from place to place. Whatever the causes, the effect is felt by the farmer. The data also suggest that the problems can be traced to either lack of or inadequate water control planning, design, and operation and management of the irrigation systems. This paper deals with the water control and management aspects of gravity flow irrigation systems.

TYPES OF GRAVITY FLOW IRRIGATION SYSTEMS

Gravity flow irrigation systems can be classified into:

- individually owned irrigation systems
- community-managed irrigation systems
- large-scale, governmentally operated irrigation systems.

Individually Owned Systems

In individually owned irrigation systems, performance on the farm depends upon the farmer's ability to manipulate (or control) the water on the farm. The supply of water is usually from a well. There is usually no interference from other farmers, and the farmer-owner can irrigate at his convenience.

Community-Managed Systems

In community irrigation systems, performance on the farm depends upon the farmer's ability to control the water on his farm and to share water with the other farmers in the command. Overall project performance depends upon the performance on the individual fields and the performance of the water distribution system. The farmer depends upon the other farmers in that command area for his water.

Large-Scale Irrigation Systems

A large-scale gravity flow system is made up of three different components, as shown in Figure 4.1: main system, unit command area (similar to community-managed

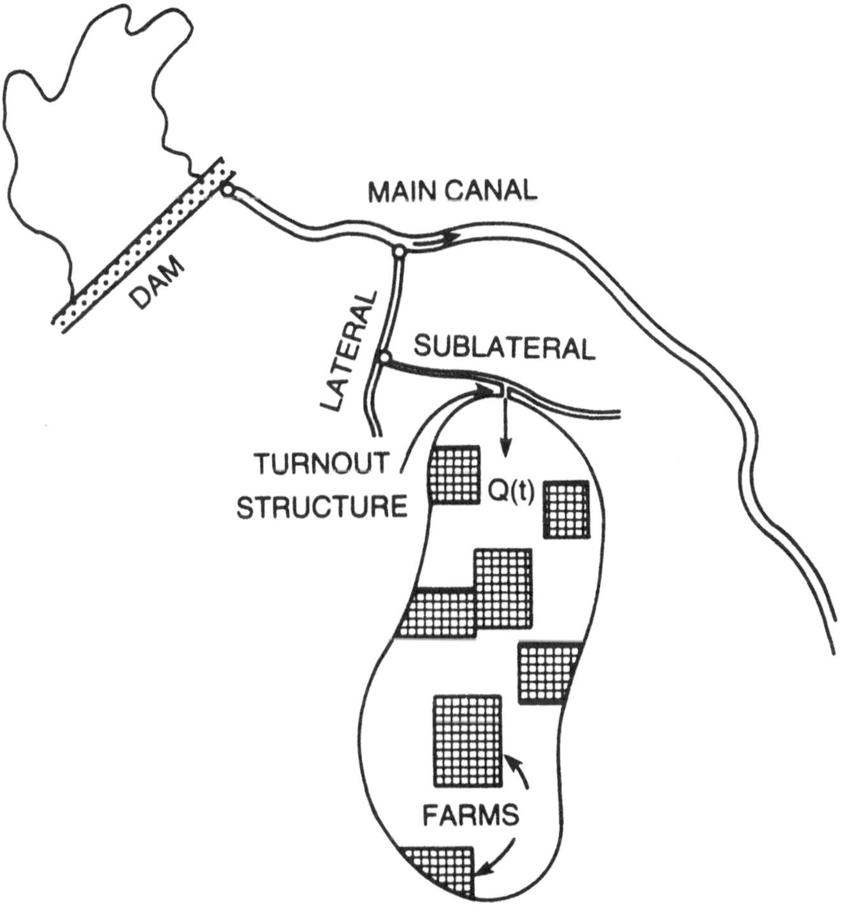

Figure 4.1 Layout of a typical irrigation system

system), and farm system (individual ownership), each with distinct properties. They are discussed below.

Main System extends from the dam to the various outlets which supply water to a group of farmers. It draws large flow rates (50 to 20,000 cfs) from the reservoir and distributes the water to the outlets. The length of canals varies from 10-200 miles long, excluding the distribution network, depending upon the size of the project area and topography. Hence, the travel times are significant. The irrigation bureaucracy operates the system.

Unit Command Area refers to the piece of land commanded by an outlet. Usually, a group of farmers is supposed to share the water available at the outlet among themselves and apply the water to their fields. The water supply at the outlet depends upon the operation of the main system. The group of farmers does not have any control of the flow rate, duration, or timing of water received at the outlet.

Farm System refers to the individual farm in the command of an outlet. The farmer operates and manages the farm system. System performance is dictated by the flow rate characteristics at the farm and the farmer's ability to manage the available water supply at the farm. The flow rate and its timing and duration at an individual farm are influenced by the operation of the main system and the system below the outlet.

The operation of large-scale irrigation systems is complex because of the diversity in the crops grown and soil types. The performance depends upon the integrated effort of the irrigation bureaucracy and the farmers in the command area, and the rules and tools (structural control) provided for operation of the irrigation system. This paper deals with the structural control aspects of large-scale gravity flow irrigation systems.

MAIN SYSTEM VERSUS ON-FARM EMPHASIS

The overall performance of any irrigation system depends upon the performance of individual components (fields, unit command area, and the main system). The relative performance of one component may be lower or higher than the performance of the other components. Wade and Chambers (1980) state that main system management is "canal irrigation's blind spot," and large increases in production may be achieved, with equity benefits to the tail enders as well as head enders, by managing the main system. In addition, any attempts to improve management below the turnout, including farmer involvement (Jayaraman, 1982), will come to naught if no effort is made elsewhere on the system to provide a reliable water supply to meet farmer needs. Although this may be true on some projects, in the author's view, the main system is managed better than the on-farm system in general.

There is ample evidence to support that the irrigation network below the outlet (unit command area) has received little attention (Kathpalia, 1982), and performance of the systems is much below the expected level. The area below the turnout has been and still is, largely, a "no man's land." Only recently have there been any efforts to improve the system performance below the outlet. While the government's responsibility and control ends at the outlet, the farmer thinks that the government should design, construct, operate, and maintain the distribution system below the outlet, similarly, the on-farm application system. Currently, these systems are not designed but only constructed and operated by the farmer. And thus, the performance of these systems has not been satisfactory. In Pakistan, for example, the average application efficiency of traditionally leveled fields was only about 35 percent (Clyma and Ali, 1977).

From the above, it is clear that the performance of all three components of the irrigation system has been unsatisfactory and needs improvement. The relative magnitude of performance of the three components varies from irrigation project to irrigation project, and for increased agricultural production, good water control and management of all three components of the irrigation system is necessary.

WATER CONTROL

The performance of an irrigation project can be measured in terms of technical efficiency with which the water is provided to the crop root zones, total agricultural production from the project, and equity. Performance of an irrigated agricultural production system is dependent upon technical, economic, and institutional and organizational support provided to the system, under a given set of constraints. The relationship between the agricultural, economic, technical, and organizational aspects is presented in Figure 4.2.

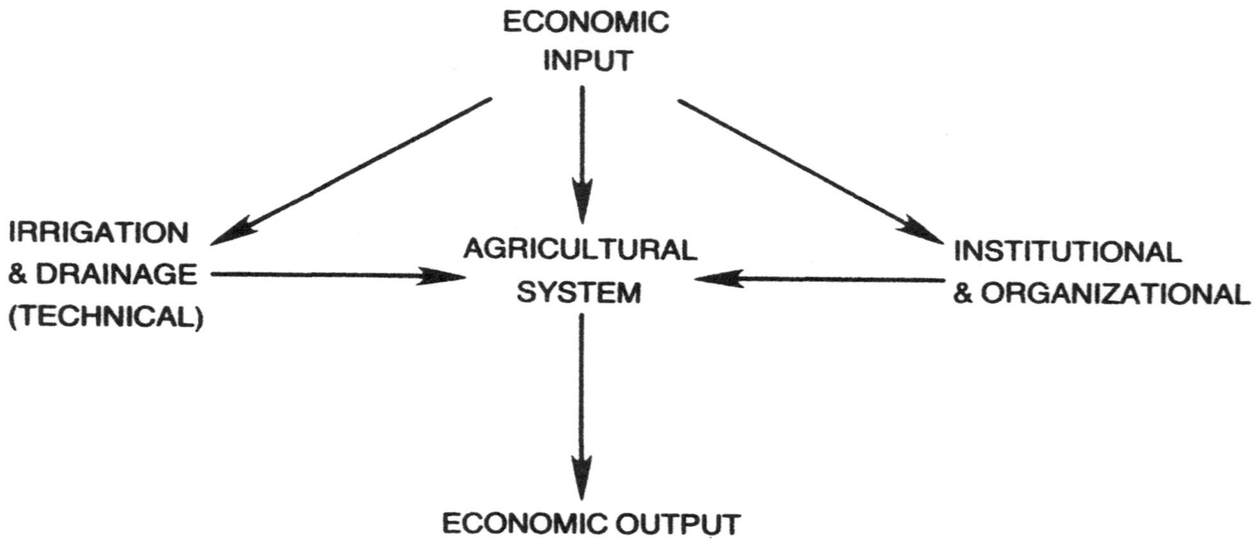

Figure 4.2 Relationships between the agricultural, economic, technical , and institutional and organizational aspects

Water control refers to the ability of the system to distribute, apply, or remove water at the right time, in the right quantity, and at the right place. Levine (1977) found that increasd control (in this case structural) would result in reduced system water requirements (Figure 4.3). This means a close match between the demand and supply. Increased water control should also result in increased reliability and equity in water distribution. However, increased water control, in general, would also require increased structural and/or management control, resulting in additional expenditures. Hence, a balance must be reached between the cost of improving water control and the additional benefits that accrue from it.

Water Control Objectives

The main objectives of water control in an irrigation project (Lowdermilk, 1981) are: reliability (temporal parameter), adequacy (volume balance, including seepage, operational, and application losses), and equity (spatial parameter).

Reliability. A reliable supply of water is crucial to successful crop production. Reliability is defined as:

$$\text{Reliability} = \frac{\text{Actual value of the given parameter}}{\text{Design value of the given parameter}}$$

The parameters of interest in irrigation distribution are the flow rate, the time of arrival, and the duration of supply. If the design values of the system are modified, the new values, rather than the original ones, must be used to estimate the reliability of the system.

Adequacy. Irrigation projects are designed to meet certain levels of irrigation water requirements. For example, many projects are supplied to meet the peak demand of the crop. The water is usually designed at the peak rate throughout the season. However, depending upon operation of the system, the flow rate delivered at any given point may or may not equal the design value. Adequacy measures the variation of the flow rate/duration around the design specifications at a given point in the system. Adequacy is defined as:

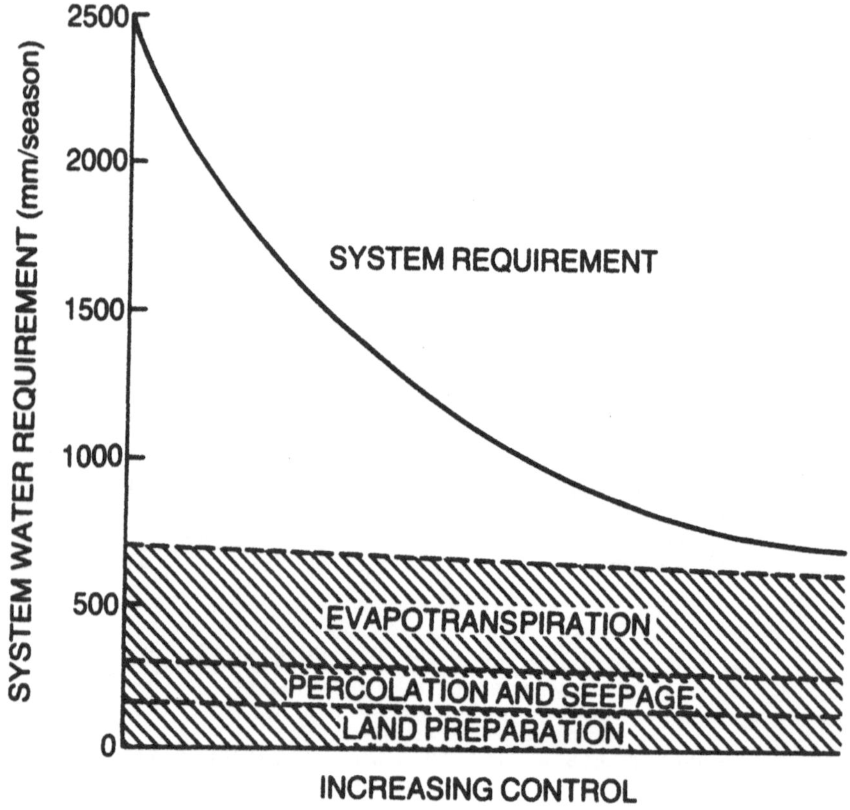

Figure 4.3 The irrigation water requirement for lowland rice as affected by the level of control inputs (Levine, 1977)

$$\text{Adequacy} = \frac{\text{Actual value of the given parameter}}{\text{Required volume of the given parameter}}$$

A value of one for the adequacy parameter is desirable. If the value is higher than one, it indicates the amount of water wasted. A value of less than one indicates water deficiency. However, if the original plan was to provide only a certain fraction of the requirement, then as long as the rate is equal to the specified fraction, the system meets the water control objective as far as adequacy is concerned. The parameters of interest here are the flow rate, the duration, and the total volume of water received (discharged) at (from) a given point in the system.

Equity. Equity refers to the spatial distribution of the above two parameters--reliability and adequacy. Whatever the basis for equity--economic, crop variety, area, soil type, water rights, or a combination of the above--equity must be incorporated into the system design and operation. A single value for adequacy throughout the project indicates that the supply is equitable; otherwise, it is inequitable. The value of adequacy can be less than one and still satisfy the equity criterion.

Reliability, adequacy, and equity are the objectives of water control for the water distribution system. A similar set of water control objectives, not elaborated here, can be defined for the water application and the water removal systems.

Prerequisites for Water Control

To have water control, the irrigation system must be planned, designed, operated, and maintained properly (Clyma and Sritharan, 1984). If any one of the above is lacking, it is difficult to achieve adequate control of water. The following discussion elaborates these aspects.

Planning. Planning should be the first phase of any action, for either designing a new project or improving an existing project. Planning starts by pooling pertinent, basic information--soils, topography, climatic conditions, hydrology, and social structure. Decisions are made regarding the layout of the main canals and distributaries; area to be covered; irrigation delivery schedule (continuous, rotational, or on-demand); extensive or intensive irrigation; equity criteria; and mode of operation (manual or automatic). Quite often, the basic information is not available, and the plan is based upon some assumptions which may not represent the site conditions. All planning decisions influence system performance.

Within the unit command area, planning refers to making decisions regarding size of the command, irrigation delivery schedule, and layout of the watercourses and drains. The farmer's concerns must be considered. As mentioned before, little planning is done below the outlet. It is usually left to the farmers to plan the distribution system.

The crops to be grown, the type of irrigation system to use (border, furrow, basin), direction of irrigation, deficit or full irrigation, the frequency of irrigation (high or low), and the layout of the system are points that must be decided during the planning stage at the farm level. But in reality, little planning takes place at the farm level except for the type of crops to grow. This is probably because farmers lack knowledge of the different types of irrigation systems and their characteristics.

Design. Based upon the decisions made during planning and the soils, topography, crop-water requirements, etc., the design of an irrigation system specifies the values for the appropriate parameters: the required flow rate; depth of flow and free-board; channel cross section and slope; location and size of regulating, controlling, and measuring structures; and size of the turnouts and outlets on the distribution network both below and above the outlet structures. In addition, design also specifies the flow rate, frequency, and duration at each control point in the distribution system.

Data from several irrigation projects suggest that, in some cases, the slopes are erosive, the location of the control structures is inappropriate, and no erosion protection structures have been provided. This resulted in systems that could not deliver water according to the requirements of plan.

Design at the field level specifies the appropriate flow rate, bund height, size of the field (size of a border, etc.), and time of irrigation. The number of irrigations required per season is also estimated. Based upon the cost of water and labor, optimal design of irrigation systems--based upon either minimal cost or maximum profit --can also be accomplished.

Generally, the concept of design on farms is nonexistent. The irrigation systems are constructed and operated. The most common method of irrigation is wild flooding, which is the most inefficient method of surface irrigation. Often, the topography is uneven, fields are irrigated on reverse slope, and flow rates do not match the field size or the soil type, resulting in nonuniform distribution (over- and under-irrigation in the same field) of the applied water (Tyagi and Narayan, 1983). The application efficiency values assumed in the design of the main system are very high for the management level that exists at the farm.

Designs within the distribution system level are generally better than on the farm. However, a lack of control structures at appropriate locations, erosive slopes, and greater seepage than originally assumed have resulted in reduced system performance.

Operation. Operation refers to manipulating the structures that convey, distribute, and apply irrigation water according to the design specifications. On large irrigation systems, travel times are significant. Therefore, changes in flow rate in the system are bound to cause delays. To deliver the right quantity of water at the right time, travel times must be included in the operational plans. Also important is that the delivery schedules or changes in the delivery schedules or flow rates must be communicated to the farmer and the personnel operating the irrigation system. Too often, communication and travel times are not considered in the operation of the irrigation system. The result is an unreliable and inequitable water supply.

Automatic control: Automatic open-channel systems are used extensively in some countries, such as France, Morocco, and Hungary. There are basically two types of automatic control: a constant-level control method and a constant-volume control method. In constant-level controlled systems, system response moves upstream from one point (gate) to another, all the way to the source. Hence, the response is slow. In constant-volume controlled systems, the system behavior (actual performance) is measured and the system parameters are modified to maintain a constant volume in a given reach. This system has the advantage of reducing the response times by rendering the flow rate variations in each reach independent of the stored-volumes. Hence, the system responds more quickly. However, these time lags cannot be eliminated by the sole use of the above-mentioned local control systems. Significant reductions in response times can be achieved, however, by resorting to centralized control.

In the fields, since there is no design to follow, the farmers irrigate the fields for as long as the supply is available, until the high spots are covered, or until the water reaches the downstream end of the field. Depending upon the field size, flow rate, topography, and infiltration characteristics at the time of irrigation, the farmer usually either underirrigates or overirrigates.

Sometimes, irrigation system operation according to the original plan may not achieve the design specifications for flow rate and duration, due to deficiencies in the design and/or construction of the system. Since it is difficult to modify the existing system design without expending additional amounts of money, which might be prohibitive, the original operational plans can be adjusted to meet the design specifications. However, in some cases, the original design might be a constraint to system operation in meeting the design specifications.

Maintenance. Regularly maintaining the irrigation system is a prerequisite for sustained performance of an irrigation project. A system that is not maintained deteriorates. Therefore, the canal cross sections, structures, fields, and drain must be checked and maintained. However, in many projects around the world, maintenance is not considered, and sufficient financial resources are not allocated for routine maintenance. Hence, several irrigation projects have deteriorated, resulting in a system performance that is significantly different from the original design performance. One should bear in mind that as long as the system condition is not close to the original condition, the actual performance of the system will not be the same as the design performance. Therefore, routine maintenance is a prerequisite for achieving good water control.

MANAGEMENT OF IRRIGATION SYSTEMS

Lowdermilk (1981) defines irrigation water management as "the process by which water is manipulated (controlled) and used in the production of food and fiber... (It) is not water resources, dams, or reservoirs to capture water; nor codes, laws, or institutions to allocate water; nor farmers organizations; nor soils or cropping systems. It is, however, the way these skills and physical, biological, chemical, and social resources are utilized to provide water for improved food and fiber production."

Improved irrigation performance depends on the management, not only of water, but of irrigation systems as a whole, including management of information and controls;

of people (farmers and those who work in irrigation organizations); and of other inputs besides water (Peterson, 1984). In essence, the management of these elements revolves around water and its control.

In addition to better water control, a set of objectives is necesary for management. There is no management without water control and objectives. The previous section discussed the aspect of water control. Management refers to the operation of the system to meet the objectives. To check whether the objectives are met or not, the system performance must be monitored and evaluated, and a feedback control mechanism is devised to meet the project objectives. The monitoring and feedback aspects of an irrigation project are discussed next.

Monitoring and Evaluation

Monitoring a system refers to the process of checking the parameters that define or indicate the performance of a given system. The aim of monitoring a project is to discover how well the project fulfills the objectives set during the planning stage of the project. Monitoring provides:

- a data base to facilitate the ongoing evaluation of project operation and performance
- a means of ensuring that scheduled tasks are executed correctly and at the appropriate frequency
- a help in postevaluation of projects to improve the planning of subsequent projects.

Based upon monitored data, performance is evaluated and a judgment made regarding the performance (or improved performance) of the irrigation system.

Monitoring is a prerequisite for good management and must be considered during the planning stage (Rydzewski, 1978). Even irrigation projects that are automated need monitoring. The reasons irrigation systems are not monitored seem to be the following:

- systems--once planned, designed, and operated--should perform as planned
- detailed monitoring systems are costly in time and personnel

- o monitoring exposes the weaknesses of performance and management.

Effective use of capital and other resources on agricultural projects will be extremely difficult to attain without monitoring the important parameters bearing on project performance (Clayton, 1981).

Once a decision is made to monitor performance, the parameters that influence system performance must be identified, and the right information to measure the performance must be collected. Data collection for the sake of collection alone is an expensive luxury (Biswas, 1984) as the cost of data collection increases more quckly than the value of information (Figure 4.4). Quality of data collected is at least as important as the quantity of data. In addition, decisions regarding the following must also be made:

- o frequency of data collection (hour, day, month, season)
- o staffing required
- o means of recording and processing the data (equipment, manual or automatic)
- o communication of data (frequency, receiver)
- o spatial distribution of the monitoring stations (head/middle/tail, large/small farmer).

Biswas (1984) presents a list of parameters that should be monitored in relation to the operation and maintenance of irrigation projects. The important thing is to monitor the performance parameters and the factors that influence the performance. Otherwise, it is difficult to identify the causes for the low performance. Table 4.1 presents the factors that influence performance of the irrigation distribution and application systems. For example, in water control, the following parameters should be monitored:

Factors influencing the performance	Performance parameters
Planning factors	
Design factors	Reliability
Operation factors	Equity yield
Maintenance factors	Losses/adequacy

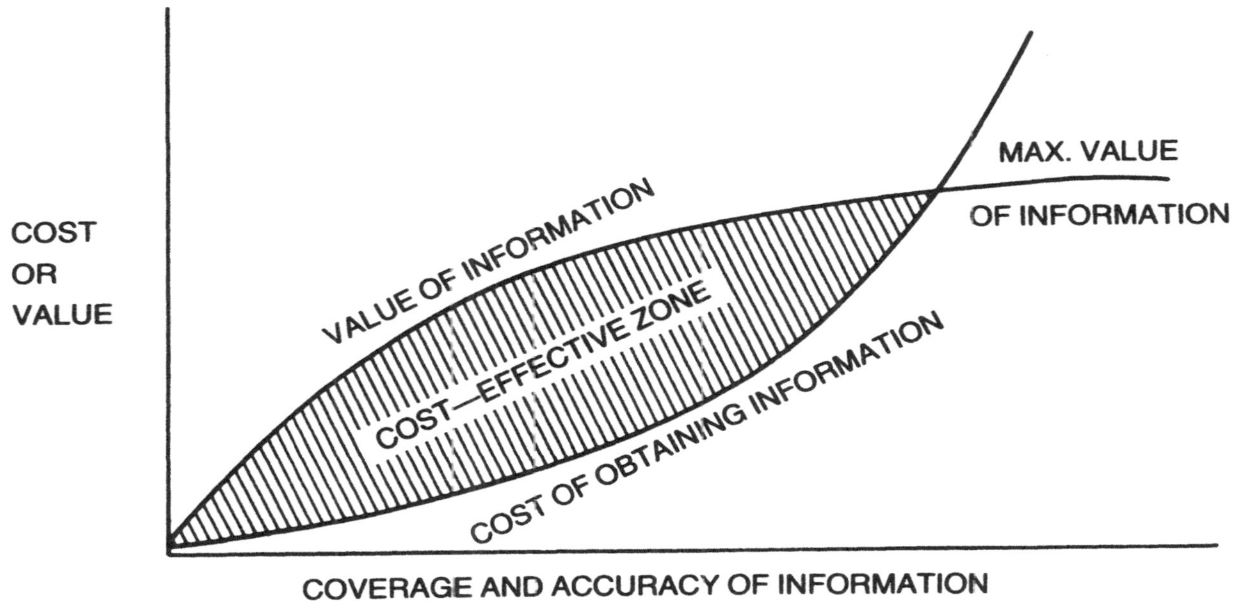

Figure 4.4 Cost – effectiveness of monitoring and evaluation information

Table 4.1 Factors influencing irrigation system performance in distribution and application of water

	Factors Influencing Performance	
	Distribution System (Main & Unit Command Area)	Application System (Field)
Planning	Cropping pattern (crop water requirements) Equity Layout of the distribution network command area Earthen or lined canal Rotation, continuous, or on-demand	Crop (rooting depth sensitivity Type of irrigation Direction of irrigation Topography
Design	Seepage losses, design discharge, depth of flow Location, type and size of structures, section parameters (Shape, size, roughness, slope)	Length & width of the field Flow rate Design depth Time of inflow frequency
Operation	Timing of operation Skills of the operator Communication related to supply and demand	Timing of operation Skills of the operator Communication related to supply and demand
Maintenance	Structures Canals (weeds, sediment, erosion)	Maintain field surface, bunds, furrow geometry

Automatic monitoring: Depending upon the volume of data and the availability and cost of qualified personnel, it may be economical and timesaving to use automatic recording equipment, such as water level recorders, flow rate recorders, and flow volume recorders (Dedrick et al., 1983; Dedrick and Pettit, 1983). The likelihood that these structures would be vandalized seems high in some countries. In addition, when deciding about automation equipment, factors such as its sturdiness, portability, cost, accuracy, and longevity should be kept in mind.

Some form of monitoring, at least qualitatively, does take place in almost all irrigation projects. Tail enders complaining that they don't get any water and low crop yields are also indicators. If there is a system response mechanism, system performance might be improved without any quantitative data. However, the level of improvement can be increased more by quantitative data (measurement). Hence, measurement structures are a necessary, but not a sufficient, factor for management. Many irrigation projects do not have any measurement structures. Even if they are present, the measurements are not taken. And finally, even if the measurements are taken, the data are not used for the intended purpose of improving the system performance.

Feedback and Control (Response)

Monitoring an irrigation project--its performance parameters and the factors influencing it--will not, in itself, result in improved system performance, even though it is a valuable data base for planning future irrigation systems. The information gathered during monitoring must be processed and communicated to the decision-making authority. As mentioned earlier, very few projects are monitored, and some are monitored for the sake of monitoring. The information gathered during monitoring is not used for the intended purpose of improved management. Information management (communication) is an important aspect of irrigation water management (Rao, 1982; Ritchie et al., 1978).

Next, utilizing the data to improve system performance by modifying one or more factors is required. Modifying the system parameters to achieve improved system performance is called the "control" or "response" mechanism of the system. If a control mechanism is not present, then improved system performance cannot be attained.

There are two types of feedback control. In the first type, information gathered during an irrigation event (or a particular day) is used to improve system performance during the next irrigation event or season. In this type of feedback control, the reasons for the deviation in system performance might not be the same during the next season, and modifying one or more of the system parameters that influence system performance sometimes might not result in improved system performance. In the second type of feedback control, the monitoring information gathered during an irrigation event is used simultaneously to modify the system performance by adjusting some of the operational parameters. This type of monitoring and feedback control usually calls for automation.

MANAGEMENT PLAN

Management starts by defining objectives, for there is no management without objectives. Next, it delineates procedures to achieve the objectives and monitors the performance parameters to see whether the objectives are achieved or not. If the objectives are met, then the system is well managed. Otherwise, appropriate changes in the system parameters must be made in the planning, design, construction, operation, and maintenance of the system. The cycle outlined in Figure 4.5 continues until the project objectives are met.

If it is a new irrigation project area, one can plan, design, and operate the system for management. In an existing irrigation system, however, the management plan starts with monitoring the system performance and comparing it with the expected (or design) performance. If there is a difference between the actual and expected system performance, we go back to the problem identification stage and delineate the factors (planning, design, construction, operation, and maintenance) responsible for the low system performance and prescribe solutions for testing and implementation. The solution may be in the form of modified planning, design, construction, operation, and/or maintenance that would result in improved system performance.

In addition to having a good management plan, the operating staff needs to be committed to management pro-

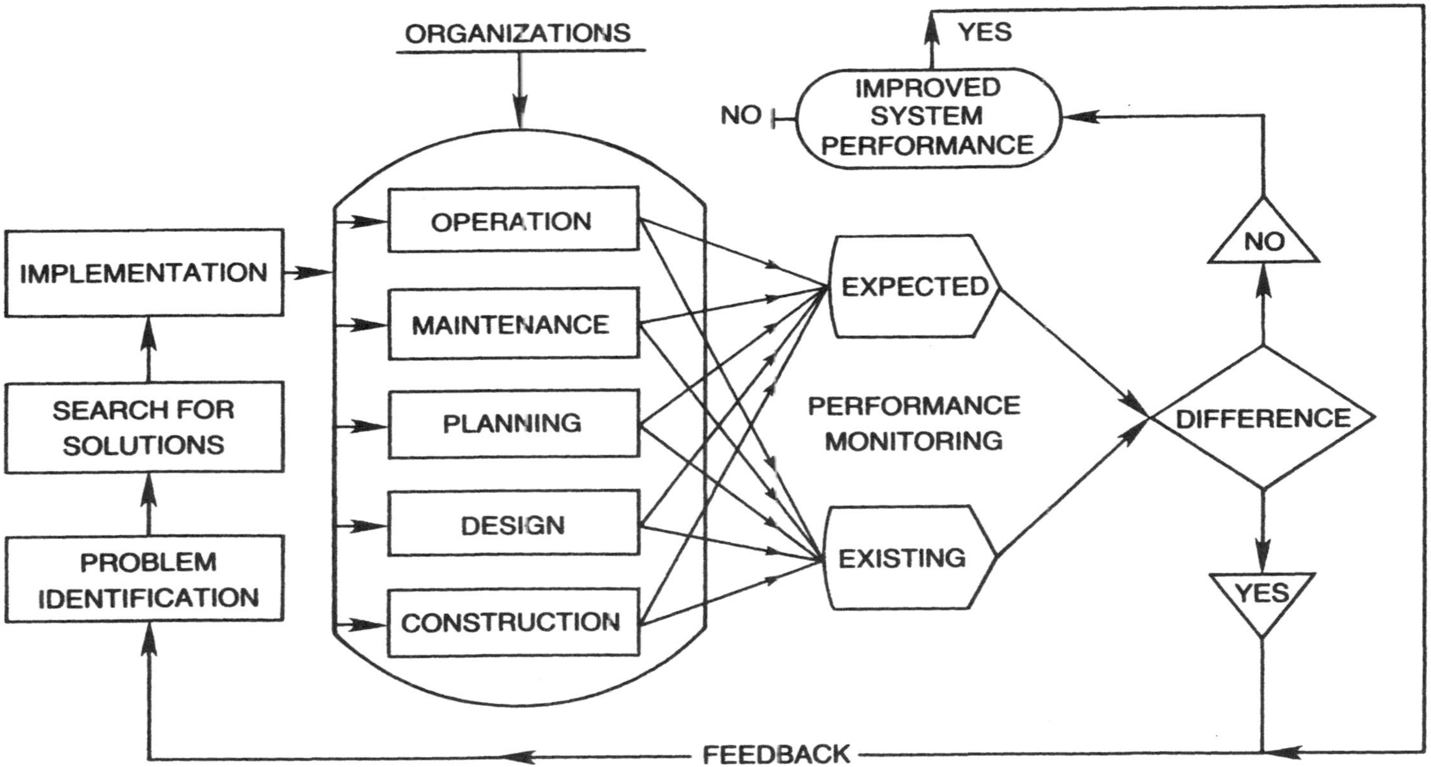

Figure 4.5 Flow chart of management plan

cedure. Sundar (1984) presents several examples of a lack of commitment in management and says:

> Management is based on the premise that things can be done better, which in turn means that one wants better performance. In a socio-political situation where what is legitimate is what one can get away with, can there be any concern for public system performance? And, if there is no desire to manage, what can management techniques do?...'In the land of nudists, what can a washerman do?' --(Panchatantra).

REFERENCES

Biswas, A.K. "Monitoring and Evaluation of an Irrigation System," Water Resources Development, 2(1):3-25, 1984.

Bottrall, A. "The Management and Operation of Irrigation Schemes in Less Developed Countries," Water Supply and Management, 2:309-332, 1978.

Clayton, E. "Monitoring, Management and Control of Irrigation Projects: the Examples of Mwea, Kenya," Water Supply and Management, 5:107-115, 1981.

Clyma, W., and A. Ali. "Traditional and Improved Irrigation Practices in Pakistan," Proceedings of Water Management for Irrigation and Drainage, American Society of Civil Engineers, Reno, Nevada, July 20-22, 1977.

Clyma, W., and S. Sritharan. "Planning, Design, and Operational Plan for Management of Irrigation Projects," mimeo, Water Management Synthesis Project, Fort Collins, Colorado, 1984.

Dedrick, A. R., A. J. Clemmens, and J. R. Dust. "A Microcomputer-Controlled Open-Channel Flow Monitoring System," ASAE Conference on Microcomputers in Agricultural Engineering, American Society of Agricultural Engineers, Chicago, Illinois, 1983.

Dedrick, A. R., and D. E. Pettit. "Automatic Surface Irrigation Systems with Water Flow Rate Fluctuation Compensation," ASAE Conference on Microcomputers in Agricultural Engineering, American Society of Agricultural Engineers, Chicago, Illinois, 1983.

Gates, T. K., W. Clyma, and T. W. Ley. "Systems Analysis for Improvement of Surface Irrigation," ASAE Paper

#81-2073, American Society of Agricultural Engineers, St. Joseph, Michigan, 1984.

Jayaraman, T. K. "Irrigator's Organizations for Better Water Management: A Case Study of Attitudes of Irrigators From Gujarat State, India," Agricultural Administration, 10:189-212, 1982.

Kathpalia, G. N., and V. Prakash. "On-Farm Development in the Commands of Irrigation Projects," ICID Bulletin, January 31(2): 45-49, 59, 1982.

Levine, G. "Management Components in Irrigation System Design and Operation," Agricultural Administration, 4:37-48, 1977.

Lowdermilk, M. K. "Social and Organizational Aspects of Irrigation Systems," lecture for the Diagnostic Analysis Workshop, Water Management Synthesis Project, Colorado State University, Fort Collins, Colorado, 1981.

Peterson, D. F. "Systems and Technology for Improved Irrigation Water Management," paper prepared for Agriculture and Rural Development Training Workshop, Washington, D.C., June 18-22, 1984.

Rao, P. K. "Policy Objectives and Information System for Irrigation Projects, India," Water Supply and Management, 6(3):243-259, 1982.

Ritchie, I. J., J. B. Dent, and M. J. Blackie. "Irrigation Management: An Information System Approach," Agricultural Systems, 3:67-74, 1978.

Rydzewski, J. R. "Irrigation Project Monitoring: A Feedback for Planners and Management," The Civil Engineer in South Africa, 20(12):309-312, 1978.

Sundar, A. "Modern Techniques for Management of Irrigation Systems: What Can They Do in the Absence of Commitment to Manage?" Wamana, July 4(3):1, 20-22, 1984.

Tyagi, N. K., and V. V. D. Narayan. "Evaluation of Some On-Farm Water Management Constraints in a Surface Irrigation System," ICID Bulletin, 32(2):11-16, 1983.

Wade, R., and R. Chambers. "Managing the Main System: Canal Irrigation Blind Spot," Economic and Political Weekly, 15: A107-A112, 1980.

5
Some Tools and Concepts for Better Irrigation Water Use

John A. Replogle

INTRODUCTION

The world has always known famines. It has not always worried about surpluses. So many factors go into the world food production picture that the opportunities for well-intended "fixes" to go astray seems to far exceed the chances for success. The pessimistic view is that empirical data confirms the distribution of failure opportunities. The Food and Agriculture Organization (FAO, 1979) of the United Nations (UN) estimated that irrigated agriculture represents 13 percent of the global arable land but accounts for about 34 percent of the crop produced. Both percentages appear destined to increase (Jensen, 1980), with the latter likely to increase more as we learn to positively control more of the food production process.

The massive infrastructure supporting any advanced society and its advanced food production system can be quickly visualized by considering the consequences of destruction by natural disaster or by global war. Imagine the United States with no working electrical power, no telephones, no highways, no railroads, and no airports. Agriculture as we know it would cease almost at once. Food stocks, even if enormous, could not prevent mass starvation, no matter how benevolent the governmental structure.

Thus, agriculture (especially irrigated agricultural development) exists and thrives in an overall development scheme that considers farm-to-market roads, rail systems, and communications. Supporting industries, from trucks and machinery to hardware items and fertilizer chemicals, are also needed. Space here prevents even attempting a really comprehensive list. Therefore, we will concentrate on some concepts and equipment that encourage better irrigation practices.

The ideal irrigation system, or scheme, would somehow default to correct water delivery methods unless energetic efforts were made to subvert the operation. I'm not aware of the existence of such systems, but I point out that high efficiency is unlikely until it is as convenient to irrigate correctly as it is to irrigate incorrectly.

IRRIGATION SYSTEMS

Some Psychological Aspects of Irrigation Project Operation

For perspective, I will relate an incident in the United States in the early 1970s, shortly after the steep rise in world oil prices. A national television talk show comedian host was lamenting the resulting economic disruptions, saying that the whole world seemed to be on "back-order" with shortages of everything, "even toilet tissue." (This was meant to be humorous, since the supply-demand for this item is, in reality, extremely uniform.) However, the general public had recently been conditioned to some inconveniences in item shortages, and the national response was an irrational run to the nation's store shelves and the hoarding of toilet tissue to the point that, for a time, it could no longer be found for purchase.

The sudden empty supply chain certainly did not mean that the national demand pattern had changed at all, only that the distribution pattern had changed, and storage of the item was now located in millions of home cupboards rather than on store shelves. This didn't help other families that had depended on an available supply, however.

The national tissue manufacturers knew that it would be foolhardy to build new facilities and increase production because of the temporary shift in storage (not true demand for their product), and eventually, as reliability of supply became apparent, the "shortage" disappeared.

Irrigation projects are also subject to some of the hoarding psychology which can be aggravated or ameliorated by the history of the system reliability. If system reliability has been low, then the psychological incentives are to use all opportunities and all mechanical adjustments, authorized and unauthorized, to "hoard" the water as if it were the last irrigation. Thus, an otherwise adequate project irrigation supply does not meet system demands. Further complicating the social picture is that these shortages are not distributed but are usually completely borne by those farmers farther from the supply source. This means that the reliability usually becomes lower and the hoarding pressures become higher as a function of distance from the supply source. A vicious circle is created wherein hoarding causes low reliability which then increases hoarding. Legal remedies and administrative threats are generally ineffective, so that remedies through physical system changes and system management techniques are more feasible alternatives.

Some Physical Aspects of Irrigation Project Operations

A recent study of irrigation delivery system management in Egypt (Richardson et al., 1984) confirms, not surprisingly, that the operation of the delivery system was significantly constraining improved farm water management. The major problem discovered in this instance appeared to stem from basic operation of the system as a static distribution entity, when, in reality, the demands varied between daytime and nighttime. The resulting spills at night subtracted significantly from total deliveries and caused wide delivery rate changes that made on-farm handling difficult.

Constraints caused by these and other operating techniques and policies are associated with on-farm water management problems throughout the world and indicate the broad need to improve project delivery systems and methods in order to allow better farm water management. It is becoming apparent that both the physical system canals and gates (hardware) and the management, social, economic, and psychological aspects (software) need careful attention.

SCHEDULING POLICIES OF IRRIGATION PROJECTS

Let us first look at some of the "software" aspects of irrigation projects. The greatest majority of government-sponsored irrigation projects on the world scene use canals, from the large conveyance canals to the smaller field canals, in some part or all of the project. These irrigation projects have the purpose of facilitating water delivery, application, and removal. The water delivery policy established and used by project managers is called the water delivery scheduling system, or simply, the schedule. Most schedules have been in place for several decades and were chosen for reasons usually valid at that time. It does not necessarily follow that those reasons are still valid, and many projects need to closely examine their scheduling policies.

Usually, the schedule selected for the canal system is theoretically capable of delivering adequate water to match the peak seasonal crop water need. This assumes that maximum unit-land production and minimum unit-production costs coincide with adequate crop water availability in the root zone. This also assumes certain crop-yield production functions, particularly the crop-yield effects of adding another unit of water. In some countries, schedules are intentionally designed to underdeliver because of the usual reliability of rainfall and the assumption that reduced production on the larger area totals more than the improved production in a lesser area. In many cases, social considerations and labor availability may be controlling parameters.

Determining and satisfying crop-water needs require applying knowledge of soil, water, plant, and atmosphere relationships, as well as various system conveyance and application efficiencies. The volume of water needed (or depth on the area) for a specific irrigation then determines the specific flow rate and the duration combination (Jensen et al., 1970; Jensen, 1980; Merriam, 1966). The crops and soils knowledge relates closely to the farm operator problems; at some point, his applicable knowledge and the physical arrangements of his fields are limited by the ability of the project delivery system to respond.

As with most system studies, a taxonomic effort is useful to bring a sense of order to the study, frequently allowing insights into relationships otherwise overlooked. For irrigation crop production, the irrigation delivery to the field crop is a function of three basic components of an irrigation schedule. These are: (1) the delivery flow rate onto the field, which may be by surface spreading, sprinkler, or drip systems; (2) the delivery frequency to the field, or simply the times of the deliveries; and (3) the duration of the delivery. Other parameters have been listed by various authors, such as delivery volume and delivery elevation. However, the flow rate and duration result in the volume of water delivered for that irrigation, and, when combined with the frequency of delivery, they provide the total seasonal application, information for estimating efficiencies, and information for such things as leaching fraction and drainage volumes. Delivery elevation can be treated like pipeline pressure and thus is a necessary requirement for achieving the basic parameter of flow rate. Thus, we can describe the operating policy, or delivery system schedule, of an irrigation project in terms of rate, frequency, and duration (Replogle, 1984; American Society of Civil Engineers, 1984).

These three components can be restrained on several levels by administrative policies or a physical system that causes a schedule to be considered either flexible or rigid. The flexible systems are usually more elaborate and permit farm-operator participation, usually by honoring requests for water delivery on a timely basis but, in turn, require a good communication system between the water authority and the farm operator. An exception is the most flexible case of a "demand" system, which allows the farm operator to simply open or close a valve or gate as needed. The rigid schedules include what has been called the rotation system, wherein a fixed flow rate of water is delivered at a fixed duration for a predetermined frequency. Many intermediate combinations exist. The flexible systems usually enhance farm operation, while the rigid systems ease delivery system complexities and problems.

These general divisions of rigid and flexible can be further divided into categories according to the restraints placed on the three components (rate, frequency, and duration). Since many restraints are possible for each of these three components, many scheduling systems are also possible.

System Components

Let us first examine the three components and discuss the range of possible options available to the planner for project operation.

Flow Rate. For the delivery flow rate, policy may specify a constant flow at each delivery cycle throughout the season. One limit is a continuous, unchanging, uninterrupted flow. Variations would include continuous but seasonally modified flow rates. The more common cycled flow is usually in some kind of rotation system among users. The canal delivery equipment associated with constant deliveries is frequently a fixed orifice or fixed pipe size. Seasonal changes in flow rate to users can be implemented within narrow limits by adjusting the canal flow elevation. Uniform changes are difficult unless the percent changes in effective head on each delivery orifice can be controlled accurately.

Often the rate is negotiable, that is, the farm operator requests a flow rate, and the water authority usually accommodates the request. Variations on this may include a standard flow rate that is delivered if not otherwise specified; a flow rate which is not to be adjusted during delivery; a requested flow rate that can be adjusted between limits, say ± 30 percent, during delivery; or the flexibility to completely shut off the delivery at the discretion of the farm operator. Many of the latter operations may cause canal spills or require the delivery system to be automated. Another variation may be that delivery volume is specified, nominal flow rate is requested, and total time delivery adjusted to achieve the correct volume. This requires volume meters on the outlets since flow rate must be integrated with delivery time, but canal elevations can fluctuate, which may ease delivery operations.

In some cases, the canal flow can be accessed on demand without any special arrangement with the water authority. A physical limit is usually imposed by a specified size of canal gate opening, or a default limit due simply to canal capacity restrictions. In most instances, the canal runs for only a few days at a time. Thus, user control of the other two components, frequency and duration, is not necessarily present unless the canal is operated continuously.

It is important to note that fixed flow rates do not necessarily limit production, even with surface irrigation systems, except at extreme limits of very small or very large flows. On-farm hardware and one-time, field-size adjustments can usually bring operations to within a workable range.

Frequency. Delivery frequency, like flow rate, can vary widely. Water can be delivered periodically throughout the season, and this period can be changed seasonally. For example, the cycle may repeat at two-week intervals in the hot season and be changed to three- or four-week intervals in the cooler seasons.

Water may be passed from user to user as each completes his irrigation needs, so that the cycle varies as the use varies. Each user is in a fixed queue, but use time is variable, and thus frequency is tied to duration.

If, indeed, accurate volume delivery is to be achieved, then frequency will be linked to flow rate as well.

More flexible systems permit farm operators to request the delivery date. The water authority may strive to honor the request to within a margin of 24 to 48 hours of the requested time. Under the most flexible schedule, no communication or special arrangements need be made with the water authority, and the valve or delivery gate to the farm is opened by the farm operator as needed.

Shallow-rooted crops and sandy soils require shorter periods between irrigations. Coupled with climatic factors, these determine the ideal irrigation frequency. Thus, high yield potential and even the choice of crop are affected by limits on frequency of irrigation.

Duration. The length of time that the farm gate or valve is allowed to operate (the duration) can vary from continuous flow throughout the season to a few minutes or hours per irrigation cycle. In some projects, small flows are taken by all farms on a canal for the entire period the canal is operating. The canal may be filled on alternate weeks. Other projects deliver in 24-hour units of duration or in some other standard but fixed time block.

More flexible schedules permit the farm operator to specify the duration. If the canal operation can provide a steady flow rate, the duration time then allows easy and accurate volume determination. Otherwise, specified duration and poorly controlled delivery rates can cause crop production problems. Poorly controlled delivery rates can be partly compensated for with flexible durations where the farm operator controls the shutoff time. Theoretically at least, he has the opportunity to obtain the required flow volume and also to adjust for added inefficiencies that the flow rate changes may have caused.

Scheduling Terminology

It is not surprising to find that each irrigation project has a unique schedule, in that some component is handled differently from other projects. Also, it is apparent that a complete naming system would require up to three descriptors--possibly three paragraphs--one for each of the three components of the schedule, and would be cumbersome, if not impractical, to use for a naming scheme. Attempts have been made to standardize scheduling terminology (Replogle, 1984; ASCE, 1984), but this standardization is still being modified. Former terminology has included "demand," "modified demand," "rotation," and "continuous flow." However, these terms have had to describe too many situations and therefore have not always conveyed the intended operating concept.

Having outlined some possible variables in each of the scheduling components, I will select important variations of each component and discuss some of the resulting combinations for rigid and flexible groupings. At the same time, I will try to introduce meaningful terminology into these groupings.

One approach to standard terminology is to assume the extreme case for each category, rigid or flexible, and then relax or modify the necessary components. This means that for the rigid case, all three components are fixed and unadjustable during use. Conversely, for the flexible case, all three components are unrestricted and adjustable during use. The next problem is to indicate rigid or flexible without the cumbersome addition of yet another word to the name. The goal is a two-, three-, or four-word system that has some intrinsic definition value that likely will be interpreted correctly with minimum translation when first encountered.

Yielding somewhat to tradition and to the custom that most rigid schedules assign water in user rotation, we will use the term "rotation" to indicate rigid systems. Yielding again to tradition, we will use the term "demand" to indicate the more flexible systems, and introduce the term "arranged" for other flexible schedules, a term which implies communication and reply, or negotiated arrangement between the water authority and the farm operator. These latter variations have sometimes been called "modified demand."

We try to avoid long names such as: "fixed-flow-rate, requested-frequency, adjustable-duration scheduling system," but even this long name must assume that an adjustable flow rate, while not unlimited in size, is large enough for most practical operations.

Rigid Schedules. Rigid schedules are usually water-authority oriented and are predetermined by law, policy, formula, water-right shares owned, or other means. They are rigid in that the decisions contain no current farm operator inputs based on current farm crop needs. Allotments are previously determined by the water authority, perhaps in the early spring, and delivered by some type of rotation plan among the various farms. Some probable combinations of flow rate, frequency, and duration for rigid schedules are as follows:

- <u>Fixed Rotation</u>: This term implies a rigid schedule with fixed flow rate, fixed frequency, and fixed duration. The flow rate is the same from irrigation to irrigation throughout the season and is delivered at regular intervals. The word "fixed" under the original assumptions is redundant but is included to warn users that it is not necessarily their old definition of "rotation." This combination meets the word-length goal.

  Typical flow rates are from 25 to 500 liters per second; typical frequencies are one to two weeks; and typical durations are 8, 12, or 24 hours. Limiting cases include durations that last as long as the flow is in the canal, which may be one week full and two weeks empty, or for the whole season if flow is continuous.

  This schedule requires the least capital investment in canals or distribution pipelines and involves the least water-agency management and operational input. The canal flows at its maximum rate and is the minimum size for the job. However, this schedule encourages the farmer to operate at low efficiency, which wastes water, energy, and labor and increases drainage problems. It also restricts crops to those adaptable to the set frequency. Since all soils in a project cannot be expected to be similar, only part of the project area is well suited to the selected frequency.

- <u>Varied-Frequency Rotation</u>: Again, the word "rotation" indicates a rigid schedule, with only the frequency modified. The flow rate and duration components remain constant. A shortcoming here is that no information is conveyed on how the frequency is varied. A two-week interval during the hot season changed to a one-month frequency during the cool season could represent one extreme. More typically, the variable frequency would be in response to seasonal changes on a more refined increment or even in response to immediate local weather.

This schedule has potential for reducing early season and late season over-irrigations and, therefore, reduces the delivered water quantity. However, with the changing irrigation interval, the management-allowed deficiency of the soil moisture should be, ideally, the same everywhere and, ideally, always satisfied. Such a condition will not be reached on most farms because it implies that all soils and root-zone depths of all crops are the same everywhere throughout the season. This condition may be approached where a perennial crop, such as a single kind of tree crop, is grown and soils are similar. For other conditions, crops are limited and production lowered or water wasted. During the early part of the season, when the frequency is increasing, the first farms to receive water on the first round may still be quite wet from rainfall or pre-irrigating. In the late season, with decreasing frequency, unneeded water will be applied to the last farms. For the water authority, this schedule means that the same size canal is used throughout the season, but it is empty part of the time. Farm gates or valves could be of the fixed opening style.

- Varied-Rate Rotation: As before, the unnamed components (in this case, frequency and duration) are fixed. The flow rate in the canal, and hence to the farms, on each cycle is varied by the water authority to approximate seasonal crop demands. This schedule still allows the water authority to use a minimum-sized canal, but in this case the canal is always in operation. Again, perennial crops with deep root zones on uniform soils are best suited to this schedule. With short, constant intervals, annual crops requiring frequent application early in the season may be grown successfully. Farm gates or valves would need to be adjustable to obtain the prescribed discharge against a changed canal level or to receive a prescribed reduced flow against a full canal.

- Varied-Duration Rotation: This term implies that the flow rate and frequency are fixed, and the duration is varied, perhaps seasonally, again to accommodate the seasonal crop-water needs. This schedule is particularly suited to projects that have fixed orifice openings at various elevations, which operate at the intended rate only at a certain canal elevation. Otherwise, reducing the canal elevation or flow may not properly divide the flows to the various farms.

This combination, assuming the frequency is otherwise satisfactory, would allow annual crops to be grown throughout the season and would accommodate a wider range of soil types. Because the total delivered volume is somewhat matched to the seasonal crop-water requirements, opportunities exist to reduce water demand and drainage needs. The canal size again is minimum and is empty for part of each cycle.

o Varied-Frequency-and-Rate Rotation: The duration is fixed, and the other two components are varied by the water authority to approximate crop-water needs. Without changing the duration, and thus the daily schedule of the ditch attendant, the changing frequency and flow rate can accommodate a fairly wide spectrum of crops and soils. Farm outlet gates or valves would need to be adjustable. Again, the canal would be minimum-sized and may flow continuously, depending on the combination of frequency and rate prescribed.

o Varied-Duration-and-Rate Rotation: With the frequency fixed, combinations of rate changes and duration changes can again follow seasonal crop needs and be adjusted to local canal capacities. The same restrictions on crop types and soil types are shared with those of the varied-rate rotation schedule described above. This schedule is not common but could accommodate the long durations and low flow rates needed with sprinkler or drip systems, or at the other extreme, could accommodate the high flow rates and short durations that favor level-basin irrigation systems. It may be difficult to accommodate both extremes on the same canal lateral.

- <u>Varied-Duration-and-Frequency Rotation</u>: This combination is also adjusted seasonally by the water authority to approximate average crop-water requirements throughout the project. The seasonally varied frequency, if it is sufficiently frequent, can allow most annual crops to be raised, and the variable duration allows fixed delivery orifices, etc., to be operated on each farm. Another advantage of this schedule is fewer seasonal delivery cycles and presumably lower seasonal labor costs to the water authority than the varied-duration rotation schedule would require.

<u>Flexible Schedules</u>. Starting with the flexible schedule having all components unrestricted, we will name and discuss combinations with various restrictions on one or more components. The main feature for flexible schedules is user input for selecting one or more of the three components. Compromises between the needs of the water delivery authority and the farmer will create restrictions on each that should result in an optimal solution. However, increasing water, energy, and labor costs have made formerly acceptable schedules too restrictive for present conditions, and present upgrading must consider future changes that could affect the optimal economic solution. For example, pressure systems constructed for only sprinkler applications now face high energy costs, with no easy way to increase flow rates to facilitate other irrigation methods that operate best with short-duration, high-rate deliveries.

The term "demand" has been widely used for the flexible schedules. To be precise, the term is restricted to mean that only the user operates the turnout controls, without restraint on frequency, rate, or duration, and that no communication with the water authority is required. Of course, unlimited flow rates are not practical, but as long as flow rates meet the optimum needs of the farm and do not restrict the farm operation, then the demand is met and the name could apply. Where the water authority, outlet size, or system capacity limits rate, then the term "limited-rate demand" is more precise.

For the condition for which "demand" has sometimes been used, the term "arranged" will be applied. The latter term more correctly implies that the farm operator and the water authority are both involved in the rate and timing decisions. It further implies that some sort of communication system is available, and that the water authority has some control over the delivery, usually by a ditch attendant or the equivalent.

Again, all variations and ranges of restrictions on rate, frequency, and duration could produce a long list of possible variations in the flexible schedules. The most common of these can be summarized as:

- Demand: This implies, as described above, that no limits exist on rate, frequency, or duration, and that there is no external control by the water authority. This is usually available when the access is directly to a large lake, river, pumps, or main canal, and the farm unit is reasonably small. Totalizing flow meters, as opposed to rate-type meters, are needed for billing purposes, if billing is to be based on water used. (Examples of totalizing meters include propeller and turbine meters, and weirs or flumes with integrating devices.)

- Limited-Rate Demand: This is a practical schedule in which the valve used or system capacity restricts the user flow rate but not to a level that seriously limits the choice of irrigation methods or creates problems for efficient labor use. The limit should be quite large, so that it does not limit foreseeable future needs. In practice, pressures often vary as flow rates vary in small capacity systems, so that set flow rates may not be constant at a constant outlet setting unless additional controls (Merriam, 1973) are included. It should be noted that increasing pipeline diameter from 200 to 250 and from 250 to 300 mm doubles and then triples capacity, respectively. Thus, irrigation systems that are not highly sensitive to flow rate (only to total delivered volumes, such as level-basin irrigation systems) (Dedrick et al., 1982) could tap the system directly. Other systems, such as sprinklers, may require pressure regulators or booster pumps for optimum operation. Again, no communication system is needed, but totalizing flow meters would be necessary.

- Arranged-Frequency Demand: In practice, this implies that the flow rate has an upper limit established by practical considerations but that the rate and the duration can be adjusted during delivery by the farm operator. However, he must negotiate with the water authority as to the date when water will be present in the supply canal. This implies that the water authority automates the system or closely estimates the total volume needed and the demand rate, so that the user is satisfied, but the canal is not overtopped. This may produce considerable operational spillage or the application of accurate input parameters to a suitable computer model; otherwise, the use of canal storage, automatic controls in level reaches of canal, oversized canals, and off-line storage are usually needed in some combination to make the demand or limited-rate demand and the limited-rate, arranged-frequency schedules workable (Merriam, 1973, 1977; Replogle et al., 1980; Replogle and Merriam, 1980). The "arrangement" requirement by the water authority allows a method of "decoupling" synchronized demands that would naturally occur after a general rainstorm over the delivery area. Again, totalizing meters would be needed. High on-farm efficiencies are possible with this schedule as with demand systems.

- Restricted/Arranged: This implies that the three components are negotiated between the water authority and the farmers but are restricted in that they are not further adjustable during the delivery by the farm operators. If, indeed, the rate remains constant, the meter can be a rate meter only, such as a weir or flume, with the time duration serving as the totalizing parameter.

- Fixed-Rate/Restricted/Arranged: The duration and frequency are negotiated between the water authority and the farm operators. Again, access to the gate controls are restricted, and no changes are made during the delivery. As explained previously, the fixed-rate feature is usually not a severe restriction and is very usable with level-basin design. If, indeed, the fixed rate is accurately maintained and measured, flow duration provides the total volume accounting.

- <u>Fixed-Duration/Restricted/Arranged</u>: This schedule has a duration (usually 24 hours) that is fixed by policy, but the flow rate and the date of delivery are arranged. As before, the gate controls are restricted, and the flow rate is not changeable during delivery. Again, if constant flow exists, then rate meters such as weirs or flumes can be used to verify the delivered volume. The 24-hour duration almost always causes difficulty in the use of farm labor. It usually means that water must be accepted for too long a period, and the arranged rate, which is then fixed, is selected to be small, resulting in poor field distribution and the need to over-irrigate to be sure of an adequate volume. This schedule is not conducive to efficient on-farm use of water and labor, but yields are not greatly affected.

- <u>Fixed-Frequency Demand</u>: This schedule is common in areas with many small ranchettes or lawn waterings. Typically, the canal is filled one day every two weeks, and the users may access it at will during this period. If the delivery system is small, then it degrades to a rotation (formal or informal) among neighbors, requiring the current user to alert the next user.

- <u>Fixed-Frequency/Restricted/Arranged</u>: Although the frequency may be fixed, the user can arrange for the total delivered volume and the way it is delivered by requesting the flow rate and the delivery duration. As with other arranged schedules, the water authority operates the canal gates to the farm unit. This is one of the least desirable of the user-input schedules, having many of the faults of the rigid schedules. The fixed frequency limits the types of crops that can be grown and the soils that can be efficiently irrigated. However, it offers the opportunity to refill the depleted soil profile using a hydraulically efficient flow rate and thus can allow high water distribution uniformities.

Scheduling Policy and Water Conservation

The demand or arranged schedules may be considered too ideal to approach. Also, without careful, knowledgeable farm operators, they do not in themselves produce water conservation, even though (theoretically) extremely high efficiencies cannot be obtained without it. Thus, it is a two-edged sword that sometimes must be approached in evolutionary stages.

Typical case problems for demand systems are found in some developing countries. They usually stem from the long-term irrigation construction procedure. Initially, the water supplies are captured (dams), then canal mains are started and distributary canals added, etc. Meanwhile, the dams are filling and the water supply is overabundant for the small area initially serviced. These "top-end" farm operators are allowed to take water at extremely inefficient quantities--the lakes might as well be used, since otherwise the water must be released. This process is usually accompanied by a delay in providing effective on-farm irrigation management. Thus, into this vacuum, the new irrigation farmer learns low efficiency methods that, in a few seasons, border on dependency. These early users appear to establish an unintended "water right" as first beneficial users that continues even though the remaining canals may have been constructed. Because of the large number of small-farm operators usually involved and the slowness of communication and technology transfer, the farmers do not usually know that their canal should really have water in it and may assume that it is empty due to drought. This then becomes a problem entwined with educational, economic, and sociological aspects, as well as with engineering ramifications, such as high water tables and salinity problems that usually accompany the extremely low irrigation efficiencies.

One way of implementing water conservation in these cases may require an evolution of processes and some extraordinary water management education on a grand scale. For example, a rigid irrigation schedule with fixed rates, frequencies, and durations (the worst possible schedule from a cropping and water conservation viewpoint), may have to be announced and imposed, and simultaneously, those top-end farmers must be convinced and shown by demonstration that their previously learned practices were actually leaching nutrients from some fields, causing waterlogging and saline seeps in others, and, in general, depressing yields. This process can be used to reduce on-farm water applications and raise field application efficiencies to approximately 50 percent. Later, in perhaps a decade, portions of the district can return to a demand or an arranged-type schedule, provided communications systems have evolved. This would then allow efficiencies in excess of 70 percent. Alternate procedures should be developed to avoid the intermediate step, which is undesirable.

Even in the developed countries, efficient irrigation application methods available on the farm do not guarantee water conservation. But as pointed out, a rigid schedule encourages or forces early season over-applications, thus adding to possible salt loadings and drainage problems.

The total impact of over-irrigation depends on whether the area is in an upper-basin region with the possibility of downstream reuse, as exists in some areas of the United States, such as California, Colorado, and Idaho. In lower-basin areas, such as the Imperial Valley of California, recovery of over-irrigation flows is difficult. Thus, water conserved in the latter area is truly saved, while in the upper-basin case, water quality and energy conservation may be more tangible benefits (Jensen, 1980, 1982).

A strong case for conservation by improving the first use of water through increasing efficiency of distribution systems and field irrigation systems is presented by Horner et al. (1983). In particular, they point out the savings in capital costs otherwise needed for a recapture system. As long as energy is limited, it is important to improve first-use efficiency and avoid spending additional energy to recapture dispersed water lost from the system through deep percolation and field runoff.

The arranged-frequency demand schedule is attractive for many irrigation projects. It may be considered to be the desirable one in the United States for new or improved projects. It permits the farm operator to nearly optimize all his water-related operations. The water authority is able to provide acceptable service with a peak system capacity that is smaller than the demand schedule would require. It is desirable that the delivery system be automated and totalizing meters used; otherwise, frequent manual flow changes are needed, and difficult time-discharge records may be required.

CASE STUDIES

Let's look at three United States irrigation projects that have different irrigation scheduling policies and speculate on some of the implications of those schedules.

Table 5.1 Scheduling policies for three projects

Project	Command Area	Approx. Field Eff.	Schedule Details		
			Rate	Duration	Freq.
A	20,000 ha	65-85	400 L/s	Arranged	Arranged
B	100,000 ha	55-75	Arranged	Arranged	Arranged
C	200,000 ha	50-65	Arranged	24 Hrs.	Arranged

Comparing the three different ways that these projects deliver water, we see that only Project B is flexible enough to allow all three components of the schedule to be arranged. The other two projects have one item restricted to a constant value. As explained in the discussion on schedules, the restriction to a fixed 24-hour duration can be severe, except for particular matches with crop, soils, and field sizes. To irrigate a small field and make the irrigation last for the required 24 hours, the operator must accept a small flow rate that does not allow efficient use of labor or good distribution uniformities.

Notice, though, that highest efficiencies are not acheived in the most flexible system of Project B, but rather in Project A, where the flow rate is fixed. There are a number of explanations for this, including project size, urbanization near Project B, the relative young age of Project A, the special government programs in Project A, and other factors, none of which act alone.

Major factors may be that the urbanization mentioned near Project B delays interest in converting to modern systems. Besides, with the available flexible system, an operator can nearly optimize his crop water management at a sufficiently high efficiency, about 70 to 75 percent, so that further improvements are not deemed mandatory. Very few modern laser-controlled level systems (described in another section) are constructed in the command area of Project B.

Project A, on the other hand, has a relatively large proportion of innovative farmers, as a result of early efforts to attract highly qualified irrigation farmers to the area immediately after construction. Also, the large delivery flow rate could be more easily handled with large level basins that could be designed to appropriate size for a variety of crops, once and for all. Thus, the flow rate and field sizes matched well and caused little or no restraints on field efficiencies.

Also, Project A has been the recipient of special government programs and was the first project to be extensively leveled using laser-controlled equipment. One could reason that the large stream size and the difficulty of handling it predisposed those farmers to accept level basins immediately when their advantages were first generally recognized.

SOME STRUCTURES AND EQUIPMENT ASPECTS

The basic function of an irrigation system is to place water and nutrients in the crop root zone on a timely basis. Secondary requirements include the placement, with minimum deep percolation consistent with salinity control and without surface runoff. This requirement reduces to furnishing a certain volume of water at a steady flow rate for a fixed time, or at a fluctuating flow rate for a period that extends until the volume is appropriately distributed to the required area. Since there are many ways to distribute this water, it must be a primary element of irrigation system design.

Considering the field crop, water may be placed in the root profile zone by a wide variety of sprinkler systems; subsurface and surface drip irrigation systems; and surface, or flood, systems.

Contrary to popular concepts, all methods can be managed to similar degrees of high efficiency, or mismanaged to similar degrees of low efficiency. Granted, the management effort, the investment level, and the maintenance inputs may vary widely between systems for similar results. Likewise, the water requirements of crops do not materially change for the various systems, but rather, the apparent differences really represent management losses, not crop-water use changes.

The selection of a farm irrigation system includes consideration of soils, topography, water quality, water delivery mode, system construction costs, maintenance support availability, farm operator knowledge, crop requirements, and other factors.

When soils are of medium texture, deep, uniform, and on nearly level topography, almost any system can be designed and constructed for almost any crop. On the other extreme, variable soils, steep topography, and tree crops strongly favor drip system designs.

When system economics (including energy costs, labor inputs, water costs and crop yields) are a prime concern, surface systems prove to be well suited to a large portion of the irrigated lands of the world, despite an average field efficiency (water needed by the crop divided by the water applied) presently on the order to 30 to 50 percent in most of the United States (Jensen, 1980) as well as the rest of the world (Bos and Nugteren, 1978). It is these very numbers that have prejudiced politicians, planners, and laymen against surface systems because other systems are quoted as having field efficiencies higher than 70 to 80 percent, with some drip systems pushing 85 to 90 percent if intense design effort and management attention are available.

Ideally, we should start with the concept of designing a food and fiber production system rather than an irrigation system. From this standpoint, we may notice that if water is the really limiting resource, then it should be used on the most suitable land within reasonable range of energy, transport, and other aspects of economics. Also, it helps to make the choices regarding whether to upgrade present areas of command (vertical) or to increase the area of command (horizontal). Maximum production sometimes must be subordinated to social or political requirements and thus a clear decision from the engineering standpoint is not usable.

MODERN SURFACE SYSTEMS

Considering the engineering aspects, the highest production with minimum expenditure of energy and resources, limited water would be used first on the well-drained, medium-textured, deep, and level soils, where almost any irrigation system can be effectively used. Thus, surface systems would appear to be favored if the matter of general low efficiency can be satisfactorily addressed.

Let's examine what is happening to surface systems. Recently, we have seen large areas, in excess of 100,000 ha in Arizona alone, converted from traditional (for the

United States, at least) sloping border strips and sloping furrows to level border strips or basins, and level furrow systems presently graded to zero slope with laser-controlled scrapers. These precision-leveled basins are measured to have field efficiencies of 35 percent and higher (Dedrick, et al., 1982), easily exceeding the efficiencies of most other systems. Granted, topography, soil depth, and soil types pose some economic and operational restrictions, but the fact remains that a large portion of the present surface-irrigated land and some of the irrigation expansion areas are suitable for such surface system design.

The success of such systems has not been closely documented in all aspects, and some of the testimony is, indeed, anecdotal. One large farm operator near Blythe, California, reported that his irrigation labor efforts using spiles (small pipes through earthen berms) to apply flow to sloping furrows planted to lettuce, a short-season crop requiring frequent irrigations, were reduced from $100 per ha to $18 per ha. Near Delta, Utah, trial basins installed as a part of an irrigaton extension program (Heneggeler, 1983) were so successful in the first year at permitting normal to excellent wheat yields where low to marginal yields had been the rule, that some farm operators put priority on the use of time and equipment during the second farming season to construct level basins for the third year and beyond.

These experiences and the operational reality of high efficiency and increased crop yields, along with the ever-present threat of increasing energy costs that become locked to pressurized systems, can be viewed as compelling reasons to expect that surface irrigation systems will continue to be viable.

IMPROVING EASE OF CONTROL AND MANAGEMENT

One major attraction for the large center pivots and the drop systems that are independent of the system efficiencies is the built-in control and management ease that are usually installed as part of the system. Flow meters are frequently installed with the center pivots equipment and are almost always part of a drip system. These provide the farm operator with much of his water management information. Pressure regulators provide steady flow for convenient water management and general ease of operation. Granted, maintenance problems peculiar to each system exist, but the water management decisions were frequently provided by equipment preprogramming by experts or by the farm operator himself, at his leisure. The feeling of control has positive selling points and has frequently been used as a major selling point for some types of pressurized systems.

On the other hand, surface systems have often been constructed with minimum or nonexistent flow metering and rather haphazard flow control that can place severe requirements on a farm operator. To make surface systems attractive from the operational standpoint, we need economical, simple, but accurate open-channel flow measuring methods and flow controls. We will now examine some recent developments in canal measuring equipment.

FLOW MEASUREMENT

Most of the major flow metering methods for canals include variations of sharp-crested weirs, short- and intermediate-throated flumes, and long-throated flumes (Replogle and Bos, 1982).

Sharp-crested weirs have changed little over the past several decades. Their major advantages are low cost and extensive documentation. Disadvantages include the need for high head loss and poor passage of sediments. They can tolerate no backpressure from the tailwater channel, that is, their limiting submergence ratio (modular limit) is zero. Actually, most references suggest a margin of at least 50 mm (2 in).

Flumes of the short and intermediate variety are those in which major streamline bending occurs in the contraction, or throat region, of the flume. This affects the hydrostatic pressure distribution and limits the ability to accurately predict the discharge from theoretically derived relationships. It also affects the tolerance to downstream backwater. The sharp-crested weir could be considered the limiting case of short-throated flumes.

The older and more familiar short-throated flumes are the Parshall flumes, the Cutthroat flumes, and the H-flumes. All depend on laboratory-derived calibrations. All should be installed with careful attention to duplicating the calibration situation as closely as possible. These flumes have higher tolerance to downstream water levels than thin-plated weirs, commonly tolerating 60 to 65 percent of the upstream flow depth in the downstream tailwater depth before corrections are needed in the calibration. These corrections are determined by a second depth reading made on the downstream water elevation. With two head readings, submergence ratios sometimes exceeding 90 percent can still produce a flow measurement, although at reduced accuracy compared to the flow range that needs only a single depth reading.

The long-throated flume types have experienced several recent technological advances for measuring open-channel flows. In these, the throats are proportioned to be at least as long as the measured head, and preferably twice as long, so that hydrostatic pressure conditions prevail at the control section. Application of fundamental fluid flow characteristics related to fluid friction and flow velocity distribution in channels permits accurate computations of flow for a wide range of channel shapes and flume throat shapes. These "computables" are becoming the standard for most new open-channel installations and are usually the best candidates for retrofitting older canal systems because of high accuracy, very low head-loss requirements, and simplified construction details (Replogle and Bos, 1982; Bos et al., 1984).

There is virtually no limit on size or variety of cross-sectional shapes that can be designed to satisfy the relatively liberal hydraulic and geometrical requirements. These flumes, in combination with appropriate gates or valves, are seen as the basic control and measuring devices for open-channel flows for the foreseeable several decades.

The throat contraction can be made by moving the sidewalls of the channel inward or by raising the channel bottom with a sill. In all cases, a smooth transition is required between the upstream channel and the throat. This is usually made from plane surfaces converging at no faster than 3:1 with respect to the centerline of flow. A transition between the throat and downstream channel diverging at about 6:1 is sometimes used on larger sizes to obtain maximum head recovery and to produce submergence ratios exceeding 90 percent with a single upstream depth reading. Smaller sizes are usually truncated, and the flow simply dumps into the downstream channel or pool.

With side contractions, or a combination of side contractions and bottom sill, the devices are usually called "flumes" or "critical-flow flumes." Those devices with only a bottom sill, while theoretically identical, are usually called broad-crested weirs. Thus, broad-crested weirs and long-throated flumes are variations of the same device. A recent advance in the state of the art is the ability to predict flow to within about 2 percent for any mathematically describable cross-channel shape (Replogle, 1975; Bos et al., 1984).

One of the older flumes, the Palmer-Bowlus flume (Wells and Gotaas, 1958), has a configuration that allows its calibration to be computed as a long-throated flume, at least for the lower ranges of discharges when the throat length criteria are met.

Of all known flumes and weirs, the "computables" have the following major advantages:

- The weir or flume can be shaped in such a way that all practical ranges of discharge can be measured accurately
- For any weir or flume, a rating table can be calculated with an error in the listed discharge of less than 2 percent
- The required head loss over the structure is the lowest attainable
- The head loss requirement of each combination of structure and channel can be calculated
- Under similar hydraulic and other boundary conditions, these weirs and flumes are usually the most economical for accurately measured flow.

The largest broad-crested weirs installed to date can measure 50 m^3/s (1800 cfs) each and are installed on the Arizona Main Canal and the South Canal of the Salt River Project. They are 16.5 m and 18.3 m (54 and 60 ft) wide, respectively, with trapezoidal throat and channel shapes. The required head loss is only about 0.1 m (4 in). The smallest in routine use measures 2.5 L/s (40 gal/min) as its maximum capacity, and 0.125 L/s (2 gal/min) for its low rate (Replogle and Bos, 1982), with a required head loss of less than 12 mm (0.5 in).

For the usual canals with concrete linings, precomputed ratings are presented by Replogle and Bos (1982) and Bos et al. (1984). The usual configuration is a trapezoidal, broad-crested weir, with an approach flow ramp. Precomputed ratings are also available for partly full circular culverts fitted with a similar sill (Clemmens et al., 1984).

CANAL CONTROL SCHEMES

Controlling flow in canals usually follows one of two philosophies: upstream-controlled flow or downstream-controlled flow. The first implies that the flow rate is selected and released. From there it flows to its final destination, with few opportunities for changes. Lag times between release and destinations can approach days on large canal systems. The major advantage is that few electronic or computer-assisted controls are necessary. The main control processes are closely related to flood-routing procedures.

Downstream control allows the outlet user to start and stop the flow at will, much like that afforded by a household plumbing faucet. In pipe flow, the faucet simply increases the backpressure on the pipe, and this backpressure is ultimately transmitted to the source tank. Static pressures in the pipeline can be as high as that of the source. In open-channel flow, the pressure pulse is less easily used and would require canal walls to be as tall as the source reservoir level unless artificial controls are introduced. To be highly effective, the sensed information at the outlet must be interpreted quickly into control actions all the way to the supply source. Electronic transmission and computer controls are usually used on long systems. Mechanical transmission of a surface level from station to station is usually suitable for only small systems or subsystems where the response time is on the order of a few minutes. Operating a high-efficiency surface irrigation system is reasonably convenient with canal systems that can deliver water in response to the direct control of a knowledgeable farm operator (demand system). Downstream control schemes offer this option, and advances in electronics and computer controls almost assure that these will become the general practice in the next decades.

On transmission canals, flow depth is important for keeping the canal safe and intact. In distribution canals, both flow rate and depth have important meaning, while in field deliveries, flow rate is usually of main concern. Thus, the control processes in each should be tailored to the particular need. For example, in transmission canals, storage volume can generally be changed by about 10 to 20 percent with a 0.3 m (1 ft) change in depth. Thus, for downstream-control systems, depth sensing and controls that can detect the captured volume between gates, called canal reaches, and can keep it to within 10 to 20 percent, may be adequate. At the final destination, the delivered volume and the flow rate should be within 5 percent for highly efficient irrigation. These accuracies are not inconsistent, since those associated with the transmission canal represent storage changes in the canal.

Again, two control methods come to mind. One is based on flow-rate control. The sensed flow-rate demands are transmitted to all upstream gates, which quickly respond by each supplying that rate to the downstream reach. If this is done accurately and instantaneously, the water surface in the reach tilts in the direction of change (actually, waves progress up and downstream) so that the total volume in each reach remains unchanged. This requires each structure to be an accurate meter, or a slight error will eventually deplete one reach and flood another.

An alternate control procedure ignores flow rate, for the most part, and simply transmits the loss or gain in depth (volume) being experienced in the canal near a point of water use. This canal depth is then electrically transmitted to the gate at the upper end of the reach, which responds enough to achieve a rough makeup volume, displaced laterally by up to several kilometers. Upstream from this second gate, a change in volume is noted through the hydraulic connection, and again a transmission is made to the next gate which starts a makeup volume into the reach, and so on, back to the source. The transmitted information does not arrive instantaneously at the flow source reservoir but can travel from reach to reach, with no central control, at an average rate of up to several hundred kilometers per hour. The delay time then relates to the necessary storage changes that must be accommodated in the system for given operating situations. The controls are simple, in that depth sensings are converted into mechanical gate movements with feedback, so that the gates do not need stable discharge ratings for a given opening. This is ideal to compensate for partly clogged gate openings due to floating debris or sediment buildup and lends itself to a series of individual and essentially independent microprocessor-controlled gates (Burt, 1984).

While depth and volume control measurement are suitable in some situations, the farm outlets ideally need a controllable flow at a measurable rate. For this situation of canals and surface irrigation, the long-throated flumes described previously should serve well. Progress in secondary devices that convert them from rate to totalizing meters should increase their usefulness to farmers and to canal operators.

UPGRADING SCHEMES

To reiterate, the rigid schedules are water authority oriented and designed for minimum system investment and operating expense. The impact on production per unit of water--water conservation--is severe. Viewed as a total production system, the delivery system and the farm operations cannot be readily optimized with these systems. Therefore, efforts and methods for upgrading schedules need high priority and are essential to conserve both water and energy.

There are several possibilities for upgrading less flexible schedules to more flexible schedules. These concepts include total reconstruction; repair-replacement with increased capacity; adding regulating reservoirs; using automated level-top canals (Merriam, 1977) and/or closed or semiclosed pipelines (Merriam, 1973), both of which allow a no-flow condition to exist; and automating systems (Burt, 1984). There are difficulties in accurately determining limiting capacities for any of these methods because true operational data or adequate models of the peak irrigation requirements are usually lacking. Using values that later prove to be too small is undesirable and probably uneconomical. It is inexpensive to appreciably increase capacity to obtain increased efficiencies. As mentioned previously, increasing pipe diameter from 200 to 250 and from 250 to 300 mm doubles, then triples, capacity.

Many other possible combinations exist. One of these, a statistical approach to canal capacity requirements, is discussed and illustrated by Clemmens (1979). This concept envisions using freeboard capacity and canal-storage volume changes to create regulatory reservoirs. The changing canal levels would, in turn, require farm delivery canal turnouts that could compensate for changing canal surface levels to maintain stable turnout flows. It appears that low-cost microprocessors could be developed to use with available flow metering equipment, or that automated gates that control water levels downstream could effectively accomplish this (Zimbelman, 1981; Burt, 1984).

Institutional changes, computer-assisted scheduling, and accurate system response modeling are "software" items that require no construction project in order to upgrade a schedule. Most water authorities already incorporate weather and crop projections into their canal operations. If the upgraded schedule increases irrigation efficiencies, thus reducing water volumes delivered, excess canal capacity is generated to assist further upgrading.

A moderate encroachment on, or raising of, freeboard can be used to facilitate level-top canal operation. The channel bottom can be on any gradient, but the top must be level. Level-top canals can permit a zero-flow situation.

Incorporating on-channel reservoirs will reduce the magnitude of the needed increased capacities. They can reduce the need for canal automation in the upper canal reaches. The small level-top canals, only a few kilometers long, can be automatically maintained at a constant downstream level, even when the outflow rate is varying (Merriam, 1977). A 100-ha pilot project has been established in Sri Lanka by the Mahawili Development Board that has successfully demonstrated the feasibility of these practices to permit a limited-rate delivery schedule (Merriam, 1980). A number of other techniques could be added, including pumpback systems (Strongham and Hamid, 1975) on the farm, and even by the canal delivery system to recover operational spills (Jensen, 1980).

Multiple Scheduling Policies

Among institutional changes that could be considered by a water authority is to abandon project-wide uniform scheduling policies. Instead, the most flexible schedule that can be supported on each submain or lateral would be allowed. It is relatively easy to find farms that are adjacent to large supply mains. Because of the residual storage and bypass flow capacity of the large canal, these adjacent farms could be assigned limited-rate demand schedules, while other areas may be served by an arranged schedule, and yet more difficult areas not yet properly reconstructed, may be assigned one of the rigid schedules.

Usually in the name of fairness, multiple scheduling techniques have not been widely considered. This may be shortsighted, since the area that is served by a demand schedule releases management attention that can be concentrated on effective application and upgrading of the areas with less desirable schedules.

Some districts already practice a type of multiple scheduling. However, the different schedules are distributed seasonally, rather than geographically, to areas of a project. In off-peak seasons, the more flexible schedules are used, reverting to the rigid schedules during peak seasons.

COORDINATION

We should reiterate that cooperative efforts by farm unit operators and water delivery organizations to improve schedules and conserve water and energy do not create new water supplies, but water so conserved remains at the elevation of the supply reservoir (Interagency Task Force, 1979; Jensen, 1980, 1982). These improved schedules make it possible for the farmer to use water more efficiently only if he makes the effort. Some areas presently providing limited-rate demand schedules still have farm operations diverting up to three times that needed for crop water requirements. Most of this excess passes through the farm as surface runoff to be recovered lower down, or passes through the soil to be recovered by pumping. Granted, some of these savings are subject to evaporation and, perhaps, losses to saline sinks.

Utilization of the potential value of the flexible schedules requires that the farmer determine when he needs to irrigate. It is up to him to understand the soil-water-plant-atmosphere conditions so that he may optimize the irrigation frequency (Idso et al., 1977; Jensen, 1970, 1980). Techniques available to him have been improved so that the need for improved schedules is now more apparent. Irrigation management services offered by some private companies, irrigation districts, and government agencies have been effective in making the irrigator aware that he really can improve his conditions in regard to labor, production, and costs, if he can properly schedule his water (Jensen, 1980). For example, the use of deficit irrigation requires careful control of frequency and duration to make it operational. There is a growing awareness of the necessity for bringing about the attitudes, conservation consciousness, and institutional changes, as well as the agronomic and engineering applications needed for sustained food production. It now appears that with improved irrigation scheduling and water delivery on demand, with its control and measurement, that we now have the tools and concepts to bring well-managed crop production systems to within practical reach for a large proportion of irrigated agriculture.

REFERENCES

American Society of Civil Engineers (ASCE). "Recommended Irrigation Schedule Terminology Report of the On-Farm

Irrigation Committee of the Irrigation and Drainage Division," Proceedings of the Specialty Conference of the Irrigation and Drainage Division, ASCE, Flagstaff, Arizona, July 24-26, 1984, pp. 219-221.

Bos, M. G., and J. Nugteren. "On Irrigation Efficiencies," International Institute for Land Reclamation and Improvement,Publication 19, 2nd ed., Wageningen, The Netherlands, 1978.

Bos, M. G., J. A. Replogle, and A. J. Clemmens. Flow Measuring Flumes for Open Channel Systems. New York: John Wiley and Sons, 1984.

Burt, Charles M. "Canal Automation for Rapid Demand Deliveries," Proceedings, Special Conference of I & D Division, American Society of Civil Engineers, Flagstaff, Arizona, 1984, pp. 502-509.

Clemmens, A. J. "Control of Modified Demand Irrigation Distribution Systems," Proceedings of the Irrigation and Drainage Division Specialty Conference, American Society of Civil Engineers, Albuquerque, New Mexico, 1979, pp. 303-313.

Clemmens, A. J., M. G. Bos, and J. A. Replogle. "RBC Broad-Crested Weirs for Circular Sewers and Pipes," The Ven Te Chow Memorial Volume, G. E. Stout and G. H. Davis, eds. Journal of Hydrology, 68(1):349-368, 1984.

Dedrick, A. R., L. J. Erie, and A. J. Clemmens. "Level-Basin Irrigation," Advances in Irrigation, Daniel Hillel, ed. New York: Academic Press, 1982, pp. 105-145.

Food and Agricultural Organization (FAO). "Water for Agriculture," FAO of the United Nations, UN Water Conference, Mar del Plata, Argentina, March, 1977.

Heneggeler, Joseph C. "A Comprehensive Program for the Introduction of Laser-Controlled Land Leveling for Irrigation in Delta, Utah," M.S. thesis, Utah State University, Logan, Utah, 1983.

Horner, J. A., Charles V. Moore, and Richard E. Howett. "Increasing Farm Water Supply by Conservation," California Agriculture, 33 (11-12):6-7, 1983.

Idso, S. B., R. D. Jackson, and R. J. Reginato. "Remote-Sensing of Crop Yields," Science, 196:19-25, 1977.

Interagency Task Force (ITF). "Irrigation Water Use and Management," Interagency Task Force Report, U.S. Dept. of Interior, U.S. Dept. of Agriculture, and Environmental Protection Agency, Washington, D.C., 1979.

Jensen, M. E., D. C. N. Robb, and C. E. Franzoy. "Scheduling Irrigations Using Climate-Crop-Soil Data," Proceedings of the American Society of Civil Engineers, Journal of the Irrigation and Drainage Division, 96(IRI):25-28, 1970.

Jensen, M. E., ed. "Design and Operation of Farm Irrigation Systems," Monograph No. 3, American Society of Agricultural Engineers, St. Joseph, Missouri, 1980.

Jensen, M. E. "What are the Technical Possibilities for Maintaining Agriculture with Less Water?" Proceedings of the National Conference on Impacts of Limited Water for Agriculture in the Arid West, University of California, Asilomar Conference Grounds, Monterey, California, Sept. 25-Oct. 1, 1982.

Merriam, J. L. "A Management Control Concept for Determining the Economical Depth and Frequency of Irrigation," Transactions of the American Society of Civil Engineers, Vol. 9, No. 4, pp. 492-498, 1966.

Merriam, J. L. "Float Valve Provides Variable Flow Rate at Low Pressures," Proceedings of the Irrigation and Drainage Division Specialty Conference, American Society of Civil Engineers, Fort Collins, Colorado, 1973, pp. 385-402.

Merriam, J. L. "Level-Top Canals for Semi-Automation of On-Farm Irrigation and Supply Systems," Proceedings of the Irrigation and Drainage Division Specialty Conference, American Society of Civil Engineers, Reno, Nevada, 1977, pp. 217-224.

Merriam, J. L. "Demand Irrigation Schedule Concrete Pipeline Project," Mahaveli Development Board, Colombo, Sri Lanka, 1980.

Replogle, John A. "Critical-Flow Flumes with Complex Cross-section," Irrigation and Drainage in an Age of Competition for Resources, Irrigation and Drainage Division, American Society of Civil Engineers, pp. 366-388, 1975.

Replogle, John A. "Some Environmental, Engineering, and Social Impacts of Water Delivery Schedules," Proceedings of the 12th Congress of the International Commission on Irrigation and Drainage, Question 36, Report 61, 1984, pp. 965-978.

Replogle, J. A., and M. G. Bos. "Flow Measurement Flumes: Application to Irrigation Water Management," Advances in Irrigation, Daniel Hillel, ed. New York: Academic Press, 1982, pp. 147-217.

Replogle, J. A., and J. L. Merriam. "Scheduling and Management of Water Delivery Systems," Proceedings of the American Society of Agricultural Engineers Second National Irrigation Symposium, American Society of Agricultural Engineers, St. Joseph, Michigan, 1980, pp. 112-126.

Replogle, J. A., J. L. Merriam, L. R. Swarner, and J. T. Phelan. "Farm Water Delivery Systems," Design and Operation of Farm Irrigation Systems, M. E. Jensen, ed., Monograph No. 3, American Society of Agricultural Engineers, St. Joseph, Missouri, 1980.

Richardson, E. V., H. R. Horsey, and J. W. Andrew. "On-Farm and Delivery System Water Management," Proceedings of the Specialty Conference of the Irrigation and Drainage Division, American Society of Civil Engineers, Flagstaff, Arizona, July 24-26, 1984, pp. 493-501.

Wells, Edwin A., Jr., and Harold B. Gotaas. "Design of Venturi Flumes in Circular Conduits," Transactions of the American Society of Civil Engineers, 123:744-775, 1958.

Zimbelman, Darell D. "Computerized Control of an Open-Channel Water Distribution System," Ph.D. dissertation, Arizona State University, Tempe, Arizona, 1981.

Part 2

Economic Aspects

6
On the Allocation, Pricing, and Valuation of Irrigation Water

Robert A. Young

INTRODUCTION

The purpose of this paper is to review some economic concepts and evidence regarding the allocation, pricing, and valuation of irrigation water. The research program from which it derives arose from concerns with how irrigation water might best be allocated and financed in Third World countries.

The paper first discusses some general economic considerations regarding the allocation of water, touching on the role of government. Beneficiary charges in theory and practice are discussed next, with reference to the role of valuation and to the potential effects of various types of charging mechanisms. Next, alternative approaches for determining marginal value are listed and evaluated. The paper concludes with a review of research on irrigation water pricing and some suggestions for appropriate directions for research.

ECONOMIC CONCEPTS AND WATER ALLOCATION: AN OVERVIEW

Kenneth Boulding (1980) has noted that mankind employs three major mechanisms to reflect human values in the process of organizing human utilization of the earth's natural resource endowment. He labels these the "three P's"--Prices, Policemen, and Preachments. "Prices" represent the market system, operating through free exchange and a relative price structure. "Policemen"--the legitimated threat system or the political order--establish and enforce property rights and administer public regulations. "Preachments" represent the moral order, the process by which human values are learned, conveyed, modified, and employed in making choices.

Water, as with other resources, has been governed by a combination of these mechanisms. In contrast to many other natural resources, the political and moral modes have had, up to the present time, the dominant role. As Boulding puts it, water "has been the subject of sacred observance from very early times in human history...[it] becomes the object of a very complex structure of evaluations, rituals, superstitions and attitudes." Thus, water has been viewed as too important to be left to the marketplace, so that its administration falls largely in the political realm.

The Market System's Role as an Allocator of Resources and an Evaluation Mechanism

The term "market system" is used by economists in two senses. It may refer, in one sense, to an actual functioning system: the set of institutional and cultural arrangements that serves to allocate resources through the price mechanism. The term may also refer to an intellectual idealization of the system and how it performs. This idealization, or "model," has been studied to determine how apparently unrelated sets of activities achieve economic order, such that goods and services are provided to consumers at the place, time, and form desired, and capital, labor, and natural resources are organized through the productive system to provide these requirements.

The Idealized Market System. Any economic system must answer these questions: (a) What goods and services are to be produced? (b) What technologies are used in producing them? (c) Who is to enjoy the use of products? The adoption of the market system to answer these questions is based on the premise that the personal wants of individuals should decide the employment of resources in production, distribution, and exchange, and the individuals themselves are the best judges of their own wants (consumer sovereignty).

An idealized competitive market system (one that has many producers and consumers who are well informed, motivated by individual self-interest, and individually own and control resources) can be shown to have certain desirable properties. One such desirable attribute is that the system will produce the maximum-valued bundles of goods and services to consumers, given the endowment of resources, the available technology level, the preferences of consumers, and the distribution of purchasing power. Individual producers and consumers, acting within their own self-interest will, in accordance with Adam Smith's "invisible hand," arrive at an allocation of resources which cannot be improved upon. Producers, encouraged by prospective profit, buy inputs as cheaply as possible, combine them in the most efficient form, and produce those things which have the highest value relative to cost. Consumers' tastes and preferences influence their expenditure patterns, thereby encouraging firms to produce the commodities people want. Prices are bid up for the commodities most desired, and producers allocate resources in the direction of greatest profits. The firms most successful in the process (producing desired goods most efficiently) are rewarded by profit, and the unsuccessful are eliminated, so production occurs at least cost.

A second desirable property of the idealized market system is its ability to accommodate change in conditions of production and patterns of consumption. New knowledge and technology are rapidly reflected in the prices which producers are willing to accept for their products. On the consumer side, changes in income and preferences soon show up in expenditure patterns. Hence, a market system yields maximum satisfaction in not only a static but a dynamic context.

The actual market system may not always meet the precise preconditions of the idealized construct. The principal problems arise with public or collective goods (those which are nonrival in consumption), external or spillover costs (uncompensated side effects, such as pollution), and economies of large size (a precondition for monopoly). Mixed capitalistic systems are based on the presumption that for most goods and services, the allocation resulting from market processes sufficiently approximates the idealized system. Where this is not the case, regulatory processes or public production are provided to allocate resources.

Obstacles to Market Allocation of Water. Markets in water, however desirable from a conceptual point of view, are not yet common anywhere in the world as a means to a more productive use of resources. Several reasons might explain the relative lack of water markets. These are (a) physical (due to the nature of water and how it is used in production and consumption activities), (b) economic (which stems from the fact that, until recently, water has been in relatively plentiful supply), and (c) conflicting social values (in that material well-being is not the only yardstick used by society to measure success in water allocation).

The physical barriers to more extensive markets in water stem from its mobile, flowing nature, the fact that it is seldom fully "used" by the consumer, and the further fact of its potential for absorbing and carrying pollutants. As water changes from solid to liquid to gas throughout the season and the hydrologic cycle, it is relatively difficult to identify specific units of water. Hence, water presents unique problems in the establishment and enforcement of property rights, which are the essential foundation of any market allocation system. Second, most users of water only consume a part of it, even within one phase of the hydrologic cycle, the remainder being available to subsequent producers or households. The impediments to measuring portions consumed are a constraint on defining water rights and facilitating exchanges. Finally, the potential for water quality degradation is another problem difficult to deal with in market exchanges or water rights.

What may be labeled "economic" reasons for the heretofore limited development of markets stem from both the varied nature of water "use" and the relative plentitude of water (compared to demands). Water consumption is most often thought of in terms of the consumptive and diversionary uses, such as irrigation, and household and industrial uses. An important set of growing demands for water is in the class of instream, nondiversionary, and nonconsumptive uses. Recreational demands for flows (including wildlife habitat and noncontact streamside uses) constitute an important growth area. Hydroelectric power generation and waste load dilution are also increasingly in demand. A number of these instream uses represent collective consumption demands, which are partially nonrival in consumption. It is well known that such commodities are likely to be undersupplied in a market economy (Haveman, 1976; Randall, 1983). Most societies have therefore chosen nonmarket administrative mechanisms for allocation. The second economic reason for rudimentary development of markets lies in the apparently paradoxical assertion that water has not been particularly scarce, at least in the specific technical economic sense of the term. Even in exceedingly arid climates, additional supplies from mountain runoff or extensive groundwater supplies have often been relatively inexpensive. New uses did not strongly conflict with the interests of established water-consuming groups. The relative plenty implies that formal institutions for managing scarcity are only now becoming important.

Water is also a very "bulky" commodity, in that the value per unit weight tends to be relatively low. Therefore, costs of transportation and storage tend to be high relative to economic value at the point of use. Hence, only in limited cases is it economical to transport water, and the extensive rail, truck, and pipeline network that the market system has developed to transport more valuable liquids (e.g., petroleum) is absent for water.

The third major force inhibiting the adoption of market institutions for water allocation can be identified as conflicting social values. This is an example of Boulding's third "P": "Preachments." Even though it is likely that economic improvement would be best served by market allocations, several important conflicting themes emerge in opposition to the directions dictated by pure willingness to pay for water. One theme is, in Boulding's terms, "the sacredness of water as a symbol of ritual purity, exempts it in some degree from the dirty rationality of the market" (p. 302). Later in the same essay, Boulding remarks that water is "so holy and valuable to use as a symbol that we are apt to carry the production and transportation of it far beyond the point of rational economic returns" (p. 309). Religious teachings may explicitly or implicitly prescribe against market allocations of water.

Where markets are absent due to any of the above causes, government regulations may be established to provide for regularity of water use and to protect a given use against present and future competing demands. This type of protection may preclude economically efficient resource allocation, if demands for alternative uses outweigh the economic value of protected uses. Conversely, institutions designed to preserve a given use may provide an inadequate supply in the face of growing demands, but will be economically inappropriate in that they leave the impression that the problem is solved.

BENEFICIARY CHARGES FOR IRRIGATION WATER USE

This section focuses on the problems of setting rates for beneficiary charges for irrigation, a case in which water with private good characteristics is often publicly supplied. The rates or prices set have both resource allocation and equity impacts and influence the level of agency revenues. (See Seagraves and Easter, 1983, for a more general discussion.) Empirical evidence on the effect of pricing on water consumption suggests that imposition of a measuring/metering system together with volumetric charges, results in significant impacts on consumption. (Schramm and Gonzales, 1976, present a case study on irrigation in Mexico.)

Concepts for Rate Setting and Pricing

Nonspecialists often experience some confusion with the concepts and terms used in discussions of pricing. Figure 6.1 will help to sort out some of the ideas involved. The curve MB represents marginal benefits or demand for water reflecting marginal willingness to pay. Marginal cost (MC) represents the incremental cost of supply. The Pareto-optimal pricing policy, as is well known, would use MC as the price schedule:

$$p = MC \tag{1}$$

The optimal quantity to supply and consume is found by equating marginal cost with marginal benefit (or marginal value).

$$MC = MB \tag{2}$$

At that point, (labeled q^* in Figure 6.1) the willingness to pay for the marginal unit exactly equals the opportunity cost (willingness to pay for foregone opportunities). Any consumption greater than the optimal level will involve marginal units whose worth to the user is less than the incremental cost of supply. Conversely, price policies which constrain use below q^* will create a situation in which the value of additional units exceeds the cost of supplying them. A principal theorem of welfare economics shows that, in a properly functioning competitive market, price will equal marginal cost.

Returning to the problem of setting water rates, several points are noted. First, the functions MC and MB represent empirical relationships. That is, they describe relationships which exist in the practical problem setting. While these relations may, in some cases, be difficult to measure, techniques exist (discussed in the

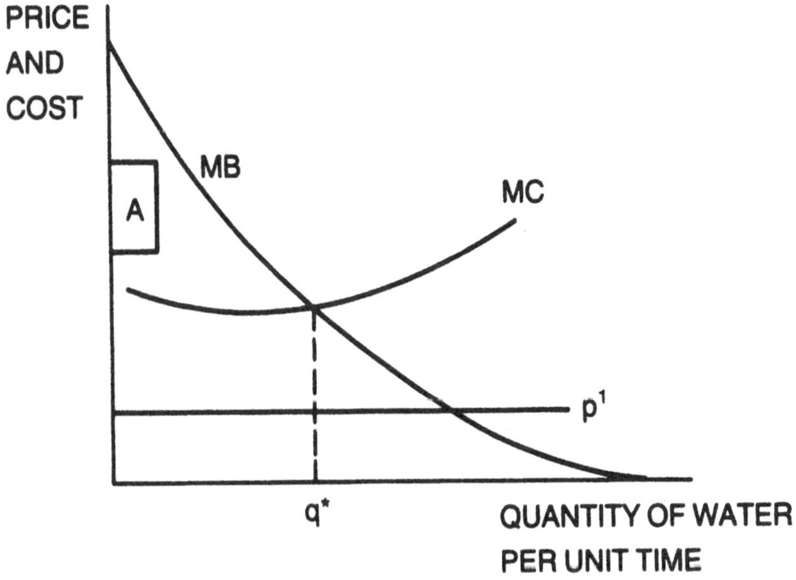

Figure 6.1 Concepts employed in analyzing water rates

following section) which provide directions by which the task may be accomplished. Price, however, is a decision variable for public water supply planners and must be established by a policy judgment.

However desirable it might be to follow a marginal cost pricing policy from an economic efficiency point of view, conflicting revenue and income redistribution objectives often dictate alternative solutions. In Figure 6.1, the allocative impacts of two broad alternative classes of rate-setting rules can be analyzed. Consider the line p^1, which represents a volumetric price set (arbitrarily) below marginal cost. Line p^1 can be set at any level, the only constraint being that it cannot exceed the maximum willingness to pay, represented by the vertical intercept of MB. Allocative efficiency losses are incurred to the degree that p^1 differs from MC.

The box labeled "A" in Figure 6.1 represents a non-volumetric rate system, which is the most commonly observed method of charging beneficiaries. This approach charges for access to water supply but does not measure or collect for incremental consumption units. Consumption of water by self-interested, fully informed water users under such a system would be predicted to occur at the horizontal demand intercept. Water use would be rationed only if the charges exceeded total willingness to pay, measured as the area under MB.

Rate Setting in a Multi-Objective Framework. Rate setting represents a choice of policies within a multiple objective framework, in which the major social objectives include (a) allocative (Pareto) efficiency, (b) equity of income distribution, and (c) "fairness" of apportioning costs (in the sense that persons in like circumstances should be treated alike). Subsidiary criteria include simplicity, administrative feasibility, and stability (Bonbright, 1961, pp. 290-292). A general principle or rule for setting rates can be associated with each major criterion. These principles can be thought of as converting one of the major social goals into a broad practical guide or formula for setting rates.

The Marginal Cost Pricing Principle is the rate-setting rule applied where allocative efficiency (maximizing net social product) is the primary objective. When rates are set according to the schedule of marginal cost of supplying water, then the user will demand the commodity as long as marginal willingness to pay exceeds incremental cost, and the optimal level of usage will result. A corollary of this principle is that the common practice of "flat rate" pricing of water, in which no marginal charge is imposed, is likely to encourage consumption beyond the optimal level.

While economists have generally endorsed the marginal cost principle, application of it is difficult because of the variety of definitions of the appropriate marginal cost concept for pricing policy. An example concerns the transactions costs associated with measuring, allocating, and monitoring a water pricing system. For example, in an irrigation system with plentiful water supplies and numerous small field units, the transactions costs of a volumetric pricing system may exceed the value of water saved (Bowen and Young, 1983). A second example is the long debate over the "short run marginal cost" principle stemming from the work of welfare economists in the 1930s. Strong objections have been voiced to the proposal to set utility prices equal to marginal costs, especially where marginal cost is below average cost (hence, requiring public subsidy). Coase (1971) emphasized the absence of a market test to determine whether users were willing to pay the total cost of supplying the commodity. He also viewed with disfavor the potential misallocation of resources stemming from the additional taxation, the redistribution of income in favor of users of products of decreasing cost industries, and the impetus toward centralization of the economy.

While most of these criticisms can be dealt with by a multipart pricing system (where marginal price is set equal to marginal cost, and an assessment is levied on users to reflect the costs which do not vary with output), establishment of such multi-part systems which accurately reflect costs is difficult. Nevertheless, multi-part rate structures are now frequently found in municipal and industrial irrigation and hydroelectric power systems. However, as it has been applied, multi-part pricing systems often fail to account for the economically correct concept of opportunity costs, focusing rather on historical or embedded costs. The opportunity costs which are relevant include both the value of water in alternative uses and the cost of securing incremental supplies in the presence of demand growth (Meier, 1983; Milliman, 1972; Davis and Hanke, 1971; Randall, 1981; Seagraves and Easter, 1983). In this view, historical costs are sunk and therefore irrelevant to establishing an efficient rate structure (Warford, 1977). Moreover, the opportunity costs of water should be determined by a market mechanism rather than by administrative procedures (Randall, 1981; Howe, 1984).

We turn next to a brief discussion of some alternative rate-setting principles which have been proposed or utilized.

The Ability-to-Pay Principle is an alternative principle for rate setting and rests heavily on the equity criterion. The rule provides the most common basis for setting rates for irrigation in the U.S. (and elsewhere) and is also regularly applied to village water supplies in developing countries. A common practice is to require only operating costs to be recovered fully, plus a small fraction of the initial investment.

The U.S. experience with federal irrigation projects is illustrative. Originally planned early in this century according to a full cost-recovery concept, three decades of unsuccessful attempts to fully recover costs ensued. In implicit recognition that costs overshadowed benefits (thus yielding zero demand if farmers were required to fully reimburse costs), an ability-to-pay procedure was authorized in 1939 (Huffman, 1953). A complex formula has been developed which limits the farmer repayment requirement to about 10-20 percent of estimated federal costs (North and Neely, 1976).

The ability-to-pay approach has little to commend it except in instances where low-income groups are to be explicitly subsidized. The concept is inherently subjective, and political pressures arise to set the formula in ways which redistribute income from taxpayers to water users. Since charges bear little relation to costs, no test of whether users would be willing to pay the total costs of supply exists.

The Net Benefit Principle, sometimes termed the "rent" principle, seeks to employ charges to capture part or all of the economic surplus accruing to the user. (In Figure 6.1, the net benefit or surplus is represented by the area under curve MB.) Net productivity of the user would govern the calculation, but neither past nor opportunity costs would enter in. The approach has been proposed for pricing public irrigation water in more centralized political systems (Ansari, 1968). The net benefit principle is consistent with the view that water and its fruits are the property of the state. However, setting rates strictly on the basis of net benefits appears to reflect a relatively deterministic view of the resource allocation process, one which ignores the incentive effects of pricing structures. Further, other producing sectors are not similarly charged for their resource inputs; this approach appears to violate the "fairness" principle.

The Average Cost Principle calls for recovery of all costs by charging for each unit received according to the average cost. It is simple and easy to understand. It is fair and equitable, in that beneficiaries pay just the resource costs incurred in their behalf. The desired signals to resource users are provided, although not in so precise a way as could be achieved by multi-part pricing. As the approach is usually applied, however, historical or "embedded" costs serve as the basis of the calculation rather than opportunity costs.

To sum up, in many places water is not yet sufficiently scarce to justify the tangible and intangible costs of establishing formal pricing systems. In such cases, flat rates will satisfactorily ration use and satisfy repayment requirements. However, when signals of scarcity of water (and of the costs of related construction capital and labor) are absent, pressures arise for structural solutions to satisfy incorrectly perceived water "needs." The expectation of increasingly scarce water supplies suggests eventual adoption of multi-part rate systems which reflect opportunity costs of water. Such systems are both efficient and fair, and have been shown to be operable in practice.

Cost Allocation

"Cost allocation" is the process of assigning an appropriate share of joint multipurpose project costs to each project purpose or user class and is a basic measurement issue in designing appropriate pricing or cost-sharing policies. User classes may be grouped according to economic sector, political subdivision, or both, and joint cost allocations among them have both allocative and distributive implications.

Given the nature of the problem, there is no ideal allocation procedure, and some degree of arbitrariness afflicts all of the suggested alternatives. Gittinger (1982, p. 233) and James and Lee (1971, p. 529) each list several guidelines for selecting allocation rules, of which three stand out. First, the method should be fair, in that the user class be charged at least the incremental cost of receiving project benefits. Second, the joint cost allocation procedure should not make infeasible any service class for which incremental benefits exceed separable costs. Third, no class of service should be assessed charges in excess of the benefits to be received.

Numerous cost allocation formulas can be identified, the most common of which are the "proportionate use of capacity" and "separable costs-remaining benefits" (SCRB) methods (James and Lee, 1971, p. 533). Because the first method assigns joint costs in proportion to the quantity utilized, expressed in terms of volumes or flow rates, it may be difficult to apply in cases where project outputs cannot be measured in volume terms, as with nonconsumptive uses, water quality, or flood control. A more significant objection to this procedure is that it can fail the second or third guidelines above (Herfindahl and Kneese, 1974, p. 291-292).

The SCRB method allocates to each user class the identifiable (or separable) costs of including that purpose or service in the project, plus a share of the joint or common costs. The joint cost share is allocated as a proportion of the benefits net of separable costs ("remaining benefits"). The SCRB method satisfies the guidelines listed above and is relatively simple to apply. Accordingly, it has been selected by federal agencies in the U.S. as the most acceptable approach.

A complication with SCRB concerns the sharing of the savings resulting from multipurpose developments as compared with single-purpose projects. Loughlin (1977) has suggested a credit to separable costs to remove the possible inequity from the SCRB procedure of crediting all savings to joint costs, an adjustment which results in a more suitable allocation of savings resulting from multipurpose projects. Riley et al., (1978) presented a detailed analysis of the problems with the various approaches in a case study of a multipurpose, multicountry project.

Some recent cost allocation proposals are based on a game theoretic framework. The theory of cooperative games provides approaches to joint cost allocation which take strategic possibilities into account. Heaney and Dickinson (1982) provide an integration of this literature with the more traditional analyses. See also H. P. Young et al. (1982) and Loehman et al. (1979) for applications. These highly formal approaches identify limitations of the traditional (i.e., SCRB) methods, but their complexity has inhibited the adoption of alternative solutions at the applied policy level.

ADAPTING THE IDEALIZED MARKET CONCEPT TO EVALUATE NON-MARKETED RESOURCES

The constructs embodied in the idealized market system discussed previously have been brought to bear on non-market resource allocation decisions in the form of the analytic system commonly known as benefit-cost analysis (Pearce and Nash, 1981). Water resource planning, in fact, represents one of the initial subjects and perhaps still the topic most widely studied with the benefit-cost evaluation mechanism (Krutilla and Eckstein, 1958).

The benefit-cost framework adopts the same principles as underlie the idealized market system, i.e., consumer sovereignty and acceptance of the existing distribution of purchasing power. The main effort in a benefit-cost analysis is the derivation of surrogate prices (usually called "shadow prices"). These are those that would emerge in the presence of a properly functioning market system and can be used in guiding resource allocation decisions. The use of techniques to shadow price water is the subject of the remainder of this section.

The process of shadow pricing can properly be understood as an attempt to establish an exchange ratio in monetary terms which would be exactly that which would emerge from a properly functioning exchange market. The basic concept is <u>willingness to pay</u> as an indicator of economic value. Willingness to pay reflects the amount that a rational, fully informed consumer would be willing to forego rather than do without the commodity in question. In accordance with the principles of diminishing marginal utility (in consumption) or diminishing marginal productivity (in production), willingness to pay falls as quantities increase. The willingness-to-pay relation is equivalent to the conventional demand function for a commodity or input, and exact shadow price estimates are points on the marginal willingness-to-pay relationship. A representative demand curve, labeled "D," is shown in Figure 6.2. Also shown in Figure 6.2 is a relationship labeled "MC" (marginal cost), representing the incremental cost of water supply.

The reader will recognize the correspondence of the relationships in Figure 6.2 with the textbook supply and demand curves of microeconomic theory. While the marginal value of water, depending on supply, may be at any point on D, the locus of most interest is the intersection of the two curves, reflecting q^* supply units, and identified on the diagram as p^*. Points not at q^* are suboptimal.

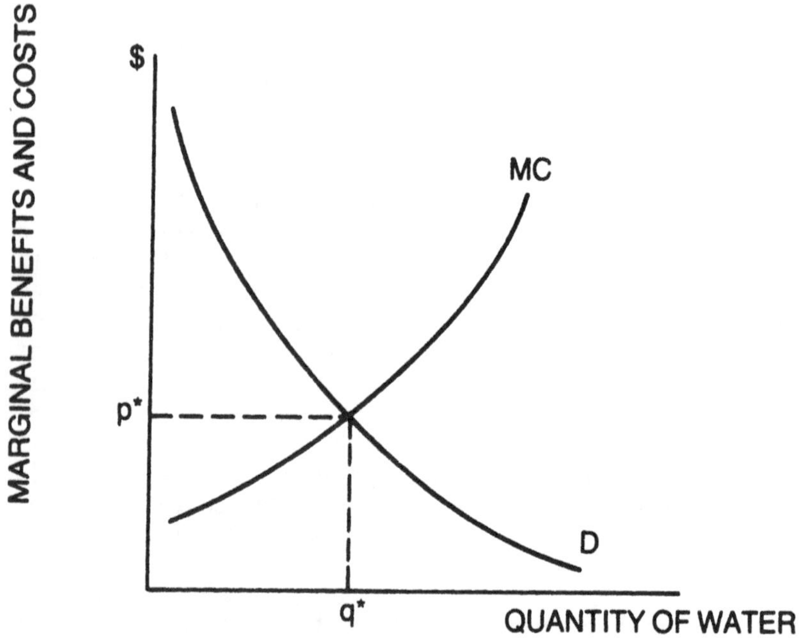

Figure 6.2 Demand (marginal value) and marginal cost curves for water supply

To the left of q^*, marginal value exceeds marginal cost, so gains can be achieved by adding q. The converse is true to the right of q^*. Many synthetic estimates aim to identify p^* when specifying a shadow price for water. (Often, however, the analyst is attempting to find what the willingness to pay would be for some specific quantity, usually in order to establish whether or not an added supply increment is valued in excess of its incremental cost, which amounts to determining whether the increment of supply in fact lies to the left or right of the optimal quantity, q^*.)

An important attribute of the demand for water is the responsiveness or willingness to pay to varying quantities. This is the inverse of the price elasticity of demand. Some types of use exhibit value which is highly responsive to quantity, so that small increases in quantity drive willingness to pay rapidly down. Industrial and household use fall in this category. The value of agricultural uses tend to be somewhat less responsive, but in all uses, significant increases in supply will negatively affect value at the margin.

Further Conceptual Distinctions

The hydrologic system must be considered in terms of its interactions with climate, land, ecosystems, and the human social and economic systems. This intricacy is further complicated by the highly variable nature of moisture supplies, the importance of sequential uses as water flows from the upper watersheds to its eventual destinations in sea or sump, and the importance of transportation costs in establishing water value. Concepts of the economic value of water can be relevant only when explicit recognition is given to quantity, location, quality, and time of supply. Put another way, the value of water is highly site specific, and varies directly with local conditions of supply and demand for the resource.

There are a number of methods and conceptual bases for generating shadow prices for water. Space limits preclude a detailed discussion here; the interested reader is referred to Gray and Young (1984) or Young and Gray (1972). Several of the more important issues are touched on briefly below.

The Residual Method of Benefit Estimation: Intermediate Goods. Benefit estimation is essentially a problem of assigning or imputing a shadow price to resources or commodities in the absence of markets to perform this function. The "residual method" is the most common of several approaches to shadow pricing producer's or intermediate goods (Young and Gray, 1972). The shadow pricing is achieved by allocating the total value of output among each of the resources used in a given productive process. The method is relatively straightforward. (See Heady, 1952, pp. 402-411, for a detailed exposition.) If appropriate prices can be assigned (presumably by market forces) to all inputs but one, the remainder of total value of product is imputed to the remaining (or "residual") resource.

Two principal postulates are required in the derivation. First is the condition that the prices of all resources are equated to returns at the margin (value of marginal product). (This is a well-known condition for competitive equilibrium.) Second, the total value of product can be divided into shares such that each resource is paid according to its marginal productivity and the total value of product is completely exhausted.

Consider a production process in which four factors of production, capital (K), labor (L), natural resources (R), and water (W), are used to produce a single output Y:

$$Y + f(K, L, R, W) \tag{2}$$

By the second postulate, we may write

$$TVP_Y = (VMP_K \cdot K) + (VMP_L \cdot L) + (VMP_R \cdot R) + (VMP_W \cdot W) \tag{3}$$

where TYP_Y represents the total value of output Y, VMP_i represents the marginal value product of the ith resource or factor of production. Substituting according to the first postulate (which asserts that $VMP_i = P_i$),

$$TVP_Y = P_K \cdot K + P_L \cdot P_R \cdot R + P_W \cdot W \tag{4}$$

Then equation (4) may be rearranged:

$$TVP_Y - (P_K \cdot K + P_L \cdot L + P_R \cdot R) = P_W \cdot W \tag{5}$$

The costs represented within the bracket on the left-hand side of equation (5) (capital, labor, and opportunity costs of other natural resources) are the "associated costs" referred to above.

The right side of expression (5) represents the contribution of water to the production process. Assuming that all variables in the expression are known except P_W, the equation (5) can be solved for that unknown to impute the shadow price of water, $\hat{P}_W$.

The question arises as to whether or not the postulates are satisfied. First, will factors paid according to their marginal value productivities just exhaust total product? The answer is provided by a principle known as Euler's Theorem, which states that, under certain conditions, resources paid according to marginal productivity will result in complete exhaustion of total product (Henderson and Quandt, 1978). The postulates cited previously are satisfied by production functions homogeneous of the first degree. The Cobb-Douglas function, subject to constant returns to scale, is one which satisfied Euler's Theorem and has been used in empirical estimation of marginal value products.

However, for only a very specific production function and where there are no fixed inputs are the conditions of residual imputation met. Thus, the method is valid so long as the requirements of the competitive model (including the equilibrium condition that marginal cost equals average cost) are satisfied. There also may be operational difficulty encountered through the use of prices as indicators of value marginal products for all resources but one. If resources are not allocated so that all factor inputs are employed to the level where prices are equated with value marginal products, the imputational process may result in either under- or overestimation of the value of the resource in question. Residual imputation can lead to erroneous shadow prices, when improperly employed. These limitations should be recognized by the user. One particular case is emphasized here. Even where the production function exhibits constant returns to scale and prices do reflect marginal value products (according to the postulates), one may encounter the problem of omitted variables. Omission of factor costs (including opportunity costs of unpriced production factors) means that the returns to such resources are being imputed to the residual resource, and, thus, the value or benefit estimate is overstated.

There is an additional technique which is closely related to the residual imputation approach and warrants discussion. It is the "change in net income" approach. This method (hereafter abbreviated CINI) defines the increment in net producer income associated with adding water to a production process as willingness to pay for the incremental water. The approach has been adopted for valuing irrigation water benefits by the U.S. Water Resources Council (1979).

Generalizing the notations used in equation (2), let

$$f(X_1, X_2 \ldots X_n; Y_1, Y_2 \ldots Y_n) = 0 \tag{6}$$

Equation (6) represents the multi-input, multi-product production function. Further, let the subscripts 0 and 1 attached to the input and output variables refer, respectively, to values without and with an investment or program adding to water supply. The water resource is designated X_1. Assuming that the increase in crop production following from the added water supplies is not so large as to influence crop prices, the change in net income associated with a discrete addition to water supply per unit of time is:

$$\Delta Z = Z_1 - Z_0 = \left(\sum_{i=1}^{m} Y_{1i} P_{y_i} - \sum_{j=2}^{n} X_{1j} P_{x_j}\right) - \left(\sum_{i=1}^{m} Y_{oi} P_{y_i} - \sum_{j=2}^{n} X_{oj} P_{x_j}\right) \tag{7}$$

The second term in (7), in effect, represents the annual net returns to the fixed land resources in the "without" project situation.

The unit value of water may be obtained by dividing the expression in equation (7) by the incremental quantity of water (i.e., ΔX_1).

The CINI approach requires the same assumptions of the residual imputation procedure, namely, that resources (including water) be optimally allocated, that there be no fixed inputs, that factor and product prices correctly reflect social values, and that all inputs be properly represented in the calculations. The CINI technique can also be interpreted as an approximation to the optimal allocation conditions expressed in equation (3) preceding for the case where the incremental water input is discrete rather than an infinitesimal amount.

Mathematical programming procedures can be employed to derive theoretically similar imputations of the value of water. Burt (1964) pioneered this approach with application to irrigation water, deriving a long-run net benefit function from parametric variation of a water supply constraint in a linear programming (L.P.) model of a California agricultural region. Depending on the formulation of the objective function of the L.P. model, long-run and short-run value estimates can be derived. Bowen and Young (1985) have applied the method to date from Egypt's Northern Delta region.

Deriving Value Estimates from the Production Function. The classical approach to estimating values of nonmarketed commodities is to estimate the demand function for the good in question. In most uses, water is an intermediate good, in which case the demand function is the marginal value product function, the first derivative of the production function in value terms. This technique has been most widely employed in valuing water in irrigation use, where numerous experiments have studied crop response to water application and other factors (for example, Hexem and Heady, 1978). The general approach is to derive a schedule representing the short-run value of the marginal product under the experimental conditions. While the technique has appeal as a means of estimating short-run private values, limitations are encountered in using it for estimating the long-run social value of water. Public intervention is often present in the market for particular irrigated crops, either through direct price control or price manipulation by supply control. Most studies employ the prices received by farmers in valuing outputs. In such cases, private willingness to pay would differ from the appropriate measure of the social value of the marginal unit at the point of use. Perhaps more important, the short-run production function, estimated with all factors but water fixed, may not provide an appropriate measure of the long-run marginal product.

Cobb-Douglas-type functions fitted to farm account data with irrigation water as an explicit variable have been employed in developing estimates of long-run marginal value productivity. A number of such studies have been done in India and Pakistan. (See Khan and Young, 1979, for an example.)

SPECIAL PROBLEMS OF ASSIGNING ECONOMIC VALUE TO IRRIGATION WATER

This section reviews and assesses some specific issues arising in the process of assigning values (shadow prices, net benefit measures) to irrigation water. With few exceptions, the technique must rest on direct observation of the response of crop yield to alternative levels and timing of water applications. The difficulties and complexities involved in measuring this relationship are touched on first. Then the general approaches which have been utilized are described and assessed.

Measuring Physical Productivity of Irrigation Water

A number of difficulties are encountered in making accurate measurements of the water application-crop yield relationships. First, crop production, with or without irrigation, is a biological process carried out in uncontrolled and highly variable environments. The process, therefore, is subject to the vagaries of diseases and pests and variations in climate (temperature, sunlight, wind, humidity, and rainfall). Output, even within a field or an experimental plot, may vary significantly with soil texture and fertility. Furthermore, irrigation decisions are made by a large number of individual farmers, each representing a small proportion of the total irrigation water utilized, and varying widely in management capability. In perusing the literature on water-crop response experiments, one is struck by the high rate of failure to achieve statistically reliable measures of response or, for that matter, any measure of response at all, even under conditions where rigorous experimental control is attempted.

Second, yield response to irrigation water application is especially sensitive to the rate at which water is combined with other inputs. Soil nutrient levels and seeding rates are of principal significance in this regard. Capital investments in land leveling and water distribution systems (ditches, pipes, sprinkler systems) are also important determinants of irrigation water productivity. Capital can substitute for water (and labor) and tends to enhance the productivity of water.

Third, in any irrigated crop producing area, there are a number of possible crops, each of which exhibits a unique value productivity with respect to applied irrigation water. Further, for each crop, there are a number of adapted varieties, and these may also respond somewhat differently to water application.

Fourth, the question of technological change should be noted. Improved crop varieties can increase the physical productivity of irrigation water, which dictates caution in utilizing productivity measures which are not of recent origin.

Fifth, crop response may be inhibited by salinity in the irrigation water. Salts in the water are concentrated in the crop root zone by the evapotranspiration process. Extra water to leach out excess salts is then required in order to maintain crop productivity levels. Some steps toward incorporating such consideration into the evaluation of irrigation water have been reported by Moore, et al., (1974), Oyarzabal and Young (1978), and Yaron and Dinar (1982).

Sixth, and finally, some discussion of the proper conceptualization of production response to irrigation water is appropriate. Application of the conventional textbook production function, which simply postulates annual yield to be related to annual water input, greatly oversimplifies the true input-output relationship facing the irrigator. A more realistic model will reflect the fact that the productivity of irrigation water varies widely over the year, depending particularly upon soil moisture level and upon stage of growth of the plant.

With respect to soil moisture content, the response of plants to an application of irrigation water when the soil in the root zone is already moistened to field capacity by rainfall or by a previous irrigation, would be zero, or in some cases, even negative. Water productivity tends to increase as the interval of time from the last moistening lengthens. As soil moisture is depleted, a point may be reached at which failure to irrigate would lead to complete loss of the crop. Water applied at this time is extremely valuable because the value is equivalent to the net income loss avoided by the application of irrigation water. Water productivity may also vary over the life cycle of the plant. To illustrate, water application near harvest time may have little impact on yield, or may even diminish productivity by adversely affecting quality of the crop. The yield of many crops whose primary economic value is in a seed or fruit are known to be highly sensitive to moisture availability during the period of flowering. Growth processes are, in a sense, irreversible, and the productivity of water application depends upon the time of application and prior history of watering.

These characteristics of crop response to irrigation water have prompted a number of analysts to discard the static economic model and to formulate the irrigation water allocation problem as a multistage or sequential decision process, which can be solved within the format of dynamic programming (Flinn and Musgrave, 1967). However, attempts to formulate a dynamic programming model which could accommodate both the multistage and the multiproduct aspects of irrigation water allocation met with difficulties, due to an excessive number of state variables.

Several attempts have been made to circumvent this difficulty. Anderson and Maass (1971) developed a simulation approach to the problem. DeLucia (1969) developed what he termed a "sequential linear program" with the desired sequential optimization feature. Young and Bredehoeft (1972) utilized a hybrid of the previous two procedures. (See Vaux and Pruitt, 1983, for a general review of these issues.)

CONCLUSION

The initial conceptual framework established that an estimate of the marginal benefit was necessary in establishing a pricing policy, no matter what pricing principle (marginal cost; multi-part; ability-to-pay; flat rate) was to be applied. The next section discussed problems of estimating marginal benefit functions. I conclude with a summary of the procedures and implications from some economic research recently performed at Colorado State University (Bowen and Young, 1983).

A linear programming model of farming in the Kafr El Sheikh region in Egypt's Nile Delta was developed and operated for the purpose of estimating irrigation net benefit functions. The model incorporated a range of potential water use levels for each of the typical crops and adjustments in water use efficiency in order to provide measures of the total, average, and marginal net benefit functions for irrigation water in the region. In the absence of more satisfactory ways of reflecting crop response to alternative water supplies, generalized response functions (from the Food and Agriculture Organization) were adapted to Egyptian conditions. The results were expressed in both social benefits (international prices) and private benefits (governmental prices).

Government revenue and pricing policies lower farmers' valuations of water. A high tax burden lowers willingness to pay for production inputs. Long-run efficiency concerns are probably much more important than the short-run distortions in allocative efficiency produced by current pricing policy. The transfer of much of the farmers' surplus out of agriculture reduces the ability of farmers to invest in productivity improvements, including investments in improving water management technologies.

The marginal social net benefit function can be interpreted as an opportunity cost function, when read from right (full supply) to left (reduced supply). This function measures the social opportunity cost of reduced water supply to the study area, which might occur due to increased scarcity. Such a function can be useful in decision-making regarding the relative desirability of the alternative policies of developing new water supply or reallocating existing supply from existing irrigated lands to lands planned for reclamation.

The results of the study show that irrigation water in the northern delta study area has a high average social benefit but a low marginal benefit. Social net benefits may differ from other areas, due to regional differences in productivity of the land and cropping patterns. For instance, returns to water in many of the newly reclaimed lands are far less than the estimates in this study.

Our research effort on water pricing was aimed at determining which of several cost-recovery instruments would be appropriate for Egyptian conditions, as evaluated under the concerns for allocative efficiency and equity in income distribution. The full range of instruments that could be considered is quite large, consisting of different combinations of allocative rules, quotas, water charges, and water markets. We have evaluated two broad types of water charges: area-based taxes and volumetric prices. Since the analytical results and price policy implications of the government and market models were similar, the subsequent discussion reports only the findings derived from the market model.

Two levels of cost recovery were evaluated. One level recovers all budget operating and capital costs of providing irrigation water, plus the estimated cost of administering a water pricing system. The other level recovers budget operating costs only. These costs are estimated to be 20 L.E. and 10 L.E. per <u>feddan</u> per crop for area-based charges and 25-37 L.E. and 10-22 L.E. per feddan per crop for volumetric charges.

Single-charge instruments (area-based taxes and flat volumetric charges) can usually only guarantee the attainment of one objective. In this analysis, cost recovery is the presumed objective.

The optimal pricing instrument, judged by the efficiency criterion, is the instrument that maximizes returns to land and water in the study area, net of social costs incurred in providing and charging for the irrigation water.

Unlike the efficiency objective, there is no agreed-upon basis for defining an "optimal" equity position. The equity concerns considered here are the distribution of farm income along the watercourses and the differences in per capita income among farms of varying size.

At the present time, aggregate irrigation water supply in Egypt is generally adequate to meet demand in the agricultural sector. It is not surprising, then, that the results of this study show pricing systems with a zero marginal charge to be most efficient under nonscarce supply.

Because of lower administrative costs, area-based charges are more efficient than volumetric charges, under current water supply. The flat land tax is the least expensive instrument and has the advantage of being allocatively neutral. Although crop taxes theoretically can produce allocative distortions, no misallocations were predicted by the model under the range of conditions tested. This result follows from the fact that demand for water in the linear programming model is a step function. In this case, the demand was perfectly inelastic with respect to price (or price proxies) within the range of water charges examined.

There were also no differences among the water charging instruments, under current water supply, according to the measure of income equity. Income equality along the branch canals was achieved by all the instruments. Under constant returns to scale, distribution of income per unit of land will be equal when water is not limiting.

In conclusion, the analysis has shown area-based water charges to be more efficient and just as equitable as volumetric charges, under the plentiful water supply conditions that have been the situation since the completion of the High Aswan Dam. In particular, the flat land tax appears to be the most satisfactory, using both efficiency and equity criteria. We make this recommendation only on the condition that new water charges would not add to the current agricultural tax burden in Egypt but would be balanced by reduced taxes on crop price penalties. Effective taxation on farmers in the northern delta study area was shown to be high, and further increases would worsen long-run distortions in agricultural incentives.

Our results show that water supply to Egypt's agricultural sector would need to decline substantially to warrant volumetric pricing. Small farm size is an important factor in the high transactions cost of measuring water and providing it on a demand basis. Land taxes would continue to be an appropriate method of raising revenue but would need to be supplemented with administrative rules for allocating water when water scarcity becomes a more important concern.

REFERENCES

Anderson, R. L., and A. Maass. "A Simulation of Irrigation Systems," Technical Bulletin 1431 (Revised), U.S. Department of Agriculture, Government Printing Office, Washington, D.C., 1978.

Ansari, Nasim. Economics of Irrigation Rates. New York: Asia Publishing House, 1968.

Bonbright, James C. Principles of Public Utility Rates. New York: Columbia University Press, 1961.

Boulding, K. E. "The Implications of Improved Water Allocation Policy," Western Water Resources: Coming Problems and Policy Alternatives, Marvin Duncan, compiler. Boulder, Colorado: Westview Press, 1980.

Bowen, Richard, and R. A. Young. "Allocative Efficiency and Equity of Alternative Methods of Charging for Irrigation Water: A Case Study in Egypt," Egypt Water Use Project Tech. Report No. 37, Colorado State University, Fort Collins, Colorado, 1983.

Bowen, Richard, and R. A. Young. "Financial and Economic Irrigation Net Benefit Functions for Egypt's Northern Delta," Water Resources Research, 1985 (in press).

Burt, O. R. "The Economics of Conjunctive Use of Ground and Surface Water," Hilgardia, 36, California Agricultural Experiment Station, Berkeley, California, 1964.

Coase, R. "The Theory os Public Utility Pricing and its Applications," Bell Journal of Economics, 1:113-128, 1971.

Davis, R. K., and Steve Hanke. "Pricing and Efficiency in Water Use Management," U.S. National Water Commission Technical Report No. PB209083, National Technical Information Service, Springfield, Virginia, 1971.

DeLucia, R. J. "Operating Policies for Irrigation Systems Under Stochastic Regimes," Ph.D. dissertation, Harvard University, Cambridge, Massachusetts, 1969.

Flinn, John, and Warren Musgrave. "Development and Analysis of Input-Output Relations for Irrigation Water," Australian Journal of Agricultural Economics, 11(1): 1-19, 1967.

Gittinger, J. P. Economic Analysis of Agicultural Projects, 2nd ed. Washington, D.C.: Johns Hopkins University Press, 1982.

Gray, S. L., and R. A. Young. "Valuation of Water on Wildlands," Valuation of Wildlands Resource Benefits, George K. Peterson and Alan Randall, eds. Boulder, Colorado: Westview Press, 1984.

Haveman, R. H. Economics of the Public Sector, 2nd ed. New York: John Wiley and Sons, 1976.

Heady, Earl O. Economics of Agricultural Production and Resource Use. New York: Prentice Hall, 1952.

Heaney, J. P., and R. E. Dickinson. "Methods for Apportioning the Cost of a Water Resource Project," Water Resources Research, 18(3):476-482, 1982.

Henderson, James M., and Richard E. Quandt. Microeconomic Theory: A Mathematical Approach, 3rd ed. New York: McGraw-Hill, 1978.

Herfindahl, Orris C., and Allen V. Kneese. Economic Theory of Natural Resources. Columbus, Ohio: Merrill Publishing Company, 1974.

Hexem, Roger W., and Earl O. Heady. Water Production Functions for Irrigated Agriculture. Ames, Iowa: Iowa State University Press, 1978.

Howe, C. W. "Innovations in Water Management: An Ex Post Appraisal of the Colorado Big Thompson Project and the Northern Colorado Water Conservancy District," monograph in preparation, University of Colorado, Boulder, Colorado, 1984.

Huffman, Roy E. Irrigation Development and Public Water Policy. New York: Ronald Press, 1953.

James, L. D., and R. R. Lee. Economics of Water Resources Planning. New York: McGraw-Hill, 1971.

Khan, M. Jameel, and R. A. Young. "Farm Resource Productivities, Allocative Efficiencies, and Development Policy in the Indus Basin, Pakistan," Land Economics, 55:388-396, 1979.

Krutilla, J. V., and O. Eckstein. Multiple Purpose River Development: Studies in Applied Economic Analysis. Baltimore, Maryland: Johns Hopkins Press, 1958.

Loehman, E., J. Orlando, J. Tschirhart, and A. Whinston. "Cost Allocation for a Regional Wastewater Treatment System," Water Resources Research, 15(2):193-202, 1979.

Loughlin, J. C. "The Efficiency and Equity of Cost Allocation Methods for Multipurpose Water Projects," Water Resources Research, 13(1):8-15, 1977.

Meier, Gerald M., ed. Pricing Policy for Development Management. Baltimore, Maryland: Johns Hopkins University Press, 1983.

Milliman, Jerome. "Beneficiary Charges: Toward a Unified Theory," Public Prices for Public Products, Selma Mushkin, ed. Washington, D.C.: Urban Institute, 1972, pp. 27-52.

Moore, Charles V., P. Sun, and J. H. Snyder. "Effects of Colorado River Water Quality and Supply on Irrigated Agriculture," Water Resources Research, 10(2):137-144, 1974.

North, Ronald K., and W. P. Neely. "A Model for Achieving Consistency for Cost-Sharing for Water Resource Programs," Water Resources Bulletin, 13:995-1005, 1976.

Oyarzabal-Tamargo, F., and R. A. Young. "International External Diseconomies: The Colorado River Salinity Problem in Mexico," Natural Resources Journal, 18:77-88, 1978.

Pearce, D. W., and C. A. Nash. The Social Appraisal of Projects: A Text in Cost-Benefit Analysis. New York: John Wiley, 1981.

Randall, Alan. "Property Entitlements and Pricing Policies for a Maturing Water Economy," Australian Journal of Agricultural Economics, 25:195-212, 1981.

Randall, Alan. "The Problem of Market Failure," Natural Resources Journal, 23:131-148, 1983.

Riley, Jay P., Herbert Fullerton, and John Keith. "Cost Allocation Alternatives for the Senegal River Development Program," Water Resources Planning Series P-78/06, Water Research Laboratory, Utah State University, Logan, Utah, 1978.

Schramm, Gunter, and F. V. Gonzales. "Pricing Irrigation Water in Mexico: Efficiency, Equity, and Revenue Considerations," Annals of Regional Science, March 11, 1976.

Seagraves, J. A., and K. W. Easter. "Pricing Irrigation Water in Developing Countries," Water Resources Bulletin, 19:663-672, 1983.

U.S. Water Resources Council. "Procedures for Evaluating National Economic Development (NED) Benefits and Costs in Water Resources Planning (Level C); Final Rule," Federal Register, 44(242):72892-72976, Government Printing Office, Washington, D.C., 1979.

Vaux, H. J., Jr., and W. O. Pruitt. "Crop-Water Production Functions," Advances in Irrigation, 2, 61-97, Daniel Miller, ed., 1983.

Warford, J. J. "Pricing as a Means of Controlling the Use of Water Resources," Proceedings, U.N. Water Conference, Water Development and Management, 1, Pt.2, 659-668, 1977.

Yaron, D., and A. Dinar. "Optimal Allocation of Irrigation Water," American Journal of Agricultural Economics, 64:681-689, 1982.

Young, H. P., N. Okada, and T. Hashimoto. "Cost Allocation in Water Resources Development," Water Resources Research, 18(3):463-475, 1982.

Young, R. A., and J. D. Bredehoeft. "Digital Computer Simulation for Solving Problems of Conjunctive Ground and Surface Water Management," Water Resources Research, 8:533-556, 1972.

Young, R. A., and S. L. Gray. "Economic Value of Water: Concepts and Empirical Estimates, U.S. National Water Commission Technical Report No. PB210-036, National Technical Information Service, Springfield, Virginia, 1972.

7
Social and Economic Impacts of Investments in Ground Water: Lessons from Pakistan and Bangladesh

Sam H. Johnson III

INTRODUCTION

Although both Pakistan and Bangladesh are cut by vast river systems, the relatively flat topography of their agricultural land limits the potential for large-scale reservoirs. In spite of this, Pakistan possesses the world's largest contiguous gravity irrigation system. On the other hand, Bangladesh has no more than 10 percent of its cultivated land served by gravity irrigation. However, in both countries, given their high population growth rates, there is a continuous need to increase agricultural production. In order to meet this requirement, these countries have been forced to develop their vast underground water supplies. Pakistan started this process in the mid-1950s, while Bangladesh, due to the War of Independence, started in the mid-1970s. Yet by 1983, Pakistan had installed more than 14,000 public deep tubewells (DTWs), while Bangladesh had installed in excess of 17,000 public DTWs.

As Pakistan and Bangladesh were once divisions of the same country, and prior to that were part of India under the British (see Figure 7.1), it is not surprising that the earlier organizational structures for groundwater development and management had some similarity. However, it is of interest to study how groundwater is presently managed in the two countries and, in particular, to notice how the management structure for groundwater in the two countries has diverged.

The purpose of this paper is to examine social, technical, and economic aspects of this massive investment in groundwater. Although it is necessary to describe many of

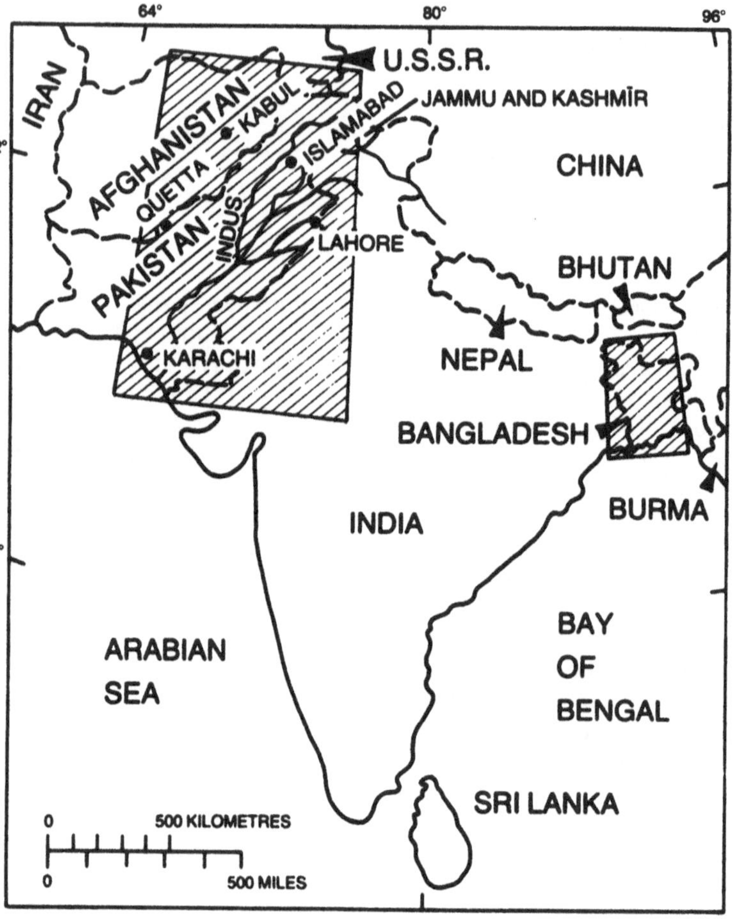

Figure 7.1 Map of Asian subcontinent showing location of studies

the physical parameters of the system, the paper focuses on government policy concerning groundwater development and documents economic and social impacts of this policy. In order to accurately present the situation in Pakistan and Bangladesh, the paper is divided into two sections: one that details groundwater development and management in Pakistan and another that covers the situation in Bangladesh. The final two sections discuss the social and economic impacts and the long-term implications of current policies and draws some conclusions that may be used by the two countries, or other countries, as they expand groundwater development in the future.[1]

GROUNDWATER DEVELOPMENT IN PAKISTAN

Between the 1830s and the 1960s, the Indus Plain, which encompasses more than 207,000 sq km and stretches 1,200 km from the Himalayan foothills to the Arabian Sea, was covered with the world's largest contiguous block of irrigated land. Here the Indus and its tributaries were developed by the British to serve an irrigated area of 13 M ha (Taylor, 1965). Yet, the bounty of the irrigation system was not perfect. Given the gentle slope of the Indus Plain, 0.2 m per km, drainage soon became a major problem in many areas.

Before the development of canal irrigation in the nineteenth century, the groundwater hydraulic system in Pakistan was in a state of dynamic equilibrium. Over moderately long periods of time, recharge to the groundwater reservoir balanced discharge, and there were no long-term changes in groundwater levels. However, irrigation changed the natural hydrologic environment of the Indus Plain. The canal system introduced additional sources of recharge and caused a rise of the water table in and around the irrigated areas. Seepage losses were greatest near the bifurcation points in the upper parts of the areas between the rivers known locally as doabs because of the greater density of canals. Seepage losses were less near the rivers because the water table was already close to the surface (Mundorff et al., 1976).

Figure 7.2 illustrates the change in depth to water table from preirrigated time to the early 1960s. The water table in the middle of the doabs rose from 20 to 30 m over this 80- to 100-year time period. This rise initially was on a linear trend and maintained a constant

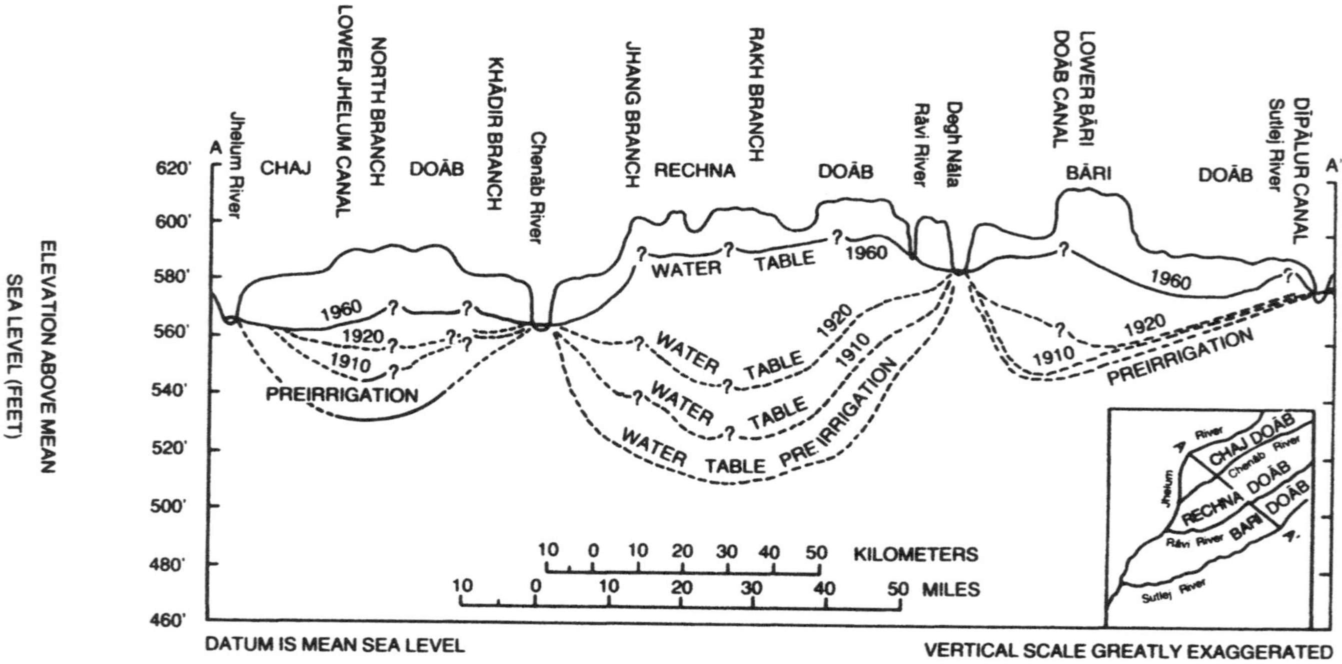

Figure 7.2 Water table profiles, Chaj, Rechna, and Bari doābs (D.W. Greenman, V.M. Swarzenski, and G.D. Bennett, 1967)

slope until the water table approached the land surface (Greenman et al., 1967).

Soon after independence in 1947, Pakistan became increasingly concerned about the growing waterlogging and soil salinity problems in the Indus Plain. By 1950 over 2.0 M ha of irrigated land had gone out of production, with additional land going out of production at the rate of 29,000 ha each year. The government of Pakistan (GOP) requested the Food and Agriculture Organization of the United Nations (FAO) for help in finding a solution to the waterlogging and salinity problems. In response, in 1950, the FAO sent a number of experts in the fields of drainage and reclamation to study the problem. In 1952, again at the request of the government of Pakistan, the United States Bureau of Reclamation (USBR) sent a drainage engineer, E. R. Maierhofer, to study the damaged areas of the Punjab and the Khairpur/Shikarpur area in the south.

After 1958, drainage and reclamation works were transferred to the Water and Power Development Authority (WAPDA). In 1961, a plan for eradicating waterlogging and salinity in the whole of Pakistan was prepared by WAPDA with the assistance of its consultants, Harza Engineering Company International, Tipton and Kalmbach, and Hunting-MacDonald. However, prior to the completion of the WAPDA plan, a project for reclamation of 490,000 ha of land, known as Salinity Control and Reclamation Project I (SCARP-I), was prepared. Using $15,200,000 made available to the government of Pakistan by the United States Development Loan Fund, work on SCARP-I was begun in 1960. The project area for construction of SCARP-I was in the center of the interfluvial area between the Ravi and Chenab rivers, known as the Rechna Doab. One of the major objectives of SCARP-I was to demonstrate the effectiveness of vertical tubewell drainage for lowering the water table over a large area and as a means of providing sufficient water for intensified irrigation and leaching of salts from saline-affected soils (Malmberg, 1975).

While SCARP-I was under construction, WAPDA and its consultants were completing their ambitious program for elimination of waterlogging and salinity throughout Pakistan. In this plan, the Upper Indus Plain was divided into 10 reclamation projects ranging from 0.4 to 1.6 M ha each, and the Lower Indus Plain was divided into 16 projects ranging from 0.3 to 0.8 M ha each. In all, the programs embodied the construction of 31,500 tubewells, 12,500 km of major drainage channels, and 42,000 km of supplemental drains serving more than 12 M ha in the northern and southern zones (Ahmad, 1974).

A panel of experts (hereafter referred to as the Panel), headed by Roger Revelle, was sent by the American President to study the problem of waterlogging and salinity in Pakistan. The Panel prepared a comprehensive report on agriculture, drainage, and reclamation in Pakistan, which examined technical, institutional, and organization solutions. Engineering aspects of the report (often called the Revelle Report or the White House Report) were generally along the lines of the WAPDA program, although the Panel used sophisticated computer models to demonstrate that the development of groundwater by public tubewells could provide an intermediate solution to waterlogging and salinity problems. These tubewells could also serve as a means of providing a much-needed additional supply of irrigated water (White House - U.S. Department of the Interior Panel, 1964).

A major issue that continues to be discussed is the merits of publicly installed deep tubewells (DTWs) contrasted to privately installed shallow tubewells (STWs). The almost unanimous recommendation for public tubewells in the early 1960s can be partially explained by the fact that many of the original studies were completed before there was any significant degree of private tubewell development in the Indus Plain. However, a small number of Pakistani and foreign consultants, most notably Dr. Ghulam Mohammad from the Pakistan Institute of Development Economics, argued that public tubewells should be installed in areas where groundwater was too saline to be applied to lands without dilution with canal water. In areas of non-saline, good quality groundwater, development should be left to private users, with the government facilitating development by providing the electrical grid and credit schemes for purchase of pumps and motors (Mohammad, 1965). More recent studies by groups like the World Bank's Indus Basin Review Mission, the Punjab Government Special Committee on the Working of SCARPs, Mundorff and the WAPDA Master Planning Division (1979), all have the benefit of hindsight. Thus, their stronger argument for private tubewell development can be explained in light of actual changes since earlier recommendations.

SCARP Design

SCARP-I was completed in 1963, and, after it demonstrated that the water table could be successfully lowered by tubewells uniformly distributed over a large area, additional public tubewell projects were implemented in both the northern and southern zones. Over 8,000 public tubewells, covering more than 2.3 M ha, were built between 1959-1977 (Table 7.1). More than 14,000 tubewells, covering 3 M ha, had been completed by 1983, and construction is still under way. Total costs are estimated to have exceeded U.S.$1 B.

In SCARP-I and some areas of SCARP-II (e.g., Lalian, Khadir, and Mona), capacities of the tubewells were fixed so that the combined water supply from surface and groundwater at the watercourse head was one cubic meter per second for 2,144 ha (one cubic foot per second for 150 acres). In subsequent SCARPs, a cropping intensity for the area was projected, and the tubewell capacities were determined to provide the necessary water supply to meet this requirement, either with or without canal supplies, depending on the area. Table 7.2 illustrates the projected changes in cropping intensity expected after the SCARPs were in operation. The larger projected increases in the more recent SCARPs reflect the change in design criteria discussed above.

In general, the capacities of the tubewells ranged from 56 to 142 liters per second. The choice of tubewell capacity was made by considering the tubewell requirements of one or more than one adjoining watercourse command or chak.[2] As chaks vary from 80 to 400 ha, this often resulted in one tubewell serving up to three chaks. Distribution works for each tubewell required structures for proportional allocation of tubewell supplies to watercourses to be served. At first it was thought that link watercourses, which connected the tubewell to the main watercourse channel for each chak, would be excavated by the farmers. However, in SCARP-I and parts of SCARP-II farmers were unable (or unwilling) to dig these link watercourses. (The link watercourses were usually completed by the contractors in the more recent SCARPs.) No provisions were made for enlarging the main watercourse channel and distribution systems, even though they were expected to carry two to three times their previous flow quantity.

Table 7.1 Implementation of public tubewell projects

Project	Zone	Gross Area (million hectares)	Tubewells (number)	Period	Installed Capacity (m/sec)	Costs[a]
SCARP-I	N	.49	2,069	1959-63[b]	180	25
SCARP-II	N	.67	2,205	1963-73	298	90
SCARP-III	N	.43	1,635	1966-73	203	40
SCARP-IV	N	.23	935	1967-73	127	20
Khairpur	S	.18	540	1969-70	48	10
North Rhori	S	.32	1,192	1973-77	69	50
Karkana Sukkur Shikarpur	S	.01	87	1973-75	8	--
Total		2.33	8,663		933	235

[a]These figures do not reflect all associated costs (1977 million dollars).

[b]256 tubewells installed from 1954-58.

Source: Central Monitoring Organization - WAPDA (1971), Review of Completed Salinity Control and Reclamation Projects. WAPDA Press, Lahore, Pakistan.

Table 7.2 Cropping intensities in SCARP projects (percent)

Project	Culturable Cultivated Area (million ha)	Intensity Pre-Project	Projected Intensity	Intensity 1975-76
SCARP-I	.46	89	150	116
SCARP-II	.6-	83	130	102
SCARP-III	.37	54	120	97
SCARP-IV	.22	63	150	91
Khairpur	.13	106[a]	135	109[a]
Rohri North	.28	98	150	

[a]Questionable value

Source: Master Planning and Review Division, Water and Power Development Authority (1979), <u>Revised Action Programme for Irrigated Agriculture Report</u> (3 volumes). <u>WAPDA Press</u>, Lahore, Pakistan.

SCARP-II Project Performance

SCARP-II is in the Upper Chaj Doab, between the Chenab and Jhelum rivers. Most of the tubewells have fiberglass screens, but about 25 percent were initially installed with mild steel screens. As mentioned earlier, some schemes within SCARP-II were designed with fixed water duty of about one cubic meter per second for every 2,144 ha, while other schemes were designed to meet a projected cropping intensity which, in general, meant that they had a higher water duty. Phalia, Bhusal, and Sohawa were designed to meet a projected cropping intensity, rather than an arbitrary fixed water duty. However, due to decline over time of the pumping capacity of the tubewells, submergence of watercourse channel inlets, improperly designed and constructed link watercourses, and under-capacity watercourse channels, actual flows were often much less than the designed supplies (Table 7.3).

The measured reduction in flow results partially from a decline over time of the output capacity of the tubewells. The decline of tubewell capacity in a sample of 81 tubewells in SCARP-II/A was 21 percent. WAPDA records indicate that the overall decline of tubewell capacity in all SCARP-II/A is over 30 percent (Master Planning and Review Division, 1979). Other reasons watercourses do not receive their full design flow are submergence of watercourse channel inlets (either by tubewell flow and/or by limited capacity of the channel to carry the canal water combined with the water from the tubewell), low flow in the distributary, and poor design and condition of link watercourses. In a sample of 22 tubewells in the Phalia section that were operating at designed pumping capacity, actual water flow (tubewell water plus canal water) in the main watercourse channel was only 67 percent of designed capacity. The rest of the flow was either lost in the tubewell link watercourses, and/or the proper amount of canal water entering the watercourse channel was restricted by submergence of the inlet.

Link watercourse channels. Connecting watercourse channels to the tubewell outlet in SCARP-II was more difficult than in SCARP-I because the wells in general had higher output capacity and usually served two or more watercourses. The project plan assumed that the farmers would construct the link connections between the tubewells and the watercourses. In practice, this has not worked out. All of the tubewells in SCARP-II are connected to watercourses, but many of the connections, especially in

Table 7.3 Actual delivery compared to design delivery: SCARP-II

Section	Chaks Per Section	Average Design (liter/second)	Actual Delivered (liter/second)	Percent Delivered
Mona	11	128	98	77
Lower Hujjan	7	113	92	81
Phalia	10	95	63	67
Bhusal	19	115	89	78
Sohawa	25	99	82	82

Source: Data collected in SCARP-II by author and researchers during 1977.

SCARP-II/A, are unsatisfactory. The high-capacity tubewells were designed to flow through a sophisticated diversion box which allocated the water to two or more watercourses. In actual practice, many of the diversion boxes are being bypassed, and the tubewell water is serving only one watercourse (U.S. Agency for International Development, 1970). Measurements taken in 21 link watercourses in the Phalia section during canal closure, when the only water entering the watercourse was tubewell water, showed an average loss of 19.6 percent of the tubewell discharge.

Once the water enters the watercourse channel, it comes under the control of farmers served by that channel, who are supposed to maintain the watercourse channel and distribution system. Often this is not done because of neglect, ignorance, and village conflicts.

Average watercourse channel losses on unimproved water channels varied from 10 percent to 15.9 percent per 300 m of length and averaged 13.5 percent. In terms of farmers' needs for water, these losses represent a critical shortage, especially at middle and tail sections of the chak. With losses of this magnitude, by the time the water reaches 1,500 m from the head of the watercourse, the users have lost half of the initial flow entering the system. Assuming an average delivery of 79 percent of the design flow entering the system and losses of 13.5 percent per 300 m, the users 1,500 m down the watercourse channel are receiving only 40 percent of their design allocation. A sample of measurements in the Sohawa section indicates that, at 1,000 m, farmers are only receiving 44 percent of the design flow. Similarly, a sample from the Bhusal section indicates that farmers 1,000 m from the junction of the main channel and the link watercourse channel are receiving only 38 percent of design flow.

<u>Operating Schedules</u>. SCARP tubewells are supposed to be operated on schedules developed by the Irrigation Department. These vary from wells in perennial canal areas, nonperennial canal areas, and uncommanded areas. Schedules do not allow for rainfall, power failures, or personnel problems, and therefore must be considered as no more than general guidelines. The Irrigation Department has two guidelines for the interagency scheduling committees, which meet biannually to schedule tubewell operations in SCARP-II:

- over the year, pumps should run at 40 percent of annual capacity

- on days when pumps are operated, they should run continuously from 12:01 a.m. until 12:00 noon, with scheduled rest periods between 12:00 noon and 4:00 p.m.

Given these guidelines, the main area of choice is the number of days per month the tubewell should be operated. These schedules should take into account plant-water relationships, rainfall, and expected canal water availability. In fact, the proposed Lalian pumping schedule varies little from month to month. It bears little relationship to that proposed by the Land Reclamation Department (LRD), a schedule that attempts to match expected water supplies with expected demand.[3] Nor do actual pumping schedules resemble either the proposed LRD schedule or that followed by private pump operators. More flexible groundwater pumping, closer to the schedule proposed by LRD, could prevent both over- and under-pumping and, potentially, could support a higher cropping intensity.

Maintenance Problems. According to WAPDA data, SCARP-II has seen a decline in the utilization rate during recent years, from an average of 49.7 percent installed capacity in 1974-75 to 37 percent in 1976-77. As electricity charges have increased at a rate exceeding 12 percent per year for the last four years and budget allocations have not kept pace, the utilization rate is expected to continue to decline. Over the same time period, the allocation of funds for maintenance and repair work has decreased by 14-15 percent, with consequent impairment of operation.

Public DTW Program Performance

The entire SCARP program has been affected by the poor operating records of individual SCARP projects. However, factors such as unforeseen increases in energy costs, shortened tubewell life, rapid development of private tubewells, and failure to achieve desired cropping intensities have all combined to make SCARPs an economic and financial burden.

Economics. Depending upon the various consultants' assumptions and mandates, their estimated costs for relieving waterlogging and salinity problems throughout the Indus Plain ranged from $1.2 to $2.7 B. Predicted benefit/cost ratios for these plans were as high as 7.5:1 and as low as 2.25:1. As vertical drainage projects of this magnitude had never before been tried, all of these ratios were very dependent upon the underlying assumptions.

One assumption that was clearly incorrect in almost all of the proposed programs was the significant underestimation of the number of private tubewells that would be developed, even with the implementation of the public tubewell schemes. Ghulam Mohammad's 1964 survey of 23,000 private tubewells in 16 districts of the northern zone of West Pakistan established that tubewells were very profitable and the number installed would continue to increase (Mohammad, 1964). His findings were validated--between 1965 and 1975 the number of private tubewells increased by four times (Table 7.4). Yet, even the Lieftinck Report, written in 1967, still failed to appreciate the fact that private tubewells had the potential to replace public tubewells in most of the nonsaline groundwater areas.

Another assumption that was also proven wrong concerned the length of life of the public tubewells. Most consultants originally predicted 40- or even 50-year service lives. When it became apparent that the pumping capacity was quickly declining in almost all of SCARP-I and that a number of wells were facing critical problems with encrustation and corrosion, the consultants first tried to change from mild steel to stainless steel and fiberglass strainers. It was soon obvious, however, that even these materials were seriously affected by minerals in the groundwater. Therefore, the consultants reduced their estimates of tubewell life to 20 or 25 years. In 1971, the Special Committee on the Working of SCARPs (Land and Water Development Board, 1971) set 12 years as the average life of a SCARP tubewell. Depending upon the acceptable degree of decline in pumping capacity and the amount public agencies are willing to pay for repairs, "life" is a relative term, but, in general, it seems that approximately 15 years is going to be the practical life for most SCARP tubewells.

A third assumption that has not proven correct was that concerning increased cropping intensity. Almost all early studies planned to double cropping intensities from 75 percent to 150 percent. This has clearly not happened; in a few areas, cropping intensity rose to as high as 135 percent, but even there it was not settled at 125 percent. In most areas, cropping intensity has stabilized at around 115 percent, with increased crop yields and more area under higher-valued crops.[4]

Unfortunately, while changes to higher-valued crops do increase revenue from water charges, higher yields do

Table 7.4 Number of private tubewells in Pakistan

Year	Punjab and N.W. Frontier Provinces	Baluchistan Sind and Provinces	Total	Annual Increase
1965	29,007	3,447	32,524	
1966	36,663	3,806	40,469	7,945
1967	45,103	4,250	49,353	8,884
1968	54,570	4,751	59,321	9,968
1969	63,000	5,267	68,267	8,96
1970	76,509	59,420	82,451	14,184
1971	83,337	6,665	90,002	7,551
1972	92,298	7,442	99,740	9,738
1973	101,425	8,050	109,475	9,735
1974	112,002	8,415	120,417	10,942
1975	122,702	9,694	132,396	11,979
1976	133,807	10,193	144,000	11,604
1977	143,355	10,675	154,030	10.030

Source: Master Planning and Review Division, Water and Power Development Authority (1979), Revised Action Programme for Irrigated Agriculture (3 volumes). WAPDA Press, Lahore, Pakistan.

not. In SCARP areas, as the water supply has theoretically been doubled, double water charges are supposed to be assessed. In fact, many farmers refuse to pay double charges because they claim that by increasing acreage they are already, in effect, paying double water charges. However, water charges have not changed since 1969 and are not very significant.

Public Compared to Private Tubewells. The rationale underlying the recommended public sector role in groundwater development was that private development (1) would be inequitable and, therefore, not benefit most small farmers, (2) would be haphazard and probably not accomplish the desired drainage function, (3) could deteriorate the groundwater aquifer through uncontrolled pumpage, and (4) could not be expected to proceed at the rapid rate desired. In the early 1960s, this rationale seemed logical--Pakistan had limited experience with private or public development of groundwater. Yet by the mid-1960s, there were over 30,000 private tubewells, and some experts (both local and international) urged that private, rather than public, development be stressed in areas overlying fresh groundwater (Eaton, 1965). While this advice was noted by the World Bank report, it was not strongly supported and was, therefore, rejected (in effect). By 1978, Pakistan had acquired substantial groundwater development experience in both sectors. Results of private tubewell development have been demonstrated to serve the needed drainage function and also improve cropping intensities. The public sector program has lagged far behind its original and revised goals and has only partially performed its drainage function. Private tubewell investment has continued in SCARP areas, as centralized management has been unable to meet the flexible needs of the water users (Hussain et al., 1976).

While a number of postproject, benefit-cost-type analyses of SCARP-I have been made, only a few have attempted to compare SCARP-I to an equivalent private tubewell area; that is, one that has developed with private tubewells supplementing canal supplies. While this type of study closely resembles "with-and-without" analysis, it is not exactly the same, for there are private tubewells within SCARP-I. As part of the development of the Revised Action Programme for Irrigated Agriculture by Master Planning and Revision Division-WAPDA (1979), SCARP-I was compared with both the perennial commanded area in Upper Rechna Doab (162,000 ha) which borders SCARP-I, and the adjacent Lower Rechna Doab (Tandlianwala) area (110,000 ha). In the Upper Rechna Doab, there was one private tubewell per 33 ha, and in the Tandlianwala, one per 55-61 ha in 1975.

Growth in private tubewells and increases in cropping intensity have been faster in both areas than in SCARP-I. From this it can be inferred that the development of public tubewells slowed investment in private tubewells in the SCARP-I area. Assuming that if SCARP-I had not been built, private tubewells in that area would have developed to a density of one tubewell for every 67 ha, Master Planning calculated a rate of return for SCARP-I of 6 percent. When Master Planning data are used but tubewell density is increased to that of Tandlianwala (i.e., one tubewell for every 50 ha), the rate of return on SCARP-I is less than 3 percent. Even with a density of one tubewell per 67 ha, the predicted cropping intensity in 1976 would have been 122 percent, compared to an existing intensity of 117 percent, and groundwater withdrawals would have increased by over 22 percent.

GROUNDWATER DEVELOPMENT IN BANGLADESH

In certain areas of Bangladesh, there is a long history of small-scale irrigation using traditional, manual methods. Most common of these has been the dhone (pivoted boat-like devices capable of lifting up to three meters and discharging about 1.13 liters/second) and swing baskets (lifting about one meter and discharging up to 1.7 liters/second). Though accurate information is not available, traditional methods still account for about half the area irrigated from surface sources and about one-third of the total irrigated area.

The postpartition era marked the beginning of major water resource investments in Bangladesh.[5] Like earlier Indian efforts, planning focused on large-scale, public undertakings to stabilize water regimes associated with rainy season rice production. Investment took two forms: improvement in flood control and drainage and the development of supplementary irrigation during the monsoon season. Emphasis was on flood control and drainage, not irrigation, to increase agricultural productivity (Bottrall, 1983).

International observers visiting Bangladesh in the late 1950s and 1960s focused attention on the country's unique flood problems and consistently recommended water resources investment strategies which relied heavily on the construction of embankments and channel improvements (Planning Commission, 1973). This approach was institutionalized in 1959 with the creation of the East Pakistan Water and Power Development Authority (now the Bangladesh Water Development Board) and the development of the nation's first Water Resources Master Plan in 1964. The plan recommended implementation of 50 major projects, which would embrace large areas of the country, provide flood protection and drainage to 4.9 M ha, and supply irrigation facilities to another 3.2 M ha in 1985. Irrigation services were to be provided by gravity canals, with secondary pumping only in areas not serviceable by the gravity system. The plan assumed groundwater development to be costly and largely ignored pumps and wells, except in small areas in Dinajpur, Mymensingh, and Comilla districts. A major review of the plan, requested by the government and implemented by the World Bank in 1966, questioned the plan's basic assumptions and concluded that smaller pump-based surface and subsurface systems, if introduced with a high-yielding input package developed at the International Rice Research Institute (IRRI), could produce the rice needed to feed the country's growing population by 1985 (Hanratty, 1983).

Although never accepted by the government, highly criticized by donors, and subsequently modified to include only 20 "core projects," the plan had a significant impact on water resource policy decision in Bangladesh. The Bangladesh Water Development Board (BWDB) adopted it as its major operating document and organized its staff and support activities in anticipation of full implementation. The original project portfolio served as the basis for the water resources investment strategies outlined in the 1965-70, 1970-75, and 1973-78 national plans. Long after large-scale projects had proven to be of dubious merit, the Water Board continued to commit scarce human and financial resources to each of these projects annually. Yet, in Bangladesh, the entire area irrigated by large-scale, BWDB-managed schemes is less than 85,000 ha, or only about 5 percent of the total national irrigated area. These systems are thus relatively unimportant in purely physical terms, although they do have historical interest as they are a product of the centralized planning and management focus of BWDB (Bottrall, 1983).

Public LLPs and DTWs

Although major emphasis fell on flood control projects, smaller-scale surface and subsurface systems were experimented with throughout the 1950s and 60s. For example, 56.6 liters/second (two cusec) low lift pumps (LLPs), using surface water, were introduced through the Mechanized Cultivation and Power Pump Irrigation Program (MCPPI) beginning in 1956. A total of 3,990 LLPs were fielded, first under the auspices of the Water Board and then the East Pakistan Agricultural Development Corporation (BADC), a semi-autonomous government agency established in 1962 to improve the distribution of agricultural inputs.

Shortcomings led to the replacement of the MCPPI scheme in 1968 by the Thana Irrigation Program (TIP). The scheme proved exceptionally successful and by 1969-70, 18,000 pumps irrigating 285,000 ha were in operation. Although constrained by problems of water losses, poor maintenance, and timely pump distribution, the major constraint facing this program was the availability of surface water. The number of LLPs rose from about 1,300 in 1960-61 to almost 33,000 in 1972-73 and somewhat more than 40,000 during the past five years. This flattening of growth of LLPs reflects the relative scarcity of surface water supplies in relation to demand. LLPs currently serve about one-third of the total irrigated area.

While experiments with minor surface irrigation were under way, early pilot programs to tap underground water were also commencing. In 1961, the German government, in cooperation with the BWDB, installed 380 113 liters/second (four cusec) electrically powered wells at Thakurgaon in the northwest part of Bangladesh. From the outset, the project had problems. Although engineering and installation work was completed in two years, construction of an electrical generating plant and transmission system delayed operation until 1965. Costs were prohibitive, averaging $58,000 per well, approximately half of which was for electrical generation and transmission facilities. Coverage per tubewell was limited because of high seepage losses and failure to train and organize farmers. Subsequent training of farmers and the formation of cooperatives led to some improvements, but, as is illustrated in Table 7.5, commanded area is still far below potential, even after extensive efforts by BWDB and IRRI. It can be seen in this table that, even though the DTWs were designed to discharge 113 liters/ second (4 cusec), present

Table 7.5 Thakurgaon Tubewell Project, Bangladesh 1981, 1982, and 1983 dry seasons

Tubewell No.	Discharge Capacity	Area Irrigated Rabi Season		Area Irrigated Per Unit of Discharge	
		1982	1983	1982	1983
63	103	8.1	16.2	.079	.157
77	53	1.2	18.2	.023	.343
89	67	8.1	24.7	.121	.369
93	99	22.7	26.7	.229	.270
117	71	6.1	13.8	.086	.194
118	57	8.5	14.2	.149	.249
119	57	8.1	11.7	.142	.205
120	79	12.9	12.6	.113	.159
125	85	2.4	21.0	.028	.258
126	106	6.1	10.9	.058	.103
138	88	6.5	10.1	.074	.115
142	85	4.4	9.3	.052	.109
Average	80	7.9	15.9	.099	.199

Source: Bangladesh Rice Research Institute, Bangladesh Water Development Board, and International Rice Research Institute (1984), "Applied Research for Increasing Irrigation Effectiveness and Crop Production," IRRI Water Management Division. Los Banos, Philippines.

discharge rates are far below this quantity (Bhuiyan, 1984).

A second pilot project, implemented by the Kotwali Thana Central Cooperative Association (KTCCA) in east-central Bangladesh, made extensive use of low-cost, manual drilling techniques, and installed 211 56.6 liters/second (two cusec) diesel-powered wells between 1962 and 1970. Using simple hand-operated drilling machinery, which relied heavily on unskilled local labor, the wells were relatively inexpensive, averaging $6,000 per well. With command areas managed by established cooperatives, the DTWs averaged .42 ha per liter, 60 percent more than wells in Thakurgaon.

The questionable depth and focus of the Master Plan, early success with surface and subsurface minor irrigation, and the introduction of a new seed technology highly adapted to dry-season irrigation led to a reevaluation of the nation's water resource development policy in 1970. Under the auspices of the World Bank, an action program focusing on food production, not flood protection, was presented. Implementation, which placed heavy emphasis on small, quick-yielding schemes, was forced to wait until after the War of Independence in 1972 (World Bank, 1970, 1972).

World Bank studies completed in 1970 and 1972 emphasized small, quick-yielding irrigation schemes. The studies projected foodgrain self-sufficiency by 1983 (18.8 M tons) through the implementation of a multifaceted program including high-yielding seed production and distribution; use of input packages comprising seeds, fertilizer, plant protection, and improved draft animal power; greater availability of low-lift pumps and small drainage improvements; completion of minor to medium size drainage works; and the rapid expansion of tubewell irrigation.

Five-Year Plans

The First Five-Year Plan (1973-78). Components in the Bank's study became the basic building blocks of the First Five Year Plan. In part paralleling the study, the plan recognized "the tremendous potential that could be realized with small- and intermediate-scale irrigation and drainage projects, low-lift pumps and tubewell development" (Planning Commission, 1973). It suggested investment totaling Tk 598 crores (1971 U.S.$1.26 B) to irrigate an additional 1.13 M has by 1975. However, the involvement of the Water Board in drafting the plan assured a strong continued bias to large-scale projects. Consequently, 54 percent of the total water resources budget and 46 percent of its foreign exchange requirements were earmarked for large-scale projects. Secondary emphasis was placed on deep tubewell development, which consumed an additional 30 percent of budgeted funds.

Emphasis on the expansion of deep tubewells was well founded. With fewer than 1,000 wells in 1970, the potential was obvious. Economic returns, which were presumed to begin immediately after installation, were favorable, with installation costs ranging from U.S.$5,300 to $15,800 (1971) and annual net returns averaging $8,000 per well. Also, the divisibility of the technology allowed for a number of different technologies to be tested simultaneously, and those proving most suitable were subsequently utilized. Finally, wells could be geographically dispersed, thus distributing benefits more equitably than large-scale projects, making better uses of location-specific soil and water characteristics, and targeting irrigation to those areas where farmer demand was high. The latter was an important factor in improving command area performance.

The Second Five Year Plan (1980-1985). This plan places heavy emphasis on technically simple, divisible, quick-to-plan and quick-to-implement projects. Stronger focus on minor irrigation is anticipated, with five-fold increases in investment over the First Five-Year Plan. Use of shallow tubewells is to be emphasized, while deep tubewell installation will be limited to areas where shallow tubewells are not appropriate. Again, the use of low-lift pumps is constrained, with investments focusing on the fielding of new pumps to bring the number fielded to 50,000 (close to the estimated limit of surface water supplies) and the purchase of replacement units. For the first time, improving command area performance became a major objective, and the private sale of shallow and hand tubewells, set at 30 and 200 thousand units, respectively, was officially encouraged. Finally, the plan recommended a reduction in government subsidies on water, through the gradual increase in low-lift pump and deep tubewell rental and shallow and hand tubewell sale prices. Sale of low-lift pumps and deep tubewells was also to begin.

Tubewell Expansion

As of early 1984, over 120,000 STWs had been installed and were irrigating 1.7 M ha. The total number of DTWs sunk under various schemes is about 18,000, with around 16,500 commissioned as of January 1984. By early 1984, approximately 220,000 manual tubewells (MTWs) had been purchased and installed and were serving in excess of 33,000 ha. Under the Medium-Term Food Production Plan (MTFPP), which is designed to attain food self-sufficiency by 1985, Bangladesh plans to increase land irrigated by mechanized lift devices from a 1979-80 level of 1.0 M to 2.0 M ha by 1984-85 (Planning Commission, 1980). As illustrated in Table 7.6, this plan involves a significant increase in the number of tubewells (of all types) in use in the country.

With a wide variety of physical, economic, and social conditions, no single lifting device is superior in every way. However, as seen in Table 7.6 in terms of total area served, LLPs, STWs, and DTWs are by far the most important lifting devices in Bangladesh, with STWs and DTWs being of immediate concern for groundwater development.

PLANNING AND MANAGEMENT

In both of the two countries, planning and management for groundwater development has been erratic, at best. The process has clearly reflected the large-scale surface water bias of WAPDA (now BWDB in Bangladesh) and has failed to recognize the unique aspects of separable tubewells compared to continuous large-scale surface systems.

Local Level Organization

In Pakistan, selection of DTWs that potentially could serve over 500 ha and as many as 200 farmers reflects the fact that planners gave little thought to local-level conditions. Even the most cursory investigation would have revealed that farmers along a single watercourse had difficulties organizing for operation and maintenance. The bulk of court cases originating from rural areas concerned conflicts over water and associated land. Profusion of large, publicly owned and operated tubewells that were designed to serve two or more watercourses immediately

Table 7.6 Present and projected status of mechanical irrigation lift devices in Bangladesh

	Approximate Number in Operation (000)		Approximate Area Irrigated (000 hectares)		Additional No. Proposed Under MTFFP (000)	Approximate Area Proposed to be Irrigated (000 hectares)
Device	1980-81[a]	1983-84[b]	1980-81[a]	1983-84[b]	1984-85	1984-85
LLP	36.0	36.0	567	567	51.0	810
DTW	11.5	16.2	259	389	18.0	437
STW	24.0	120.0	109	648	90.0	437
MTW	100.0	200.0	16	32	180.0	29
Total			951	1636		1713

[a]World Bank (1982)

[b]Estimated by IADS Water Management Office

created a potential for all sorts of new conflicts. Investigations of farmers' organizational capacities, as well as their technical ability to deal with larger flows of water, would have indicated that smaller-capacity, more localized tubewells were better suited to existing conditions. The argument that larger public wells are more "economic" than smaller private wells rests on the unproven assumption that management under both systems would be the same. Planners failed to recognize, or ignored, farmers' limited capacity to cooperate at the watercourse level, as well as technical difficulties they faced in trying to redesign watercourse channels to carry higher flows. This was plainly a gross error of planning and goes far toward explaining failures of SCARPS to be properly utilized at the local level.

Similarly, in Bangladesh, the initial DTWs installed by BWDB were too big (113 liters/second) and, hence, required an extensive effort to educate farmers. Failure to properly train farmers in irrigated agriculture techniques and provide management expertise to operate the DTWs resulted in very low returns to groundwater investment in schemes such as that at Thakurgaon. In light of this experience, Bangladesh's decision to invest in smaller 56.6 liters/ second (2 cusec) DTWs and STWs has been correct. However, even here, area commanded has been far below potential. In one pilot effort, 20 DTWs were selected by staff from the Integrated Rural Development Program (IRDP), BADC, and the Department of Extension and Management to improve the system performance. At the end of the year (1979-80), the coverage per DTW increased from an average of 21.1 ha to 32.2 ha, and grain production per ha increased by 57 percent (World Bank, 1982). In another pilot scheme, CARE and the Bangladesh Krishi Bank (BKB) joined forces with BADC to form the Deep Tubewell Irrigation and Credit Program (DTICP) to improve the performance of 10 DTWs. In the first year (1979), they were able to increase the irrigation coverage by 55 percent from the preproject level, and the average farmer's yield increased by more than 56 percent. Encouraged by this result, the scheme was expanded significantly, such that by 1983, the project was working with more than 700 DTWs in six administrative units (Johnson, 1984). Table 7.7 details data from DTICP for 101 DTWs in three administrative units.

These studies and pilot projects have proven that investment in human capital and organizational structure in DTW areas has high potential return. The formation by the GOB of the Irrigation Management Program (IMP) under

Table 7.7 Participation in DTICP schemes: averages for 1982-83 data

	Commanded Area (hectares)		Participating Farmers		Yield/Ha. (tons/hectare)	
Location	Pre-Project	DTICP	Pre-Project	DTICP	Pre-Project	DTICP
Kaligoni	5.7	15.8	24	64	3.3	5.4
Parbatipur	23.1	27.5	44	63	4.4	5.4
Dhamrai	15.4	23.5	88	144	4.1	6.5
Weighted Means	16.4	23.8	69	96	4.1	6.1

Source: Johnson (1984)

IRDP is a step in this direction, but it has yet to develop an effective method of working with farmer groups.

Both in Pakistan and Bangladesh, relative to DTWs, much less information is available on the potential and actual use of STWs. These smaller tubewells are generally owned and operated by small farmer groups, extended families, or individual farmers with no government subsidy and, effectively, no government control. There is a general impression that the potential of STWs is better utilized than that of DTWs. However, available records from Bangladesh indicate that STWs are irrigating around 0.25-0.3 ha/liter/ second, which is no better than present DTW performance (Biswas et al., 1978). A similar type of study from the Punjab in Pakistan indicates that private STWs there irrigate about the same as those in Bangladesh, i.e., 0.35 ha/liter/second (Ashraf, 1978). Yet, both countries have seen a phenomenal increase in STWs, which indicates that even with their relatively small command areas, they are viewed as being a good investment. This is primarily due to the small number of users and, hence, the effectiveness of the informal users group.[6]

Economic Costs

In both Bangladesh and Pakistan, public groundwater development has been heavily subsidized. These subsidies have led to significant inefficiencies and have also eventually grown to be a major financial burden on the government treasury.

In Bangladesh, except for manual tubewells, all of the mechanical lifting devices have been sold with a subsidy. These subsidies have been particularly large in the case of LLPs and DTWs. In the late 1970s, BADC accounts indicated that actual payments by farmers for LLPs amounted to 12 percent of the cost to BADC, while for DTWs the figure was less than 10 percent. Current arrangements for selling tubewells (compared to past policies of renting them) result in prices for STWs that are nominally unsubsidized. However, DTWs continue to be heavily subsidized, with a selling price amounting to only 43 percent of the cost to BADC (World Bank, 1978). In addition to the nominal price of tubewells, provisions for the sale of all tubewells involve subsidized credit arrangements which further reduce the effective price paid. For example, for an STW sold for Tk 30,000 (in 1984, 23 Taka = U.S.$1.00) a farmer only has to pay Tk 2,000 as a down payment and receives a 6-year loan for the balance, paying a reduced rate of interest (12 or 13 percent, depending upon the source of the loan) per year. Furthermore, repayment of agricultural credit has generally not been satisfactory, with repayment rates of less than 70 percent (World Bank, 1983).

The most obvious impact of the subsidies has been the lack of incentive to increase command area to a national-level optimum. This is particularly acute in the case of the DTWs, where, after about 16 ha, there is effectively no private incentive to expand the command area, as the rental subsidy has significantly reduced the benefits of spreading capital costs over a larger number of acres (Small, 1983). Yet, as documented in Table 7.8, there are potentially substantial per-ha public returns to be generated if command area can be increased.

As indicated earlier, Pakistan has also subsidized their public DTWs. On the average, the subsidy exceeds $12 per ha for more than 3 M ha. With current water fees and collection rates, total revenue from water charges is insufficient to cover even operation and maintenance charges. Addition of SCARP operation and maintenance expenses to the already overburdened Irrigation department operation and maintenance budget has further increased the Department's deficit. For example, in 1975-76, the Punjab Irrigation Department spent approximately $1.3 M on ordinary operation, maintenance, and staffing on 628,400 ha served by the Lower Jhelum Canal. An additional $3.2 M was spent on operations and maintenance, as well as staffing to provide tubewell drainage to 360,000 ha served by SCARP-II within the Lower Jhelum Canal Command. The combined operation, maintenance, and staff budget for 1975-76 in the Lower Jhelum Canal Circle was therefore $4.5 M. With recovered water charges of approximately $2.9 M, the deficit was $1.6 M. If emergency capital charges (including emergency operation and maintenance costs) are included, the deficit increases to about $2.2 M. This deficit does not take into account capital repayment costs for the public tubewell system and also assumes that all capital costs for the irrigation system are already sunk costs. For the entire Punjab, the deficit in the Punjab Irrigation Department budget was $17.0 M in 1978-79 and is estimated to be more than $30 M for 1983. For all of Pakistan, the annual deficit may exceed $60 M, again not taking into account past capital expenditures (Johnson, 1982).

Table 7.8 Effects of increasing command area size on per acre net present economic values (NPEV) associated with selected irrigation technologies[a]

Technology	NPEV With			Change			
				Low to Medium		Medium to High	
	Low[b] Coverage	Medium Coverage	High Coverage	Absolute	Percent	Absolute	Percent
Deep Tubewells	2306	3909	4616	1630	71	707	18
Shallow Tubewells	4389	6049	6916	1660	38	867	14
Low-Lift Pumps (20 lit/sec)	7954	8870	9273	916	12	403	5
Low-Lift Pumps (56 lit/sec)	8141	8960	9335	819	10	375	4

[a]Average net present economic value across land types

[b]Low, medium, and high coverage (in hectares) varies by technology: with DTWs = 16, 24, and 32 Ha.; with STWs = 4, 6, and 8 Ha.; with LLP (28 lit./sec.) = 8, 12, and 16 Ha., and; with LLP (56 lit./sec.) = 16, 24, and 32 Ha.

Source: Hanratty (1983).

These deficits are the responsibility of individual provinces, but the provinces' abilities to raise revenue have not increased enough to meet them. In the short run, the provinces have subsidized tubewell operation by underfunding required canal system maintenance, agricultural extension, and crop and livestock research. They have also gone into debt to WAPDA for SCARP electric charges. The provinces must either increase their revenue from water charges or reduce their costs of operating and maintaining tubewells, or both. They have already restricted funds for SCARP operation and maintenance, but this forces a reduction in utilization rate and slows the rate at which tubewells are repaired. The result is a reduction in total pumpage and an increase in per-unit water and drainage costs.

STWs, which are nominally unsubsidized, have still done very well. This is primarily, as indicated earlier, a result of the ease of organization, but it also reflects the fact that STWs are easier to maintain and, therefore, are seen as more dependable. This can be clearly seen in Table 7.9, which shows private tubewells in Pakistan are 95 percent operational, compared to 57 percent for public tubewells. Given the lower installation costs and more flexible operation arrangements, it is obvious that STWs will continue to expand rapidly.

LESSONS LEARNED

Massive debts for public tubewell development have persuaded both Pakistan and Bangladesh that the private sector has to play a major role in the process. For Pakistan, this has resulted in a decision to shift toward more private management and, eventually, to private ownership. Likewise, Bangladesh has started to encourage private ownership.[7]

Private Ownership

Over 3 M ha of land in Pakistan are served by SCARP tubewells, with a sunk cost estimated at more than $1.0 B. After 15 years of SCARP operations, waterlogging and soil salinization within the SCARPs appear to have improved marginally, at least in the less salt-affected areas. Yet SCARP tubewells are becoming older and less efficient; they must be pumped more hours each month just to hold

Table 7.9 Operational status of public and private tubewells

	Government Tubewells		Private Tubewells	
	Total	Operational (percent)	Total	Operational (percent)
Rawalpindi	868	81	2,300	96
Sargodha	1,527	67	10,700	93
Lahore	3,202	66	20,400	97
Multan	1,586	17	26,510	94
Bahwalpur	174	49	4,060	93
Punjab Total	7,357	57	69,030	95

Source: Land and Water Development Board (1971)

their own, while the price of energy is rapidly increasing. WAPDA's Master Planning Division has recommended a phased replacement of existing public tubewells in freshwater zones with private tubewells, as SCARP tubewells are exhausted. In conjunction with this, they have recommended increased operating funds to permit higher utilization factors, use of private workshops to reduce duration of breakdowns, and distinct efforts to better integrate operation of surface water and groundwater supplies. They have also suggested that pilot studies might be made of replacement of public tubewell operators by farmer groups that have a stronger incentive to keep the well operating.

Unless the government can locate and invest vast sums of money to replace and rehabilitate SCARP systems, the decline of those systems is inevitable. Private tubewells will be built where the groundwater is of good quality and markets are available for increased output. There is not any justification for continuing to subsidize SCARP systems in areas where private tubewells have already started to be installed and SCARP tubewells are in their final years. Farmer groups could be given the option of paying energy costs, establishing their own schedules, and operating the tubewells until the group decided this was no longer economic. However, SCARP tubewells are located at the head of the watercourse channels, while private tubewells are located down the channel close to the owners' fields, making distribution losses considerably higher for SCARP tubewells. Therefore, only farmers in the head end of the watercourse command will normally be willing to pay to continue to operate these large-scale public tubewells. Farmers located away from the tubewell, given increasing maintenance costs and excessive energy costs per unit of water delivered to their fields, will quickly find that owning their own private tubewell or sharing a tubewell with close neighbors is more economical.

In areas where tubewells are newer and there has been less private development, more effort could be made to form farmers' groups to operate SCARP tubewells until private tubewells become a better alternative. Giving farmers the freedom to operate the public tubewells on demand or install their own private tubewells should lead to a significant increase in total pumping from groundwater in the freshwater areas. This would accomplish desired drainage goals of the SCARP program at a mere fraction of the cost to the government, as has been demonstrated in non-SCARP areas of the Punjab and across the border in the Indian Punjab.

Poor performance of large DTWs, strong farmer demand for smaller, simpler equipment, and the growing burden of irrigation subsidies helped facilitate policy changes in Bangladesh in the late 1970s and early 1980s. These changes resulted from three major factors: (1) the need to increase the rate of minor irrigation development while simultaneously increasing the equity and efficiency of resource use; (2) the importance of private-sector ownership of equipment and of its involvement in the supply of equipment, spares, and repair and maintenance services; and (3) the encouragement of domestic irrigation equipment manufacturers (World Bank, 1982).

Recognizing that subsidies have been too high, the GOB, in consultation with the World Bank, has altered the sale and rental prices of minor irrigation equipment. Under the new system, all equipment prices are based on a shallow tubewell equivalent formula. The formula is designed to equate costs per unit of water discharged with those of STWs. This requires that DTWs still be sold at 57 percent subsidy (down from 80+ percent). In order to ensure that DTWs are not installed in areas where STWs have a comparative advantage and to further reduce the overall drain of subsidies, it has also been recommended that DTW development be further curtailed (Hanratty, 1983). This would result in the further development of an additional 100,000 STWs.

Water User Training

Initially, neither Bangladesh nor Pakistan put any emphasis on training. This reflects both the fact that little emphasis on training of water users is provided within the large-scale irrigation systems, and that there is no clearly designated organization that has the mandate and expertise to accomplish this task. However, Bangladesh and Pakistan have slowly come to realize that this type of training is important and offers a very high economic payoff. Bangladesh, perhaps, has invested the most in training and management with its Irrigation Management Program (IMP) under the IRDP (1980), but Pakistan has also invested in its On-Farm Water Management Program (World Bank, 1981) under the Ministry of Agriculture (although this program does not focus solely on DTWs, as does IMP). Bangladesh is far ahead in recognition of the need to train DTW managers and has organized specialized training under the Rural Development Academy. These programs have helped to increase technical efficiency but, more importantly, by absorbing some of the "transaction costs," have facilitated expanded utilization and, hence, increased social returns.

Technology Selection

A question that is often asked is: What type of tubewell technology is the most appropriate? This question can be answered from a number of viewpoints, but, in terms of technology, all of the alternatives have a role to play. For geohydrologically suitable areas, STWs are preferable over DTWs and can even serve as technological advances for farmers who first start with manual tubewells or open wells. Per-acre water costs are much less with STWs than with DTWs, and, as usually happens, they require far less subsidy as well. Thus, it can logically be argued that no DTWs should be installed until DTWs are the only technological option. In both Pakistan and Bangladesh, DTWs have been installed in areas where STWs are more appropriate, which has proven to be extremely expensive. In areas where the aquifer is too deep or not available for STWs, or the groundwater quality is not fit for direct agricultural application, it may be necessary to install DTWs.

However, even with the STWs, and particularly with the DTWs, there is a critical need to expand the utilization rate. In order to do this, it is necessary to encourage such actions as:

- removal of subsidies, particularly on DTWs, but also on STWs, as this provides a major incentive to expand commanded area
- exploration of means of reducing conveyance losses, such as improved conveyance systems or lining
- expansion of programs such as the Deep Tubewell Irrigation and Credit Program (DTICP) and the Irrigation Management Program (IMP) which have, by absorbing some of the "transaction costs," demonstrated another means of expanding utilization
- ensuring proper siting of tubewells
- development of cohesive water user organizations to provide proper system management and maintenance.

Together, these actions will significantly increase returns to irrigation investment and, therefore, facilitate additional groundwater development.

NOTES

1. An excellent publication covering many of the issues discussed in this chapter is the recent World Bank publication: Gerald T. O'Mara. "Issues in the Efficient Use of Surface and Groundwater in Irrigation," World Bank Staff Working Paper No.707, World Bank, Washington, D.C, 1984.

2. In Pakistan, the canal water is distributed through the minor canals and flows out of the turnout (mogha) to a village level watercourse command (80-400 ha). There are no headgates at the moghas, and if a particular minor canal has water flowing in it, there is water in every watercourse command served by that canal. There are in excess of 88,000 watercourse commands in the Indus Basin.

3. However, even the LRD schedule ignores equitable distribution of water throughout the seven-day fixed irrigation water rotation schedule that is in operation on almost every chak.

4. How much of the increase in crop yields is a function of the use of new high-yielding varieties (HYVs) and what percentage is a function of additional groundwater supplies is unknown.

5. The publications by Bottrall (1983) and Hanratty (1983) provide an excellent review of this topic; no attempt is made to repeat their work, although the author acknowledges drawing heavily from their respective documents. Another excellent source of reference concerning the larger set of issues related to irrigation development and management is: Anthony F. Bottrall. "Comparative Study of the Management and Organization of Irrigation Projects," World Bank Staff Working Paper No. 458, World Bank, Washington, D.C., 1981.

6. Individuals interested in institutional questions should read the following two publications: (a) Daniel W. Bromley. "Improving Irrigated Agriculture: Institutional Reform and the Small Farmer," World Bank Staff Working Paper No. 531, World Bank, Washington, D.C., 1982 and (b) George E. Radosevich. "Groundwater Development and Management in Bangladesh: Institutionalizing a Strategy,"

Bangladesh Agricultural Research Council, Dhaka, Bangladesh, 1983.

7. Although not discussed in this paper, the phenomenal increase (more than 300,000 installed) in manually operated shallow tubewells for irrigation (MOSTIs) in Bangladesh is a private enterprise that has taken off with very little public encouragement and almost no assistance. This is significant in terms of equity, even if the total hectarage served is not that large.

REFERENCES

Ahmad, Nazir. Waterlogging and Salinity Problems in Pakistan (Parts One and Two). Lahore, Pakistan: Irrigation, Drainage, and Flood Control Research Council, 1974.

Ashraf, Malik. "Group Owned Private Tubewells in Pakistan's Punjab," mimeo, presented at the Wheat Research and Production Seminar, August 12-13, 1978, Islamabad, Pakistan, 1978.

Bhuiyan, S. E. "Groundwater Use for Irrigation in Bangladesh: The Prospects and Some Emerging Issues," Agricultural Administration, 16:181-207, 1984.

Biswas, M. R., et al. "An Investigation into the Factors Affecting the Command Area of Different Irrigation Facilities in Bangladesh," Department of Water Management, Bangladesh Agricultural University, Mymensing, Bangladesh, 1978.

Bottrall, A. Review of Irrigation Management Practices in Bangladesh. Centre for Development Science, Bangladesh Unnayan Parishad, Dhaka, Bangladesh, 1983.

Eaton, Frank M. "Waterlogging and Salinity in the Indus Plain: Comment," Pakistan Development Review, Vol. 5, No. 3, Karachi, Pakistan, 1965.

Greenman, D. W., V. W. Swarzenski, and G. D. Bennett. "Groundwater Hydrology of the Punjab, West Pakistan, with Emphasis on Problems Caused by Canal Irrigation," U.S. Geological Survey Water Supply Paper 1608-H, Washington, D.C., 1967.

Hanratty, M. Minor Irrigation Development in Bangladesh. Dhaka, Bangladesh: U.S. Agency for International Development, 1983.

Hussain, Muhammad, Barkat Ali, and S. H. Johnson, III. "Cost of Water Per Acre-Foot and Utilization of Private Tubewells in Mona Project SCARP-II," Publication

No. 62, Directorate of Mona Reclamation Project, Bhawal, Pakistan, 1976.

Integrated Rural Development Project (IRDP). "Report of Irrigation Management Pilot Programme (1979-1980)," RD-1 Project Series-1, Ministry of Local Government, Rural Development and Cooperatives, Government of Bangladesh, Dhaka, Bangladesh, 1980.

Johnson, Sam H., III. "Controlling Waterlogging and Soil Salinity: Analysis of Salinity Control and Reclamation Projects (SCARPs) in Pakistan," Stanford Food Research Institute Studies, Vol. 18, No. 2: 149-180, 1982.

Johnson, Sam H., III. Consultants Report: Economic and Technical Operation of Deep Tubewells in Bangladesh. Dhaka, Bangladesh: Agricultural Research Council, 1984.

Land and Water Development Board, Government of the Punjab. Report of the Special Committee on the Working of the SCARPs. Lahore, Pakistan: WAPDA Press, 1971.

Malmberg, Glenn T. "Reclamation of Tubewell Drainage in Rechna Doab and Adjacent Areas, Punjab Region, Pakistan," U.S. Geological Survey Water Supply Paper 1608-O, Washington, D.C., 1975.

Master Planning and Review Division, Water and Power Development Authority. Revised Action Programme for Irrigated Agriculture (3 volumes). Lahore, Pakistan: WAPDA Press, 1979.

Mohammad, Ghulam. "Waterlogging and Salinity in the Indus Plain: A Critical Analysis of Some of the Major Conclusions of the Revelle Report," Pakistan Development Review, Vol. 4, No. 3, Karachi, Pakistan, 1964.

Mohammad, Ghulam. "Private Tubewell Development and Cropping Patterns in West Pakistan," Pakistan Development Review, Vol. 5, No. 1. Karachi, Pakistan, 1965.

Mundorff, M. J., P. H. Carrigan, T. D. Steele, and A. D. Randall. "Hydrologic Evaluation of Salinity Control and Reclamation Projects in the Indus Plain, Pakistan - A Summary," U.S. Geological Survey Water Supply Paper 1608-Q, Washington, D.C., 1976.

Planning Commission. The First Five-Year Plan, 1973-78. Dhaka, Bangladesh: The Planning Commission, Government of Bangladesh, 1973.

Planning Commission. The Second Five-Year Plan, 1980-85. Dhaka, Bangladesh: The Planning Commission, Government of Bangladesh, 1980.

Small, L. E. Economic Aspects of Minor Pump Irrigation Development in Bangladesh. Dhaka, Bangladesh: Bangladesh Agricultural Research Council, 1983.

Taylor, George C. "Water, History and the Indus Plain," Natural History Magazine, American Museum of Natural History, New York, 1965.

U. S. Agency for International Development (USAID). "Salinity Control and Reclamation Projects: Management, Operation and Maintenance," mimeo, USAID Provincial Office, Lahore, Pakistan, 1970.

White House - U.S. Department of the Interior Panel on Waterlogging and Salinity in West Pakistan. "Report on Land and Water Development in the Indus Plain," U.S. Government Printing Office, Washington, D.C, 1964.

World Bank. Action Program for Agricultural and Water Development in East Pakistan. Washington, D.C.: The World Bank, 1970.

World Bank. Bangladesh: Land and Water Sector Study. Washington, D.C.: The World Bank, 1972.

World Bank. Bangladesh: Irrigation Water Charges. Washington, D.C.: The World Bank, 1978.

World Bank. Staff Appraisal Report: Pakistan On-Farm Water Management Project. Washington, D.C.: South Asia Projects Division, The World Bank, 1981.

World Bank. Bangladesh: Minor Irrigation Sector: A Joint Review by Government and the World Bank. Washington, D.C.: The World Bank, 1982.

World Bank. Bangladesh: Recent Economic Trends and Medium-Term Development in Bangladesh. Dhaka, Bangladesh: Bangladesh Agricultural Research Council, The World Bank, 1983.

8
Economic Aspects of Irrigation with Saline Water

Dan Yaron

INTRODUCTION

The paper[1] reviews the economic dimensions of irrigation with water of varying salinity levels, with emphasis on on-farm irrigation problems. The farm-region interactions are dealt with only briefly.

The paper commences with a short review of the underlying physical water-soil-crop yields relationships and of the sources of information regarding them. In the next section, empirical estimates of farms' income losses under selected situations are reviewed. The next section discusses the alternatives open to farms to reduce salinity-induced losses and the agro-economic models designed to evaluate them. Several aspects of the farm-region interactions are then reviewed, and the final section points to some hopeful frontier-changing innovations currently under study, which, if successfully developed, may drastically change the frame of reference for salinity problems in agriculture.

UNDERLYING PHYSICAL RELATIONSHIPS

The number of studies carried out per year on different aspects of crop response to salinity in a variety of countries is enormous. The U.S. Salinity Laboratory in Riverside, California, alone has published an average of about 35 scientific publications per year in the years 1979-1981.[2] Considerable scientific activity, both theoretical and empirical, is being carried out in the USSR,

Holland, Hungary, India, Pakistan, Tunisia, Egypt, and Israel.

These studies deal with a diversity of salinity problems; they provide understanding of the mechanism of crop response to salinity and guides to water resource and irrigation management under salinity-affected conditions. However, the number of studies involved in formal modeling and quantification of the physical relationships relating soil and weather conditions, water use (timing, water quantity, and salinity), and crop yields, is surprisingly small.

In view of the complexity of these relationships, it is convenient to refer to them within a systems framework, comprising two subsystems (Yaron, 1974): Subsystem I, involving the relationship between irrigation decision variables (timing, quantity, and salinity of water) and the soil state variables (soil salinity, soil moisture, etc.) and Subsystem II, in which the soil state variables are related to the target variables (such as quantity and quality of yield). Schematically, these two subsystems can be represented by the following functions:

Subsystem I

$$SMI = f_1(IDV, \theta | K) \tag{1}$$

$$SSI = f_2(IDV, \theta | K) \tag{2}$$

where:

SMI = soil moisture index
SSI = soil salinity index
IDV = vector of decision variables
θ = rainfall
K = all other factors considered as constant for a given crop, under given agro-climatic conditions.

Subsystem II

$$Y = g(SMI, SSI | K) \tag{3}$$

where:

Y = crop yield per land unit area.

Upon substituting (1) and (2) into (3), the yield can be expressed as a function of the irrigation decision variables, rainfall, and the constant variables:

$$Y = h(IDV, \theta | K) \quad (4)$$

The study of (1), (2), and (3) separately is apt to provide a better understanding of the system; however, it is complex and implies the estimation of numerous parameters. The reference to function (4) only is a less ambitious black box approach; the benefit is the need to estimate fewer parameters. Whenever interpolations--and perhaps extrapolations--from one location to another are needed, the reference to the fully designed system, (1)-(3), is appropriate.

Relationships relating soil salinity to the irrigation decision variables (Subsystem I) have been studied and modeled by Bresler (1967, 1973), Childs and Hanks (1975), Hanks (1974), Feddes et al. (1974), Neuman et al. (1975), and others. For convenience in economic and management-oriented analyses, a relatively simple model has been suggested by Bresler (1972) for conditions characterized by: (i) the absence of drainage problems; (ii) deep or confined aquifers where the effect of return flows can be ignored; (iii) the adsorption of the relevant ions being negligible, and (iv) sprinkler irrigation. Such conditions prevail in most of the regions in Israel, with respect to which the model has been calibrated, tested, and successfully used. The model outline is presented in Appendix A; more details can be found in Yaron et al. (1979, Hebrew).

Using this model and a simulation program[3] the process of salt accumulation and leaching in soils can be simulated with reference to a variety of conditions for which the model was calibrated [(i)-(iv) above]. For illustration, Table 8.1 presents results from selected simulation runs.[4] It is important to note that, regardless of the initial soil salinity, continuously using the same irrigation practices and water quality, soil salinity converges within 3-5 years to a steady-state average level, with between-year fluctuations depending on the rainfall in any particular year.

Major sources of compiled information on the relationships between soil salinity and crop yields (Subsystem II) are in Bernstein (e.g., 1964, 1965, 1973, 1981), and a relatively recent compilation of worldwide data published

Table 8.1 Selected results from simulations of salt accumulation and leaching, sprinkler irrigation, fruit crops, southern Israel

Region	Soil Type (SP)	Annual Rainfall[a] (mm)	Irrigation Water Applied (mm)	Winter[b] Supplementary Irrigation (mm)	Salinity of Irrigation Water Applied (ppm/Cl): 200	300	400	450
					Electrical Conductivity of Soil Solution[c] (mmhos/cm)			
South	66	339	615	53	1.47	1.87	2.25	2.46
	66	398	800	49	1.50	1.92	2.35	2.55
	66	365	1050	56	1.62	2.10	2.58	2.83
Northwestern Negev	52	264	615	84	1.47	1.85	2.24	2.43
	51	254	800	88	1.52	1.94	2.35	2.55
	52	264	1050	85	1.59	2.04	2.50	2.73
Negev	39							
	39	219	800	108	1.46	1.83	2.20	2.37
	39	290	1050	72	1.43	1.82	2.20	2.39
Eshkol Region	25	248	615	92	1.36	1.70	2.30	2.20
	25	248	800	92	1.42	1.77	2.14	2.32
	25	240	1050	99	1.46	1.85	2.24	2.43

[a]Average of 10 random simulation runs

[b]Average of 20 simulation runs and random rainfall. It was assumed that the salinity of irrigation water in winter was 200 ppm/Cl.

[c]Mean spring-fall soil salinity at the end of a series of 10 years; average of 10 simulation runs with random rainfall for each run. Root zone average - 90 cm.

by Maas and Hoffman (1977). On the basis of these compilations, Ayers and Westcot (1976) prepared irrigation management-oriented guidelines "that would allow the man-in-the-field to evaluate the quality of a given water supply for agricultural use" (Maas and Hoffman, 1977). Their guidelines present the evaluated yield decrement for selected crops in response to the salinity of irrigation water expressed in terms of (EC_wY) and soil salinity-electrical conductivity of soil water extract (EC_eY)--when common surface irrigation methods are used. They also discuss the possible effects of other irrigation systems (sprinkler and trickle) and of other factors which may affect the assumed relationships (e.g., leaching, increased frequency of water applications).

The functional specification of the relationship between crop yields and soil salinity following Maas and Hoffman (1977) is shown in Appendix A.

It is important to note that the crop response compilations by Bernstein and Maas and Hoffman are dominated by data from experimental plots and containers with ample moisture supply, which do not necessarily reflect the situation(s) prevailing in actual field practice. A notable example of widely diverging results is the comparison of experimental data compiled by Shalhevet (1983, Hebrew) on citrus response to salinity, with the response estimated on the basis of field survey data (Hausenberg et al., 1973; Shalhevet et al., 1974). The experiment-based estimated response is (Shalhevet):

$$Y = 100 - 12.9\ (EC_e - 1.28) \qquad (5)$$

with

Y = the relative yield (%)

EC_e = the electrical conductivity of the saturated soil solution in the root zone (mmho/cm).

The survey-based estimate (Hausenberg, Shalhevet), adapted by Yaron et al. (1979) to the same functional form, is:

$$Y = 100 - 30\ (EC_e - 1.3) \qquad (6)$$

The comparison suggests that the loss threshold level is practically the same in both functions (1.28 versus

1.3), but the percentage-wise loss per one mmhos/cm above the threshold is 2.3 higher in (6) than in (5).

The divergence between (5) and (6) is attributable to the difference in the background variables, specifically in the soil moisture regime.

The system-oriented and modeling-based approach still faces considerable difficulties, which are due to the complexity of the relevant relationships, on the one hand, and the scarcity of the data needed, on the other. Accordingly, workers tend to present salinity-induced yield loss within the framework of equation (4) and with reference to well-defined localities. A notable example is Robinson (1978), who tabulated evaluated yields of major crops in selected locations in the southwestern U.S., under varying conditions with respect to the irrigation system, number of irrigations per year, and water salinity. It is unfortunate that many similar evaluations are published in media with restricted circulation for ad hoc applications only.

THE MAGNITUDE OF POTENTIAL LOSSES TO FARMS AND REGIONS DUE TO SALINITY

In this section, we present several empirical estimates of potential monetary losses to farms and regions due to increasing salinity of irrigation water. Note that the economic losses are, as a rule, more severe than the physical yield losses; with a $\frac{\text{value added}}{\text{gross output value}}$ ratio of of 50 percent, which is common in modern farming systems, a 10 percent loss of physical yields amounts to 20 percent loss in terms of value added; 20 percent loss of yields leads to 40 percent (!) loss of value added, which may be critical to farms' viability. In many cases, a 10 percent loss of yields is not even observed unless a special study involving soil salinity measurements and yield records is undertaken. Farmers and extension workers feel alarmed when visible symptoms, such as leaf injuries, are evident, but these are generally observed when higher percentage-wise losses have occurred.

In the estimation of salinity-induced income losses to farms and regions, distinction should be made between three situations from the point of view of adjustments to salinity on behalf of the farm:

(a) currently practiced crop mix and irrigation technology
(b) crop mix adapted to salinity and currently practiced irrigation technology
(c) both crop mix and irrigation technology adapted to salinity.

Hypothetical curves representing farm income losses induced by salinity of irrigation water under the above situations are shown in Figure 8.1. Note that the adaptation of irrigation technology to salinity is a continuously progressing process as additional knowledge is being gained and irrigation management is improved.

Moore et al. (1974) estimated the income losses to farms in the Imperial Valley of California, with reference to assumed adjustments of crop mix and technology (see also section on agro-economic efficiency in the use of saline water). They found that the increase in irrigation water salinity from EC 1.5 (current) to 2.00 (projected for the year 2000, if no countermeasures were undertaken) would cause a reduction of 12-14 percent to the returns to land and water. Salinity increase up to EC = 3.00 will result in the reduction of land and water in returns in the range of 20-29 percent. It appears that reduction in the net income would be higher due to the lower base for computing the percentage-wise loss.

Oyarzabal-Tamargo and Young (1978) estimated the losses accrued to the Colorado River Irrigation District in northern Mexico, which includes the Mexicali Valley and the San Luis Valley. This district obtains water from the lower Colorado River. In 1960, the quality of water was about 800 ppm TDS; after 1961, due to the execution of a drainage project in Arizona and diversion of its water to the Colorado River (see also section on farm-region and interregional interactions, externalities, and cooperative solutions), the salinity level was in excess of 2000 ppm TDS and, again, was reduced by countermeasures to 1200 ppm TDS by 1970. The estimated loss in net returns to the Mexicali and San Luis Valley farms in terms of 1975 prices are presented in Table 8.2.

Note that Oyarzabal-Tamargo and Young referred to situation (c), i.e., they assumed the adaptation of the crop mix and the irrigation technology to increasing salinity. Since, in practice, such adaptations are taking place rather slowly, except for farms with a very responsive management, their estimates of the losses might be biased downwards. The crops common to the region were

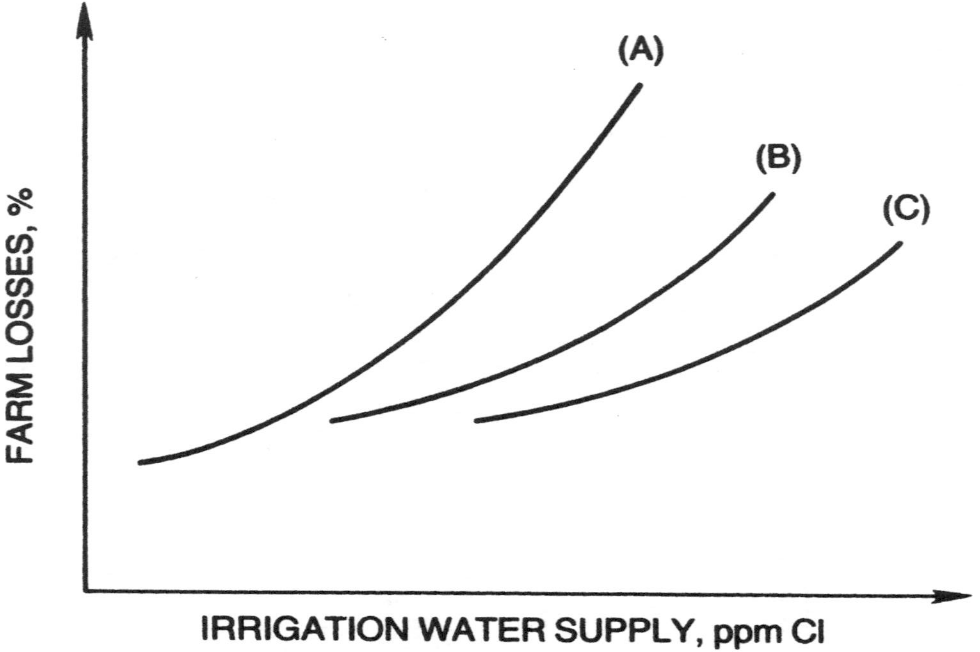

(A) CURRENTLY PRACTICED CROP MIX AND IRRIGATION TECHNOLOGY
(B) CROP MIX ADAPTED TO SALINITY AND CURRENTLY PRACTICED IRRIGATION TECHNOLOGY
(C) BOTH CROP MIX AND IRRIGATION TECHNOLOGY ADAPTED TO SALINITY

Figure 8.1 Hypothetical salinity-induced farm losses under three situations

Table 8.2 Estimated losses to net returns for alternative water salinity levels in Mexicali and San Luis valleys, North Mexico, 1975

Water Salinity (ppm TDS)	Losses (%)
800	0
1000	5
1200	12
1400	19
1600	27
1800	36
2000	46

Source: Adapted from Oyarzabal-Tamargo and Young, 1978.

cotton, wheat, alfalfa, safflower, barley, ryegrass, and grain sorghum, either tolerant or moderately sensitive to salinity.

Yaron et al. (1979) estimated potential losses to agriculture in the south and the Negev areas of Israel under conditions of deterioration of water quality. Estimates of potential income losses for kibbutz and moshav farms,[5] with reference to situation (a)--current crop mix and irrigation practices--are shown in Table 8.3.

As Table 8.3 shows, the salinity-induced (percentage-wise) losses in moshav farms are about double those of kibbutz farms. This is due to the larger share of salinity-sensitive fruit crops on moshav farms. In a later study, Yaron et al. (1982) estimated the potential losses due to increased salinity in fruit crops only on moshav farms (Table 8.4), using the same approach as that underlying Table 8.3.

Yaron et al. (1982) also studied the response of kibbutz farms and income losses induced by increased salinity. They found that the income loss accrued to kibbutz farms by the rise of salinity from 220 to 300 ppm Cl, and even up to 400 ppm Cl, is not very significant. More significant are the structural changes induced by salinity, namely, the tendency to eradicate fruit crops (sensitive

Table 8.3 Estimated income losses (%) for alternative salinity levels of irrigation water in the South and the Negev regions in Israel[a]

Salinity Level (ppm Cl)	Kibbutz Farms	Moshav Farms
200	5	10
300	14	30
400	23	45

[a]Income defined as gross revenue less variable and labor cost, spring 1978 price level.

Source: Adapted from Yaron et al., 1979.

Table 8.4 Estimated relative income losses (%) accrued to fruit crops on moshav farms for alternative salinity levels of irrigation water in the South and the Negev regions in Israel[a]

Salinity Level (ppm Cl)	Income loss (%)	
	Range[b]	Average[b]
200	11-26	18
300	21-42	34
400	33-36	51

[a]See footnote a, Table 8.3.

[b]Referring to a sample of 10 moshav villages (out of 76).

Source: Adapted from Yaron and Ratner, 1982.

to salinity) and the increase of the area of cotton which is salinity insensitive. This structural change, which is likely to happen under conditions of increased salinity, is contrary to the sound management rules in favor of diversification which guided the kibbutz farms in the past.

In summary, the extent of salinity damages is highly dependent on the types of farms and the composition of their crops. Farms with a high share of sensitive crops, such as the moshavim in the south and the Negev regions of Israel, are highly vulnerable. Furthermore, it should be noted that the loss estimates refer to averages with likely upward (and downward) deviations.

An important question is: To what extent can proper irrigation technology and management cope with the problem? This subject is discussed in the following section.

AGRO-ECONOMIC EFFICIENCY IN THE USE OF SALINE WATER

The Dimensions of the Problem

In the economic analysis of irrigation with saline water, a distinction is suggested among groups of problems according to: (1) the level of activity referred to in the farm-region-nation hierarchy; (2) the problem of external effects ("externalities," in economic semantics); and (3) the range of time referred to.

This paper deals primarily with the farm level and discusses the externalities involved in the farm-region (water basin) relationships in a partial way only.

From the point of view of time, a distinction is made between:

1. A short-run approach, which refers to relationships confined to a single irrigation season and does not take into consideration the long-run effects of salt accumulation over time. The short-run approach pertains with respect to individual crops and/or whole farms.

2. A long-run approach which does take into account the effect of salt accumulation over time in soils, river flows, and aquifers. It comprises a succession of short-run processes, the initial conditions of which are affected by salt accumulation in previous periods; the decisions over

any single season take into account the possible terminal conditions that may result from alternative decisions, and their effects on succeeding periods. Similarly, as in the short-run, the long-run approach is relevant to both individual crops and whole farms.

In the following, we review and discuss several studies addressing economic efficiency in irrigation with water of varying salinity within the framework of short-run and long-run approaches.

A Dynamic, Responsive-Type Approach to Leaching and Irrigation

The USDA Salinity Laboratory (1954, p. 37) has defined the concept of leaching requirement (LR) as: The fraction of the irrigation water that must be leached through the root zone to control soil salinity at any specified level. Under the assumptions of (a) uniform application of irrigation water per unit area; (b) no rainfall; (c) no removal of salt by the harvested crop and no precipitation of soluble constituents in the soil; and (d) steady-state water flow rates, the leaching requirement is:

$$LR = \frac{QD}{Q} = \frac{CW}{CD} \qquad (7)$$

where:

QD and Q = the depth of the drainage and irrigation water, respectively

CD and CW = their salt concentrations.

For a predetermined value of CD, the amount of water required for leaching (LR) is a linear function of the salt concentration in the water (CW). The line representing this function passes through the origin, its slope being the reciprocal of CD. As explained by Bresler (1967), Ayers and Westcot (1976), and others, this equation is most useful when applied to conditions of steady-state water flow and maintenance of the balance in the soil, which seldom occur under field conditions. It does not take into account the dynamic changes in salinity during the irrigation season, and does not apply in the case of leaching by a varying water flow or by rain. (See Ayers and Westcot, 1976, pp. 34-35, for a discussion of the above leaching formula and its modification.)

Bresler (1967) developed a more comprehensive model for tracing salt distribution in the soil profile in response to various irrigation regimes. This model (and its modifications, e.g., Appendix A), which specifically refers to the quantity and quality of water applied in each irrigation and the evapotranspiration between successive irrigations, was later used for the analysis of irrigation management with saline water.

Bresler and Yaron (1972), following Wadleigh and Ayers (1945), assumed that the relationship between crop yield and the combined water regime-soil salinity variables may be expressed in a general functional form as:

$$Y = Y(S|K) \qquad \frac{\partial y}{\partial S} < 0 \tag{8}$$

where:

Y = the crop yield

S = the index of the total water suction

K = all other growth factors, assumed to be constant.

They showed that S, the average soil water suction weighted over the main root zone Z throughout the irrigation season T,[6] increases with the amount of irrigation water Q at low values of Q, but decreases with Q at relatively high values of Q. (As is well known, S increases with each of the variables: the irrigation intended, water salinity, and initial soil salinity.) They also showed that, under the conditions studied, it was more efficient (from the standpoint of salinity) to use a given amount of water Q for changing the soil water regime rather than for leaching.

An application of the dynamic concept of leaching combined with the irrigation of a single crop was presented by Yaron et al. (1980). They presented a dynamic model for optimal scheduling of irrigation with water of varying salinity levels and with soil moisture parameters explicitly considered. The system underlying their model was characterized by two discrete state variables, soil moisture θ, and the salinity level of the soil solution c, updated in the model on a day-to-day basis.

The model provides answers to two questions arising under conditions of irrigation with saline water: (1) Given the initial soil salinity, should a preplanting leaching be applied and, if so, at what quantity? (2) What is the optimal irrigation and leaching scheduling during the entire irrigation season?

The growing season is subdivided into J subperiods, in accordance with the stages of growth of the crop. The yield of the crop is expressed by a function of the following type:

$$Y = A \prod_{j=1}^{J} (F_j)^{x_j} \tag{9}$$

where:

Y = yield, kg/ha

A = maximal yield obtained, when all $x_j = 0$

x_j = number of "critical days" in subperiod j (a critical day is defined as one during which the total soil suction s exceeds a critical level)

F_j = coefficient of yield reduction per each critical day during subperiod j, $0 < F_j < 1$.

The total soil suction in day t is a function of soil moisture θ_t, and soil salinity c_t,

$$s_t = s(\theta_t, c_t)$$

The core of the model is the following recursive relationship:

$$\Lambda_t(\theta_t^p, c_t^r) = \max_{t,k} [f_t(Q_t^k, \theta_t^k, c_t^p) + \tag{10}$$

$$+ \Lambda_{t+1}(\theta_{t+1}^{p*}, c_{t+1}^{r*})] \quad t=0,1,2,\ldots,T-1$$

with

$$f_t^k(\cdot) \leqq 0 \tag{11}$$

$$\theta_{t+1}^{p*} = g(\theta_t^p, Q_t^k | G) \tag{12}$$

$$c_{t+1}^{r*} = h(\theta_t^p, c_t^r, Q_t^k | G) \tag{13}$$

$$\Lambda_T(\theta_T^p, c_T^r) = A \cdot P_y - FC \tag{14}$$

for all p and r, where:

p = the price per yield unit (net of harvest cost)

FC = fixed cost per land unit area, with cost of irrigation and harvest excluded. Note that FC (a constant) can be ignored in the recursive maximization process

g,h = the transformation functions of soil moisture and soil salinity from day t to (t+1)

G = all other factors considered as constant

t = the number of days from the beginning of the growing season with t=0 being the planting day and t=T the end of the growing season. Note that the direction of change in t is opposite to the conventional notation in dynamic programming.

The objective is to maximize the cumulative net income $\Lambda_0(\theta_0^p, c_0^r)$ for every p and r, subject to (12) and (13), by applying a dynamic programming backward induction procedure to (10) for t=T, T-1, T-2, and so on, with t=0 denoting the beginning of the growing season.

The model has been applied to the analysis of optimal irrigation policy of grain sorghum under several situations in the Gilat area. The adjustments in the optimal irrigation policy and the changes in the yield and the income per land unit area were studied in response to two variables: the salinity of the irrigation water throughout the irrigation season and the initial soil salinity at the beginning of the growing season (t=0). Selected results of these analyses are presented in Figure 8.2. More details and analyses of additional situations are given by Yaron et al. (1980).

The results of Figure 8.2 suggest several policy rules for irrigation with saline water under different initial salinity regimes:

- Generally, frequent applications of small quantities of water are preferable to applications of large quantities at extended intervals.
- Under relatively high saline conditions (i.e., high values of either initial soil salinity or water salinity, or both) an extra amount of irrigation water for leaching is generally justified at the beginning of the growing season (for example, situations A.5 and B.4 in Figure 8.2) or in the middle of the irrigation season (C.3 and D.3). Some combinations of the above are also recommended (B.5 and C.4).
- Under relatively low-saline conditions, it is worthwhile to extend the irrigation over a longer period, as compared to the high salinity-affected situations (for example, B.1 versus B.2). The rationale underlying this rule is that when the yield potential is high, it pays to preserve the yield potential by extending the application of irrigation.
- Under the most saline conditions referred to in the analysis, it is not worthwhile to irrigate at all.

Note that the first two rules have been recently set forth in general terms (e.g., Goldberg et al., 1971; Bernstein and Francois, 1973; and others). The present model quantifies these rules for specific situations.

The model might be useful in two major applications. First, in testing and screening irrigation policies for detailed examinations by field experiments. Since computer simulation of the water-soil-crop system is considerably cheaper than field experiments, it can be used as a means for screening irrigation policies to help decide upon more expensive and more reliable field experiments to follow. Second, the model can be used for the

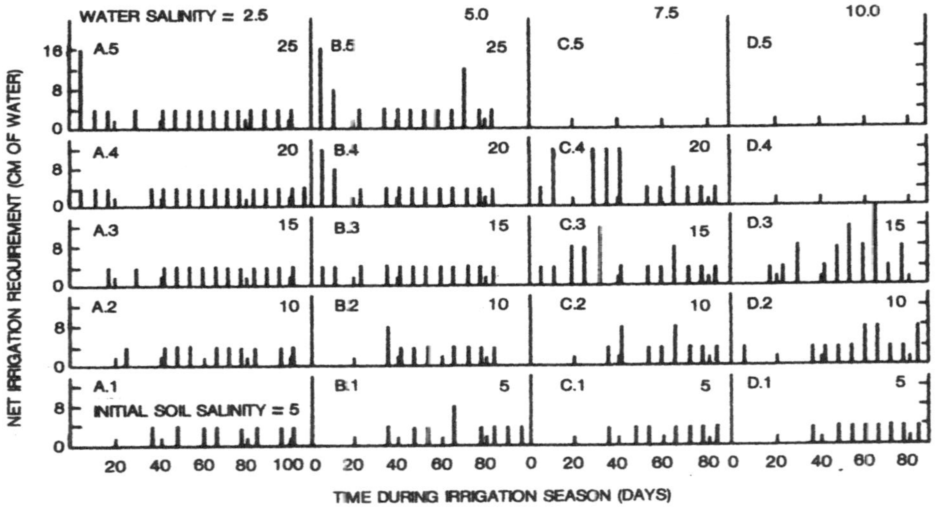

Figure 8.2 Net irrigation water requirement as a function of time during the irrigation season for five levels of initial soil salinity and four levels of salinity of irrigation water

detailed analyses of optimal irrigation with saline water under well-defined situations without field experiments. For such applications, however, refinement and calibration of the model for specific situations is needed.

Long-run Analyses of Farm Adjustments and Income Losses Due to Salinity

Analyses of irrigation with water of varying quality within a farm framework were performed by several workers e.g., Parkinson et al. (1970), Hanks and Andersen (1981), Moore et al. (1974), Feinerman (1980), Feinerman and Yaron (1983), and others.

Commonly used analytical framework is one of a linear programming model, the core of which can be schematically presented as follows:

$$\text{Maximize} \quad f = \underline{C}_1\underline{X}_1 + \underline{C}_2\underline{X}_2 + \underline{C}_3\underline{X}_3 \tag{15}$$

subject to:

$$\begin{aligned} A_1\underline{X}_1 &\leqq \underline{b}_1 \\ A_2\underline{X}_2 &\leqq \underline{b}_2 \\ D_1\underline{X}_1 + D_2\underline{X}_2 + D_3\underline{X}_3 &\leqq b_3 \\ \underline{X}_1 , \underline{X}_2 , \underline{X}_3 &\leqq \underline{0} \end{aligned} \tag{16}$$

with:

$\underline{X}_1,\underline{X}_2$ = vectors representing activity levels of crops irrigated with "good" and "low" quality (saline) water, respectively

$\underline{X}_3$ = vector of activity levels representing unirrigated crops

$\underline{C}_1,\underline{C}_2$ = vectors representing net income coefficients per activity unit or crops irrigated with "good" and "low" quality water, respectively

$\underline{C}_3$ = vector of income coefficients per activity unit of crops

$\underline{b}_1, \underline{b}_2$ = water restrictions of "good" and "low" quality, respectively

A_1, A_2 = water input coefficients related to crops irrigated with "good" and "low" quality water, respectively

$\underline{b}_3$ = vector of restriction levels other than water

D_1, D_2, D_3 = technological coefficients related to restrictions other than water.

Parkinson et al. (1970) designed a linear programming model aimed at determining the optimal crop mix in reference to three levels of water salinity and the losses induced to crops. The option to mix water from different sources was also included.

Hanks and Andersen (1981) presented an agro-economic model relating the crop mix and irrigation practices with the salt content of the return flow. The physical relationships involved are analyzed very comprehensively, with almost all relevant functional relationships regarded as endogenous to the model, including irrigation methods (sprinkling and flooding), irrigation frequency, uniformity of water distribution, transpiration and evapotranspiration, drainage, salt distribution in the soil profile, salt outflow to groundwater, and, finally, predicted yields of the relevant crops under selected conditions.

Their economic model was designed to maximize the farm income over a single year (a Vernal, Utah, farm provided the framework for the economic analysis), with parametrically varying restriction levels on salt outflows resulting from irrigation. The real cost to the farm of restricting the salt outflows was estimated. The economic analysis of Hanks and Andersen's paper was static, in that it referred to a single year; however, it can be adapted to a long-run analysis. In effect, the analyses of the physical processes are extended over a series of years.

The results of Hanks and Andersen's study show the change in the optimal crop mix in response to initial soil salinity conditions and the permissible salt outflow per farm. It amounts to varying the acreage of alfalfa, corn, and oats; under the most extreme conditions (high initial soil salinity and low salt outflow), most of the land should be left idle (Figure 8.3). Another result is the estimated relationship between the farm's income and the

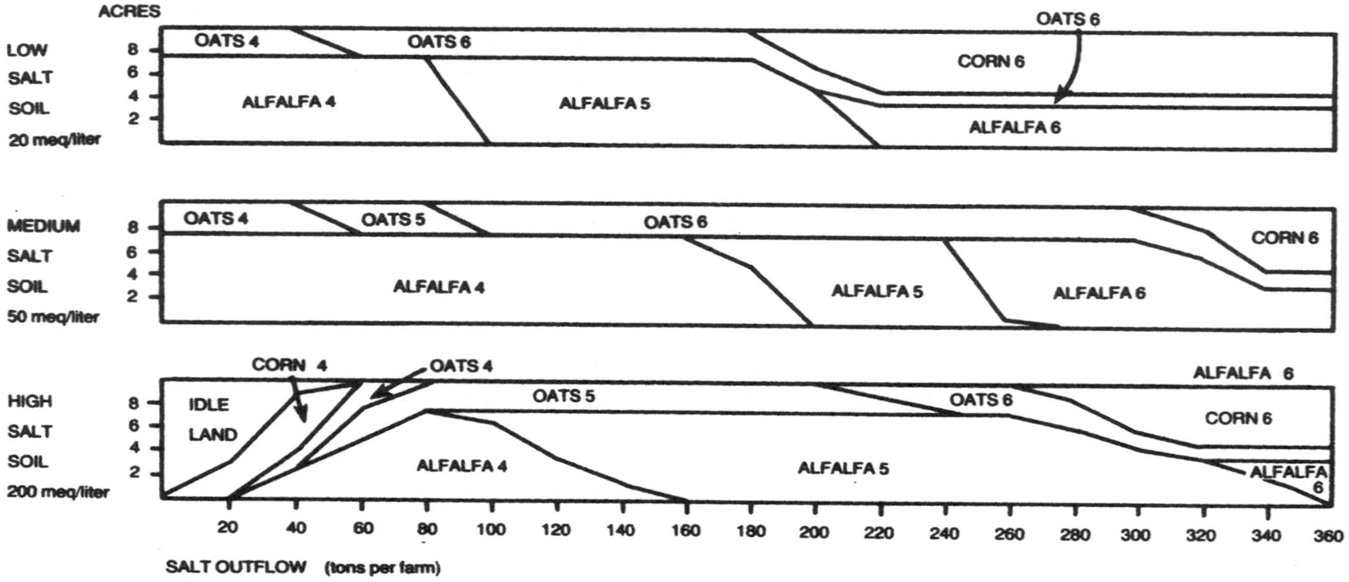

Figure 8.3 Optimal cropping and irrigation pattern for high, medium, and low initial soil salt conditions where corn roots are shallow and where flood irrigation only is allowed, a Utah farm (Hanks and Andersen, 1981)

salt outflow--a concrete monetary expression of externalities.

Moore et al. (1974) presented an economic analysis of irrigation with saline water in the Imperial Valley of California, with reference to three sizes of farms with varying quantities and salinity levels of water supplied to them. Utilizing information on the response of crops to soil moisture and salinity, alternative efficient combinations of water quantity and salinity were formulated and incorporated into farm planning, with the aid of linear programming. The objective of the analysis for each size of farm was to maximize its income under the above conditions. In response to increasing salinity, the results indicate (1) adjustments in the relative share of certain crops; (2) the tendency to reduce both the total irrigated acreage of the crops and the cropping intensity

$\left(\frac{\text{total crop average}}{\text{irrigable acreage}}\right.$ ratio); (3) reduction in the farm's income in terms of return to water and land; and (4) reduction in the average and marginal return to water of different salinities. In particular, the acreage of alfalfa (a medium-sensitive crop) is reduced to zero in response to higher water salinities, while the acreage of lettuce (a salinity-sensitive crop) remains unchanged up to water salinity of EC = 2.0 mmho/ cm, and drops to zero at only EC = 3.0. This order of acreage reduction is apparently due to the higher income of lettuce per acre. The relative reduction in the annual farm income in terms of returns to water and land is shown in Table 8.5.

Feinerman (1980) studied the optimal mixing of water of varying salinity levels on a farm with three sources of water (EC = 1.3, 2, and 4 mmho/cm, respectively), simultaneously with the crop mix determination and the optimal allocation of the mixed water to the crops. The point in Feinerman's results essential to our discussion is that the priority of crops in the allocation of good quality water should be assigned according to both income potential and sensitivity to salinity; the sensitivity to salinity is not the major criterion. For the farm studied (in the Negev region of Israel), the priorities were ranked as shown in Table 8.6.

In summary, the short-run analyses here reviewed indicate (1) what adjustments should be undertaken by the farms in terms of crops acreage and irrigation/leaching management in order to reduce salinity losses; (2) the unavoidable farm income losses under optimal management in

Table 8.5 Estimates of the relative return to water and land on small, medium, and large farms in response to increasing salinity, Imperial Valley, California

	Water Quality (EC mmho/cm)			
Farm Size	0.75	1.50	2.00	3.00
	- - - - - - - - - - (%) - - - - - - - - - -			
Small	114	100	86	71
Medium	112	100	87	74
Large	112	100	88	74

Source: Following Moore et al., 1974.

Table 8.6 Priorities in allocation of high-quality water on a farm with three sources and varying water salinity[a]

Priority Ranking	Crop	Income Potential	Sensitivity to salinity[b]
1	Potatoes	High	MS
2	Citrus	Moderate	S
3	Carrots	Low	S
4	Cotton	Medium High	T

[a]A farm in the Negev area of Israel provided the empirical framework for the analysis.

[b]Following Maas and Hoffman, 1977: S = sensitive, MS = moderately sensitive, T = tolerant.

Source: Adapted from Feinerman, 1980.

response to increased salinity. These should be compared with salinity-induced losses under the current management practices. As the number of the adaptive response measures increases, the income losses decrease.

Long-Run Analyses

Irrigation with saline water is a dynamic stochastic process. Salt accumulates in the soil during irrigation and is periodically leached by rainfall and/or irrigation. The major natural stochastic element is rainfall; other stochastic phenomena are related to uncertainty (or insufficient knowledge) regarding the physical relationships involved. A dynamic approach and stochastic elements have been introduced into the system by Yaron and Olian (1973) and others.

The dynamic process of irrigation with saline water of a single plot can be characterized by one state variable representing variations of the soil salinity of the plot over time (Yaron and Olian (1973); see Figure 8.4 for schematic presentation of the process. For several plots, the process can be characterized by the corresponding number of state variables, thus leading to a multi-state dynamic problem. However, the solutions of multi-state dynamic problems are technically difficult (curse of dimensionality) and sometimes impossible, with far-reaching simplifications or a heavy computational burden.

In some situations, the adaptive response approach (inherent to dynamic programming or optimal control) is not necessary, and a long-run, steady-state situation can be justifiably referred to. Such situations prevail whenever leaching in response to salt accumulation in the soil profile is either not needed (e.g., low salt accumulation and periodical leaching by precipitation) or not justified economically because the benefit derived from leaching is lower than the income derivable from the allocation of water to other uses. A study by Yaron and Voet (1982) may provide an illustration.

They refer to a farm in the south of Israel with two fruit groves and field crops, mainly cotton. The fruit groves (avocado and tangerine) are sensitive to salinity, while the field crops are not. The farm has at its disposal a given quota of water of low quality. In the case of excessive salt accumulation in the soil of the fruit groves due to irrigation, leaching irrigation may be applied. If justified, high quality water for leaching

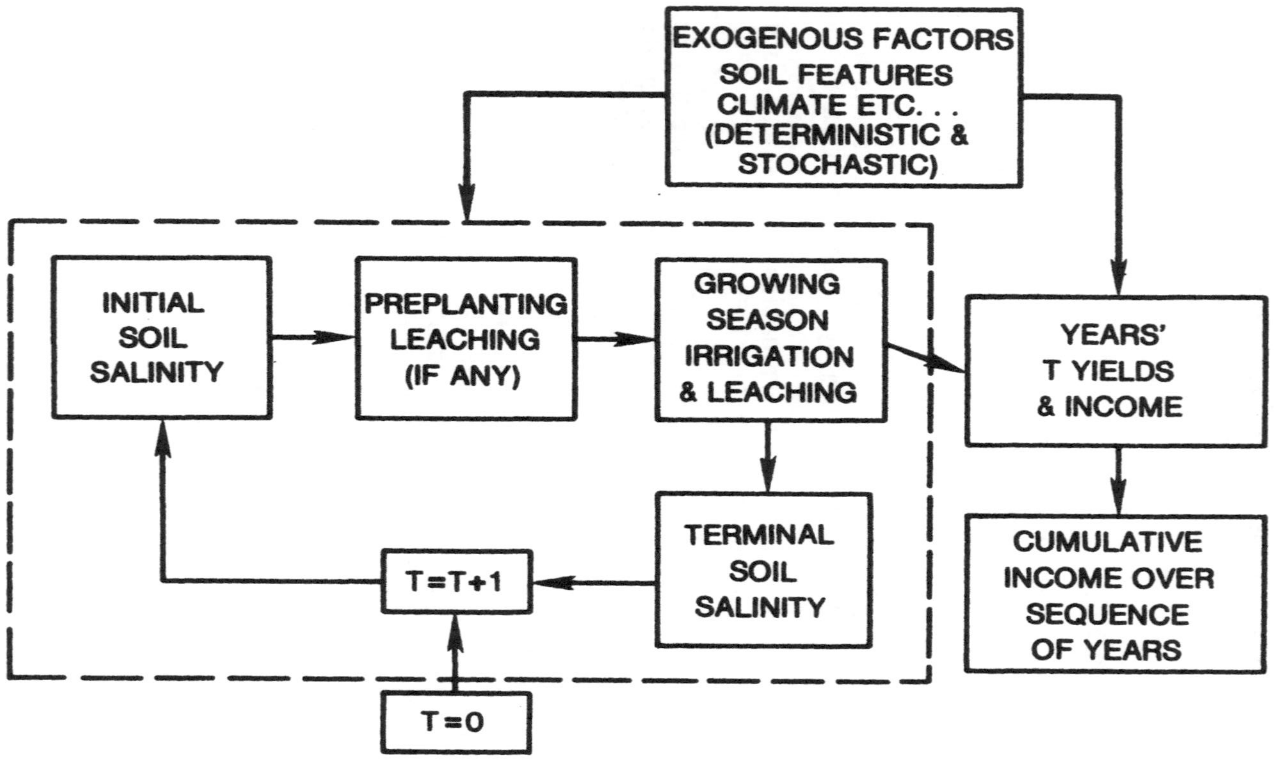

Figure 8.4 A schematic presentation of irrigation with saline water over a sequence of years

might be allocated to the farm by the regional water authority without changing the farm's annual quota (implying substitution of some of the farm's poor quality by high quality water at a 1:1 ratio). The farm's problem is the allocation of its irrigation water to the two groves and the field crops, with emphasis on the justification of leaching irrigation. The analysis, performed with the aid of an integrated dynamic and linear programming model, suggested that the optimal policy is to apply the conventional quantities of water to the irrigation of the groves, not to leach the groves over the conventional irrigation quantities, but rather to use all the water remaining for the field crops.

To support the above conclusion, the returns to salt leaching (over and above the conventional irrigation norms) in selected situations were estimated (Table 8.7). As Table 8.7 shows the returns to leaching under the condition referred to are in the range of 0.6-1/7 IL/m^3 at spring 1978 price levels (1 IL = 6 US cents), while the return from 1 m^3 allocated to cotton as the competitive alternative falls within the range of 3.5-5 IL/m^3.

The point of the above example is that in the long as in the short run, leaching should be evaluated within an intrafarm competition with alternative uses of water.

The operational conclusion, important to workers modeling and studying the long-run responses to and the effects of salinity on farms, is that in certain situations, adaptive response-type models are not necessary. Recalling that under constant irrigation policy, soil salinity converges to a steady state within 3-5 years (Figure 8.1), it is possible to address long-run, steady-state conditions and search for the long-run, steady-state optimal irrigation policies. Whenever adaptive-type decisions in response to salt accumulation are needed, the problem becomes computationally difficult.

Feinerman (1980) designed a long-run model which refers to a farm system over a sequence of several irrigation seasons and incorporates rainfall uncertainty. Conceptually, it is an extension of the two-state linear programming model under uncertainty (Dantzig and Madansky, 1961). The objective function is to maximize the present value of the expected net profits from the yields of crops over the time horizon, subject to total water and land supplies, acreage quotas for certain crops, and linear balance equations which describe the evolution of the soil-related state variables over time.

Table 8.7 Estimated returns to salt leaching under selected situations in Israel[a]

Situation No.	Region and Climate	Soil Type (SP, %)	Crop	Water Quantity Used in Leaching (m^3/ha)	Return to Leaching[b] (IL/m^3)
1	South	47	Avocado	1000	1.7
2	South	47	Avocado	1500	1.0
3	South	47	Citrus (Valencia)	1000	1.0
4	South	47	Citrus (Valencia)	1500	0.6
5	Negev	30	Avocado	1000	1.6
6	Negev	30	Avocado	1500	1.1
7	Negev	30	Citrus (Valencia)	1000	0.8
8	Negev	30	Citrus (Valencia)	1500	0.6

[a]At spring 1978 price level. 1 IL = 6 U.S. cents, approximately.

[b]In all situations, a steady state was simulated; continuous irrigation with water containing 300 ppm Cl, and leaching with water containing 220 ppm Cl were assumed. Program SALIN B (1982) was used.

Source: Adapted from Yaron and Voet, 1983

Obviously, the optimal solution of each season of the linear programming model depends on all future parameters of the system representation. As we progress over the planning horizon, however, additional data and information become available and can be used to update the model's parameters. The revised parameters are then employed as a priori information for the next model's solution (typically, an agricultural production system is relatively flexible and can accommodate itself to changing conditions at a relatively low cost). The main goal of Feinerman's linear programming model presented here is to provide a framework for decision-making in the short run, taking into account the future.

The model has been applied to the analysis of long-run irrigation on a farm with three water sources of varying salinity levels. Its results indicate the priorities in water and land allocation to the farms crops, which, in some cases, differ from those derived within a short-run analysis. For more details, the reader is referred to Feinerman (1980) and Feinerman and Yaron (1983).

While Feinerman's model and similar analytical long-run planning models incorporating uncertainty provide a much better understanding of the farm's system and its complex interrelationships, they are computationally too costly for routine work in planning or extension. Simulation provides a more practical approach for such purposes. An example is provided by Matanga and Marino (1979), who combined stochastic dynamic programming and simulation to determine irrigation schedules with brackish water for several crops, and then applied this information to the allocation of the farm's land and water among the crops. Simulation of irrigation with saline water on a farm was applied, too, by Polovin (1974). However, simulation, due to its amorphic structure, does not clearly point to the relationships among the relevant variables and the other elements of the system.

In conclusion, a combination of the computationally difficult analytical approaches with simulation is needed, with the first ones providing an understanding of the system, as well as ideas for testing by the more practical simulations. This judgment may change in the future, with development of more efficient computational algorithms.

Coming back from the phase of models to the real world which those models are expected to serve, one cannot overemphasize the importance of long-run planning of water resources and irrigation under conditions of water salinity. Water scarcity dictates the efficient use of water, which reduces built-in salt leaching. With no natural leaching by rainfall, the salinity build-up in the soil may eventually reach catastrophic levels. Ancient history abounds with examples of such events.

FARM REGION AND INTERREGIONAL INTERACTIONS, EXTERNALITIES, AND COOPERATIVE SOLUTIONS

Within the context of a region or a water basin, a distinction is useful among three phases of irrigation projects: the delivery phase, the farm irrigation phase, and the water removal phase. The farm irrigation phase has been reviewed in the previous section. Here, some problems related to the delivery and water removal phases will be discussed.

Compensation for Increased Water Salinity

Water rights, if firmly practiced under conditions of increasing water salinity, may lead in some regions to the question of what is fair compensation for the deterioration of water quality. A well-known international claim for compensation relates to the Colorado River, which crosses the U.S.-Mexico border. The 1944 U.S.-Mexico Treaty guaranteed Mexico an annual quantity of 1.5 M acre-feet (maf) of the water of the Colorado River (Oyarzabal-Tamargo and Young, 1978). Neither the quantity nor the quality was an issue until 1961. In fact, until 1960, Mexico received, on the average, over 4 maf per year. In 1961, a drainage system was constructed in southwestern Arizona, with its highly saline water discharged into the Colorado River. At about the same time, water deliveries to Mexico were reduced to 1.5 maf, due to the need for storage in Lake Powell (the reservoir for the Glen Canyon Dam). As a result, the salinity level of the water delivered to Mexico rose to 2000 ppm TDS, as compared with about 800 ppm TDS in 1960. (See section on magnitude of potential losses for the evaluation of the income losses accrued to the Mexican farmers; recall that the estimates refer to salinity-adaptive irrigation technologies, whose actual introduction to farms is, in reality, a slow process.) At the same, time the quantities delivered to Mexico were reduced. The Mexican farmers affected by this change and the Mexican government protested vigorously.

A solution was sought and found in countermeasure efforts undertaken in the U.S., with the result being the reduction in salinity to 1200 ppm in 1971. The conflict was finally resolved by an agreement signed by the presidents of the two countries in 1973, stating that the U.S. should be responsible for delivering to Mexico waters with an average salinity of not more than 115 ppm above the salinity at the Imperial Dam, the last U.S. diversion.[7] This problem is an example of a situation in which the burden of maintaining a certain level of water quality at the delivery phase is levied on the contributing region (in this case, the U.S.).

Another aspect of this problem is the evaluation of the fair rate of substitution between quantity and quality (salinity) of water supplied to farms and/or regions. This issue was studied by Yaron et al. (1982), with respect to the south of Israel. The motivation for their study emerged from the problem confronting the south and the Negev regions of Israel, in which a rise of salinity in water supply is expected. Currently (1983), most of the water supplied to the region contains between 220-250 ppm Cl, equivalent to 550-625 TDS ppm/1. Farm water quotas are determined by water rights to which the farmers, as well as the water administrators, adhere.

The necessary information for the derivation of the marginal rate of substitution (MRS) between quantity and quality of irrigation water are estimates of the production function(s) of farms in the region with varying inputs of quantity and salinity of irrigation water:

$$Y = f(\underline{Q}, S, \underline{F} | \underline{K}) \tag{17}$$

where:

Y = is the output (value added) of the farm

$\underline{Q}$ = vector of quantities of irrigation water at different periods

S = water salinity

$\underline{F}$ = vector of other production factors

$\underline{K}$ = vector of all other factors, assumed to be constant.

In view of the scarcity of data for the statistical estimation of (17), a normative planning approach was applied, also on the assumption that the farms would attempt to maximize their incomes under varying availability of quality and salinity.

The study referred to a sample of kibbutz farms with sprinkler irrigation as the predominating irrigation technology. The option of adaptation of the crop mix to increased salinity was included in the analysis (situation b in Figure 8.1).

Irrigation with saline water is a dynamic stochastic process, with rainfall being the major stochastic element in the region studied. A system integrating a dynamic stochastic programming model with a static linear programming model has been designed. Its application has led to the conclusion that, with reference to the data relevant to the specific empirical analysis, adaptive control-type decisions were not needed, and the dynamic stochastic model could be substituted by a static approach, addressing steady-state conditions (see also section on agro-economic efficiency). Accordingly, linear programming was applied to the determination of the long-run optimal mix of crops and the optimal water allocation on the sample kibbutzim. By parametric runs with reference to various combinations of water quantity and quality and other parameters, vectors of observations indicating the income, quantity and salinity of water, and other parameters were generated.

The relationship between income and quantity and quality of water was estimated for each kibbutz farm by the least squares technique. For illustration, one of the specifications of the regressions was:

$$Y = b_0 + b_1 GW + b_2 BW + b_3 BW \cdot CL \qquad (18)$$

where:

Y = income (=value added) of the farm (000 IL, spring 1978 prices, 1 IL = 6 U.S. cents);

GW = quantity of good quality water (220 ppm Cl) at the farm's disposal (000 m^3);

BW = quantity of low quality water at the farm's disposal (000 m^3);

CL = salinity index of the low quality water, defined as CL = (450-C) with C being the chloride concentration of the low quality water (ppm Cl). Cl represents the divergence from an upper bound of 450 ppm Cl.

Regression (18) was estimated for each kibbutz farm for three scenarios with respect to fruit crops. These scenarios assumed alternative policies with respect to the eradication and replanting of fruit crops in response to increased water salinity, the policies being:

- reference to the acreage of fruit groves as fixed
- eradication of fruit groves which are not profitable
- substitution of unprofitable fruit groves by profitable ones.

See Yaron et al. (1982) for the discussion and the justification of these scenarios.

Marginal productivity values of water with differing levels of salinity content were derived from the estimated regressions. These were later used to compute the marginal rate of substitution of low quality (BW) for good quality water (GW), with the income kept constant.[8]

Selected results presenting the marginal productivity of good quality (GW) and low quality water (BW) and the marginal rate of substitution between BW and GW under conditions of the last scenario are presented in Table 8.8.

Such MRS values were estimated for all 10 sample farms in the region for water salinity ranging between 260-400 ppm Cl and three scenarios assumed with respect to fruit crop policy. The frequency distribution of these MRS values is shown in Table 8.9.

An attempt to generalize and summarize the results presented in Table 8.9 is not an easy task. The individual farms differ one from the other considerably. Furthermore, functions representing the physical relationships involved and the relevant parameter values are subject to some uncertainty, as are relative prices of inputs and outputs. Generalizations, however, are needed for policy decisions and are expected by policymakers. In view of the complexity of the overall relationships, only a subjective evaluation of the results and their generalization may be attempted. The authors' (Yaron et al., 1982) subjective summary for water salinity approaching 300 ppm Cl is an MRS of 1.10 as a conservative measure and

Table 8.8 Marginal productivity and marginal rates of substitution between low and high quality water, kibbutz farms, South and Negev regions in Israel

Kibbutz No. and Region[a]	MVP (IL/m^3)[b]			MRS	
	GW	BW			
		Range[c]	Average[d]	Range[c]	Average[d]
1,S	4.76	4.07-4.50	4.32	1.17-1.06	1.10
3,S	4.43	3.69-1.96	3.84	1.20-1.12	1.15
8,SN	3.34	2.87-3.28	3.10	1.16-1.02	1.08
10,SN	7.00	6.18-6.70	6.48	1.13-1.04	1.08

[a] S = South; SN = South Negev

[b] At spring 1978 prices (1 IL = 6 U.S. cents)

[c] With BW water salinity ranging from 400 to 260 ppm Cl

[d] With BW water salinity at the mean value

Table 8.9 Frequency distribution and means of the estimated MRS values under selected situations

Observations Set and Source	Salinity Content of Low Quality Water (ppm Cl)	MRS					Total Median
		≤1.05	1.06-1.10	1.11-1.20	1.20	F	
		(%)					
All observations	260	90	10	--	--	100	≤1.05
"Scenario 1"[a]	320	30	70	--	--	100	1.06-1.10
	400	10	45	45	--	100	1.06-1.20
All observations	260	50	10	20	20	100	≤1.10
"Scenario 3"	320	10	40	30	20	100	1.06-1.20
	400	--	--	70	30	100	1.11-1.20

[a]The MRS values derived under conditions of "Scenario 2" are very close to those of Scenario 1. See text for the explanation of the scenarios.

an MRS of 1.20 as a liberal one. Obviously, different readers will formulate their own generalizations; farmers to be affected by increased water salinity in the future will obviously be cautious and tend towards higher MRS values.

A related problem is the determination of the optimal salinity level of the water supplied to the region, whenever it can be controlled at a certain cost. The issue emerges from the fact that various types of farms grow different crops with differing levels of sensitivity to salinity; the salinity-sensitive farms will tend to demand the supply of higher quality water.

Ratner (Yaron et al., 1982) found that under conditions of farm compensation for deterioration in water quality by additional water quantities, as described in the previous paragraphs, the preferred quantity-quality combination varies among different farms in the same region. Kibbutz A and B prefer a salinity content of 350 and 300 ppm Cl, respectively, while a neighboring moshav village prefers the current 220 ppm Cl/1 salinity level with no compensation in terms of additional water quantity. As previously mentioned, this difference in the preferences is easily explained by the difference in crop mix between kibbutz and moshav farms, inherent to their structures. The kibbutzim grow large areas of cotton, while the moshavim grow large areas of salinity-sensitive fruit crops. The questions posed to the region's farms and water resources administrators are: (1) What is the optimal quantity-salinity combination for the region? and (2) What is the proper scheme for cost-benefit allocation among the region's farms?

The Problem of Externalities

The interrelationship between the farm phase and the removal phase is a typical case of externalities, defined as a situation in which one group has indirect effects (adverse or favorable) on others. The problem deserves a long discussion; due to space restrictions, it will only be addressed briefly here.

The essence of the externality problem is that return flows from irrigated farms drain to rivers or groundwaters which, in turn, constitute the sources for irrigation water of other farms. (Drainage water contains considerably higher concentrations of dissolved salts than the application water.)

The quantities of salts transferred from one phase to another (i.e., from the delivery system to the farms and from the farms to the removal system) are subject to control. Sometimes, however, there is a conflict of interest among the various parties of the overall regional system. Where the source of water supply is a river, upstream farmers contribute to the pollution of downstream waters by the return flows from their fields. This tendency is exacerbated under conditions of availability of low-cost water, with no incentives for careful irrigation and avoidance of overirrigation. The tendency is even more acute when the source supplying the water is saline, and, from the point of view of the upstream farmers, excess irrigation, with a certain amount intended for leaching of salts, is necessary. In such a case, a clear conflict involves them in a loss of income, whereas using leaching exacerbates the salinization of downstream waters and correspondingly causes a loss of income to downstream farmers. The above is also true for large-size canals.

An analogous situation exists when the return flows are drained to groundwater or aquifers. Here, however, the process may be slowed down by the mixing that occurs with large volumes of water stored in the aquifers, and the salinization may be gradual, extending over decades. Such a process of gradual salinization of groundwater is observed in Israel; in such a case, the benefits from irrigation accrue to those currently farming, whereas the damage and income loss will be borne by future generations.

Reduction of the quantity of salt drained in return flows may be achieved by increasing irrigation efficiency on farms--using optimal irrigation schedules; improving irrigation systems by such means as lining canals and using pipes; and improving the interfarm water conveyance systems, thus decreasing seepage losses.

On the whole, some degree of return flow is necessary. Complete lack of soil leaching and drainage will result in a gradual build-up of salinity in the soil. In extreme situations, waterlogging may occur. Thus, if the groundwater level reaches the soil surface and evaporates, the contained salts are left on the ground surface (this process is often exacerbated by capillary action).

If it were possible to drain the return flows and deposit the salts in the deeper soil layers, below the root zone of plants but above the groundwater level, an ideal solution would pertain, but unfortunately, nowadays this seems utopian. In some situations, drainage of return flows to the sea may prove to be a useful solution. The application of the ASTRAN method, referred to in the next section, may hopefully lead to such solutions.

Obviously, the gravity of the salt problem in return flows depends on the salt content of water at the delivery phase. Reduction of salinity at this phase can be achieved by diverting salty sources out of the system and introducing and mixing high quality and desalinated waters.

The interaction between the contributing and the receiving regions poses several questions: What measures should be undertaken by the contributing region to reduce the losses caused to the receiving region? What measures should be undertaken by the receiving region to reduce its losses? What is the fair overall solution, and how should the costs and benefits be shared?

There are two major difficulties in answering the above questions. The first is a lack of sufficient information on the physical parameters needed to quantify the relationships between the contributing and receiving parties; this difficulty can be overcome by further research.

The second difficulty is considerably harder; it is related to the conceptual and moral problems involved in the externalities discussed. The questions posed above are generally dealt with by engineers and economists, who are used to the rigors of (1) problem definition and (2) optimization of a given objective function subject to certain constraints that emerge from the subject matter of the problem. Since a conflict of interests is inherent in the system, a crucial issue arises: From whose point of view should optimization be sought? I.e., should it be Group A farms, Group B, etc., or optimization from the point of view of the society as a whole, at the cost of certain parts of the society? The reader is referred to Young and Leathers (1981), Howe and Young (1981), Suzuki and Nakayama (1976), and Young et al. (1982) for a further discussion of these issues. Game theory approaches provide a useful framework for the analysis of such conflicts; however, it is still to be seen whether the axiom-oriented, quasi-automatic solutions derived with the aid of the various game theory approaches will provide answers meeting the criteria of common sense judgment. (For a formal introduction to game theory, the reader is referred to Owen, 1968.)

NEW IRRIGATION TECHNOLOGIES - NEW PRIORITIES

A variety of relatively new irrigation technologies and cultural practices for reducing salinity losses are available. They include increased frequency of irrigations, timing of the leaching irrigations (discussed in the section on agro-economic efficiency), change of the irrigation system (e.g., drip irrigation), land grading, profile modification, placement of seed, and artificial drainage (Ayers and Westcot, 1976, pp. 38-51). While many of the above technologies are successfully practiced at various locations, economic evaluations of their cost and benefit are generally nonexistent, except for specific situations.

For example, it is well known that sprinkler irrigation with short intervals between applications maintains high soil moisture, which dilutes salts in the soil solution, on the one hand, but induces higher evapotranspiration and water use, on the other. While modeling approaches to the problem have been presented (e.g., Yaron et al., 1980), no sufficient empirical work has been done for generalized conclusions.

Another example refers to drip irrigation. It has been seen as a proper technology for irrigation of numerous vegetable crops under conditions of salinity and is, indeed, successfully practiced. However, the effect of drip irrigation under salinity conditions on citrus and avocado in Israel is debatable. Drip irrigation leads to a continuous high-moisture regime in a limited portion of the soil volume, within which salts are diluted. However, at the edge of the wetted zone, salts are highly concentrated. The overall effect of these two soil-salinity zones is not clear. Other salinity-oriented issues regarding drip irrigation, which thus far (to our knowledge) have not been properly quantified, are concerned with the process of salt accumulation in the soil over a sequence of years and the processes of salt leaching by rainfall and/or by sprinkler irrigation. Additional theoretical and empirical work is needed in order to be able to quantify these processes, as well as many other new irrigation technologies.

Speaking of new technologies, mention should be made of approaches in sight which can be classified as <u>frontier-changing innovations</u>. One is the genetic work of Epstein et al. (1980), who screened a large spectrum of barley germ for tolerance. "The best selections grown under irrigation with undiluted seawater supplemented by nitrogen and phosphorus, had an average yield of 1082 kg/ha . . . . For comparison, the average annual world yield of barley is under 2000 kg/ha" (Epstein et al., p. 401). The experiment was performed on dune sand and with water salinity of EC = 46 mmho/cm!

Another approach to the same problem is based on the use of tissue culture techniques for the selection and isolation of salt-resistant plants (e.g., Chen et al., 1980). Note that these techniques also seem to be useful tools in the study of the physiological aspects of salinity resistance of plants.

A further example of frontier-changing innovations is the ASTRAN method (Helweg and Labadie, 1976), which, by clever manipulations and well-planned utilization of water resources with emphasis on salinity, may lead to the concentration of salts in aquifers selected for that purpose, and salt reduction in the others. The extreme and most beneficial expression of the application of this method would be the creation of a sink of salts which may be disposed of in the sea.

These examples of scientific efforts intended to create far-reaching innovative technologies, and other efforts to achieve breakthrough results, should not obscure the issues confronted within the present state of knowledge. A rational allocation of efforts between the step-by-step, achievement-oriented studies and the breakthrough, hopeful studies should be kept in mind.

SUMMARY

The yield and income losses accrued to farms under conditions of irrigation with saline water depend, to a considerable extent, on the farms' crop mix and the share of salinity-sensitive crops. When the share of the latter ones is high, as on the moshav villages in Israel, increasing salinity of irrigation water becomes a serious economic problem (see Table 8.3). On modern commercial farming systems with a $\frac{\text{value added}}{\text{gross output}}$ ratio of 50 percent, loss of yield of 10-20 percent is equivalent to 20-40 percent loss of income. The problem is exacerbated by the fact that salinity-induced losses in yields within this range are not necessarily visible, and the causes for the reduced yields might be unknown without a specific study aimed at this subject.

Our ability to estimate salinity-induced yield losses is subject to two major shortcomings: the first one is due to the fact that the bulk of information on crop response to salinity originates in experiments with ample soil moisture supply, which do not necessarily represent field conditions in practice (see the section on underlying physical relationships for the comparison of experiments with survey data). Secondly, proper management of irrigation and the introduction of new technologies can reduce the losses, but the adoption and application of counter-salinity irrigation methods is a slow and often costly process. Normative estimates of salinity-induced losses are biased downwards if they refer to technologies more advanced than those actually practiced. Significant empirical survey work emphasizing realistic farm practice is needed in order to improve our knowledge on salinity-induced losses. On the whole, some losses are unavoidable. This fact should be kept in mind, remembering as well that the general trend of deterioration of the quality of water resources in the arid regions, due to excessive use, is unavoidable too.

NOTES

1. Originally published in State of the Art: Irrigation Drainage and Flood Control, No. 3, pp. 263-296, International Commission on Irrigation and Drainage, n.d..

2. Computed on the basis of the List of Publications published in June 1982 by the U.S. Salinity Laboratory, ARS, USDA, Riverside, California (mimeo).

3. SALIN B, Library Program and the Computer Center of the Hebrew University.

4. Since citrus, a major crop in Israel, is specifically sensitive to the chloride ion, salinity is often referred to in terms of chloride concentration. The relevant transformation formulae under the Israeli conditions are:

$$Cl\ (ppm) = 35.5\ Cl\ (meq/l)$$
$$TDS\ (ppm) = 2.5\ Cl\ (ppm)$$
$$EC\ (mmho/cm) = 0.62 + 0.137 Cl\ (meq/l)$$

A quick reference conversion table is shown below (rounded numbers):

Cl (ppm)	TDS (ppm)	EC (mmho/cm)
200	500	1.40
300	750	1.80
400	1000	2.15
500	1250	2.50
1000	2500	4.50

5. Kibbutz--a collective farm; moshav--a village of 50-120 family farms, with a village cooperative providing production and marketing services.

6. $$S = \frac{1}{TZ} \int_0^Z \int_0^T s(z,t)\, dz\, dt$$

and

$$s = \pi(c) + \tau(\theta)$$

where:

c = total salt concentration of the soil solution

s = total water suction

t = time

z = soil depth

θ = volumetric water content of the soil

π = osmotic component of the total water suction, a function of c

τ = metric water suction, a function of θ.

7. Following Oyarzabel-Tamargo and Young (1978).
8. Referring to:

$$Y = b_1 GW + b_2 BW + b_3 BW \cdot CL$$

and taking the total derivative of Y with respect to GW and BW, we get:

$$dY = b_1 dGW + b_2 dBW + b_3 CL \cdot dBW$$

Letting Y = constant, dY = 0, we get for a given level of CL (= CL*) the marginal rate of substitution between low and good quality water (MRS):

$$- \frac{dBW}{dGW} = \frac{b_1}{b_2 + g_3 CL^*} = \frac{MVP_{GW}}{MVP_{BW}}$$

with MVP denoting the marginal value product of water.

REFERENCES

Ayers, R. S., and D. W. Westcot. "Water Quality for Agriculture," FAO Irrigation and Drainage Paper No. 29, Food and Agriculture Organization, Rome, Italy, 1976.

Bernstein, L. "Salt Tolerance of Plants," USDA Agricultural Information Bulletin 283, U.S. Department of Agriculture, Government Printing Office, Washington, D.C., 1964.

Bernstein, L. "Salt Tolerance of Fruit Crops," USDA Agricultural Information Bulletin 292, U.S. Department of Agriculture, Government Printing Office, Washington, D.C., 1965.

Bernstein, L., and L. E. Francois. "Comparisons of Drip, Furrow, and Sprinkler Irrigation," Soil Science, 110, 1973.

Bernstein, L., and L. E. Francois. "Leaching Requirements Studies: Sensitivity of Alfalfa to Salinity of Irrigation and Drainage Waters, Soil Science Society of America Proceedings, 37, 1973.

Bernstein, L. "Effects of Salinity and Soil Water Regime on Crop Yields," Salinity in Irrigation and Water Resources, D. Yaron, ed. New York: Marcel Dekker, 1981.

Bresler, E. "A Model for Tracing Salt Distribution in the Soil Profile and Estimating the Efficient Combination of Water Quality under Varying Field Conditions," Soil Science, 104, 1967.

Bresler, E. Personal communication, 1972.

Bresler, E. "Simultaneous Transport of Solute and Water under Transient Unsaturated Flow Conditions," Water Resources, 9, 1973.

Bresler, E., and D. Yaron. "Soil Water Regime in Economic Evaluation of Salinity in Irrigation," Water Resources Research, 8, 1972.

Chen, Y., E. Zehavi, P. Barch, and N. Umial. "Effects of Salinity Stresses on Tobacco I, the Growth of Nicotiana Tabucum Callus Cultures under Seawater, NaCl and Mannitol Stresses," Z. Pflanzenphysiol., 98, 1980.

Childs, S. W., and R. J. Hanks. "Model of Soil Salinity Effects on Crop Growth," Soil Science Society of America Proceedings, 39, 1975.

Dantzig, G. B., and A. Madansky. "On the Solution of Two-Stage Linear Programs under Uncertainty," Proceedings of Fourth Berkeley Symposium on Mathematical Statistics and Problems, Vol. 1, University of California, Berkeley, California, 1961.

Epstein, E., J. D. Norlyn, D. W. Rush, R. W. Kingsbury, D. B. Kelley, G. A. Cunningham, and A. F. Wrona. "Saline Culture of Crops," Science, 210:339-404, 1980.

Feddes, R. A., E. Bresler, and S. P. Neuman. "Field Test of a Modified Numerical Model for Water Uptake by Root System," Water Resources Research, 1199-1206, 1974.

Feinerman, E. "Economic Analysis of Irrigation with Saline Water under Conditions of Uncertainty," Ph.D. thesis, Hebrew University of Jerusalem (in Hebrew, with English summary), 1980.

Feinerman, E., and D. Yaron. "Economics of Irrigation Water Mixing within a Farm Framework," Water Resources Research, 1983.

Goldberg, D. C., M. Rinot, and N. Kary. "Effect of Trickle Irrigation Intervals on Distribution and Utilization of Soil Moisture in a Vineyard," Soil Science Society of America Proceedings, 35:127-130, 1971.

Hanks, R. J. "Model for Predicting Plant Yield as Influenced by Water Use," Agronomy Journal 66:660-665, 1974.

Hanks, R. J. and J. C. Andersen. "Physical and Economic Evaluation of Irrigation Return Flow and Salinity on a Farm," Salinity in Irrigation and Water Resources, D. Yaron, ed. New York: Marcel Dekker, 1981.

Hausenberg, Y., Y. Pozin, and M. Boaz. Salinity Survey: A Summary Report, Spring 1963 - Spring 1973. Israel: Israel Ministry of Agriculture, Extension Service (in Hebrew), 1973.

Helweg, O. J., and J. W. Labadie. "Accelerated Salt Transport Method for Managing Groundwater Quality," Water Resource Bulletin, 12:681-683, 1976.

Howe, C. H., and T. J. Young. "The Measurement of Regional Economic Effects of Changes in Irrigation Water Salinity Within a River Basin Framework: The

Case of the Colorado River," Salinity in Irrigation and Water Resources, D. Yaron, ed. New York: Marcel Dekker, 1981.

Maas, E. V., and G. J. Hoffman. "Crop Salt Tolerance: Current Assessment," American Society of Civil Engineers, Journal of the Irrigation and Drainage Division IR2, 103:115-134, 1977.

Matanga, G. B., and M. A. Marino. "Irrigation Planning, 2: Water Allocation for Leaching and Irrigation Purposes," Water Resources Research, 15(3):679-683, 1979.

Moore, C. V., J. H. Snyder, and P. Sun. "Effects of Colorado River Water Quality and Supply on Irrigated Agriculture," Water Resources Research, 10(1):137-144, 1974.

Neuman, S. P., R. A. Feddes, and E. Bresler. "Finite Element Analysis of Two-Dimensional Flow in Soils Considering Water Uptake by Roots: 1, Theory," Soil Science Society of America Proceedings, 39(2):224-230, 1975.

Owen, G. Game Theory. Philadelphia, Pennsylvania: W. B. Saunders Co., 1968.

Oyarzabal-Tamargo, F., and R. A. Young. "International External Diseconomies: The Colorado River Salinity Problem in Mexico," Natural Resource Journal, 18:77-89, 1978.

Parkinson, J. K., J. T. Habbi, J. P. Wagner, and M. S. Sachs, "Desalting Saline Groundwater for Irrigation: A Case Study of Buckeye Area, Arizona," Water Resources Research, 6(5):1496-1501, 1970.

Polovin, A. "Economic Evaluation of Water Quality in the ... - An Application of Simulation," M.S. thesis (in Hebrew), Hebrew University of Jerusalem, 1974.

Robinson, F. E. "Agricultural Consequences in California," Salinity Management Options for the Colorado River, J. C. Andersen and A. P. Kleinman, eds., Utah Water Research Laboratory Water Resources Planning Series Report P-78-003, Utah State University, Logan, Utah, 1978.

SALIN-B. Hebrew University Computing Center Library, Library Program, 1982.

Shalhevet, J. "Citrus Sensitivity to Salinity," (in Hebrew) Alon Ha'notea, 1983, pp. 41-43.

Suzuki, M., and M. Nakayama. "The Cost Assignment of the Cooperative Water Resource Development: A Game Theoretic Approach," Management Science, 22:1081-1086, 1976.

USDA Salinity Laboratory Staff. "Diagnosis and Improvement of Saline and Alkali Soils," USDA Handbook 60, U. S. Department of Agriculture, Government Printing Office, Washington, D. C., 1954.

Wadleigh, C. H., and A. D. Ayers. "Growth and Biochemical Composition of Bean Plants as Conditioned by Soil Moisture Tension and Soil Concentration," Plant Physiology, 20:106-132, 1945.

Yaron, D., and A. Olian. "Application of Dynamic Programming in Markov Chanis to the Evaluation of Water Quality in Irrigation," American Agricultural Economics, 55:467-471, 1973.

Yaron, D. "Economic Analysis of Optimal Use of Saline Water in Irrigation and the Evaluation of Water Quality," Salinity in Water Resources, J. E. Flack and C. W. Howe, eds. Boulder, Colorado: Merriman Publishing Co., 1974.

Yaron, D., E. Bresler, H. Biolorai, and B. Harpinist. "A Model for Optimal Irrigation Scheduling with Saline Water, Water Resources Research, 16:257-262, 1980.

Yaron, D., A. Dinar, H. Voet, and A. Ratner. "Economic Evaluation of the Rate of Substitution between Quantity and Quality (Salinity) of Water in Irrigation," Working Paper #8211, The Center for Agricultural Economic Research, P.O. Box 12, Rehovot, Israel, 1982.

Yaron, D., and H. Voet. "Application of an Integrated Dynamic and Linear Programming Model to the Analysis of Optimal Irrigation on a Farm with Dual Quality (Salinity) Water Supply," Planning and Decisions in Agribusiness: Principles and Experiences, C. F. Hanf and G. W. Schierer, eds., Elsvier, Amsterdam, Holland, 1982.

Yaron, D., A. Dinar, and S. Shamlah. "First Empirical Estimates of Potential Losses of Agricultural Income in the South of Israel Due to Deterioration of Water Quality," (in Hebrew) research report, The Center for Agricultural Economic Research, P.O. Box 12, Rehovot, Israel, 1979.

Young, R. A., and K. L. Leathers. "Economic Impacts of Regional Saline Irrigation Return Flow Management Program," Salinity in Irrigation and Water Resources, D. Yaron, ed. New York: Marcel Dekker, 1981.

Young, H.P., N. Okada, and T. Hashimoto. "Cost Allocation in Water Resource Development," Water Resources Research, 18:463-475, 1982.

APPENDIX 8.A

THE EFFECT OF IRRIGATION-LEACHING DECISIONS ON SOIL SALINITY AND YIELDS OF CROPS

The underlying relationship is the salt balance equation (Bresler, 1979):

$$Q \cdot C - \beta \cdot Q \cdot \frac{\xi - \hat{\xi}}{2} = (\xi - \hat{\xi}) \cdot V \qquad \text{(A.1)}$$

where:

ξ = soil salinity before irrigation (meq Cl/l)

$\hat{\xi}$ = soil salinity after irrigation (meq Cl/l)

Q = depth of irrigation water applied (mm)

C = chloride concentration in the irrigation water (meq Cl/l)

V = depth of water contained in the root zone (mm)

β = empirical leaching parameter, denoting percentage of chloride leached below the root zone during irrigation.

From (A.1), a transformation function is obtained:

$$\hat{\xi} = \frac{Q \cdot C + \xi(V - \frac{\beta}{2} Q)}{(V + \frac{\beta}{2} Q)} \qquad \text{(A.2)}$$

Denote the parameters of decision d^i by Q_L^i, C_L^i, Q_I^i, C_I^i, where Q_L^i, Q_I^i are the water quantities and C_L^i, C_I^i are the water salinities stipulated by d^i. Assume that it involves: (a) preplanting leaching (Q_L); (b) irrigation and leaching during the irrigation season (Q_I).

From (A.2), we obtain:

$$\xi^i = \xi^i(\xi,d^i) = \frac{Q_L^i C_L^i + \xi(V - \frac{1}{2}\beta\cdot Q_L^i)}{(V + \frac{1}{2}\beta\cdot Q_L^i)} \tag{A.3}$$

$$\hat{\xi}^i = \hat{\xi}^i(\xi^i,d^i) = \frac{Q_I^i C_I^i + \xi^i(V - \frac{1}{2}\hat{\beta}\cdot Q_I^i)}{(V + \frac{1}{2}\hat{\beta}\cdot Q_I^i)} \tag{A.4}$$

where:

ξ = soil salinity after preplanting leaching

$\hat{\xi}^i$ = soil salinity after the irrigation season

β = preplanting leaching parameter

$\hat{\beta}$ = irrigation season leaching parameter.

A similar relationship was used for the rainy season with the salinity of rainfall taken as zero and the parameter β empirically estimated.

Salinity damage to yield is determined via the electroconductivity of the soil solution, assumed to be a function of two known parameters, A and B (Maas and Hoffman, 1977). Soil salinities ξ^i, $\hat{\xi}^i$ determine the value of the electroconductivity of the soil solution (EC_i).

$$EC_i = 0.62 + 0.137\,\frac{\xi^i + \hat{\xi}^i}{2} \tag{A.5}$$

Referring the base yield of crop j, in the absence of salinity damage as Y_0^j, the actual yield is defined by:

$$Yj(d^i,\xi) = y_0^j[1 - 0.01 \cdot B^j\{\max(0,EC_i^j - A^j)\}] \tag{A.6}$$

where EC_i^j is the electroconductivity of the soil solution in the j-th crop plot using strategy d^i.

The empirical leaching parameter β was estimated on the basis of 141 observations from the Citrus Salinity Survey in Israel (Hausenberg, 1973). The estimates relating to sprinkler citrus irrigation were β = 0.73, 0.61, 0.52, and 0.40 for soil with SP = 0-39, 40-59, 40-69, 70-84%, respectively. The coefficient for rainfall leaching was estimated to be 0.60 for all soil types.

Note that these are empirically estimated coefficients, valid for the conditions referred to (sprinkling, citrus irrigation, south of Israel). For more details, see Yaron et al. (1979, Hebrew).

9
Irrigation, Drainage, and Food Supplies

Ian Carruthers

GROWING RELIANCE UPON IRRIGATION

Irrigation is playing an increasingly important part in providing the developing world's food supply. The World Bank estimates that there are now 160 M ha of irrigated land in developing countries. This is only 20 percent of all land harvested, but it receives 60 percent of applied fertilizer and produces 40 percent of all crop output. There is $15 B invested in irrigation and it is still growing at 2 percent per year. It is a favored sector for aid donors, receiving one-fifth of all aid for food and agriculture in 1980 (Carruthers, 1983).

There is no sign of irrigation falling from favor with farmers, governments, or aid donors. Irrigation is the principal means by which man modifies climates to increase food supplies. New developments in technology, plus complementary advances in plant breeding, crop protection, and agronomy "packages" have increased the potential productivity and profitability of irrigation agriculture. This increased productivity comes from higher yields and multiple cropping, often with two or even three crops a year; therefore, it is argued (mainly by those with a technical bias) that public and private investment and aid donor interest are well founded. Without irrigation, the "Green Revolution" would founder.

But irrigation investment also has had loud critics who note such matters as the huge costs ($2,000 per 10,000 ha) which are often underestimated, the delays in construction, the yields below forecast, the poor financial performance, and the environmental damage to human health and to the soils (Hotes, 1983).

THE THREAT OF WATERLOGGING AND SALINITY

This article concentrates upon the growing problem of soil waterlogging and salinity, which threatens to destroy the food-producing capacity of the irrigated lands of the Nile, Euphrates, Indus, Ganges, and many other arid zone river basins. Drainage is also required in the humid tropics where rice is the dominant crop, in order to bring about the sound water control necessary to obtain high yields. The aid lobby, such as those responsible for the Brandt Report, specifically mention large-scale irrigation basins as a major area for agricultural investment and production expansion. But in most irrigation areas, drainage, reclamation, and water control projects are needed now. The Food and Agriculture Organization estimates that 50 percent of the world's irrigated land is salinized to the extent of affecting productivity (cited by Gilbert White, 1977). In Iran, Iraq, Egypt, and Pakistan, more than 70 percent of the farmland is so affected. India has 5-7 M ha affected. Wherever evaporation exceeds rainfall, salinity is a risk. Where high sodium content leads to alkaline soils, with a consequent toxicity, loss of structure and permeability, then reclamation is technically extremely difficult and expensive. Where alkaline conditions occur, there is virtually no economic solution and this problem is reputed to be increasing in parts of northern India, Pakistan, southern Russia, Afghanistan, and Iran (Kovda, 1977).

Drainage has not been undertaken because the effect of waterlogging and salinity is generally slow to become apparent; remedial measures are expensive; in areas already irrigated, the loss of land and disruption to existing farm structures, roads, and canals causes local opposition; and maintenance of drains is costly and requires careful management.

Drainage has been consciously neglected by irrigation advocates. In arid India and Pakistan, developers have long recognized eventual drainage needs, but they deferred expenditure on grounds of political expediency and finance. As has been noted, unlike irrigation, drainage is unpopular with farmers, taking substantial amounts of land (approaching 15 percent in the case of open drains) and giving in return a benefit that is not obvious, is delayed, and is indirect, at a relatively large cost. Johnson (1982) brilliantly reviews the irrigation experience of Pakistan and argues that now massive investment in drainage is inevitable if the Indus Plains are to sustain at targeted living standards the 130 M who will inhabit the region in the year 2000. He concludes that no alternative is available and, most depressing of all, that most of the costs must be borne by the users of irrigation. This is daunting because the Pakistan government has not managed to make even the rich, among what are mostly low-income farmers, pay more than 50 percent of the recurrent costs of irrigation supplies. We can anticipate that drainage levies will be politically and administratively more problematical, even if the present trend to pay farmers higher prices by reducing indirect taxation of agriculture continues.

Waterlogging and most forms of salinity are the direct consequences of poor water management and inadequate drainage. Various symptoms of damage from defective drainage are still regarded, all too often, as unexpected indirect costs of irrigation development. For example, in Egypt, with its long experience of irrigation, there was disappointment verging on surprise at the extent and form of the deleterious effects of the large additions of irrigation water from the Aswan Dam, first on the groundwater regime, then later on crop yields. The damage has forced the government to adopt a nationwide drainage program that has absorbed the major part of the Ministry of Irrigation's capital budget in recent years.

The best technical means of drainage and the optimum operating system are not well tested. There is a need to assist poor countries with finance, technical assistance, and pilot projects. Once this phase is over, there will be a major role for aid donors who have large resources, long-term perspective, and environmental consciousness. The economics of drainage shares some of the problems of conservation, soil erosion, and tree planting. Primarily because of delayed benefits, the rates of return are likely to appear low, but the instinct is to proceed in spite of this. On long-term, irreversible matters, the economic calculus sometimes appears fragile and deficient.

CAUSES OF DRAINAGE PROBLEMS

Drainage is going to be required whenever the groundwater equilibrium is disturbed and the water table rises to the plant root zone. This will happen when the sum of incoming vertical seepage from precipitation, rivers, canals, watercourses, and fields, plus lateral seepage exceeds the sum of evaporation from the capillary fringe of the groundwater and evapotranspiration of plants and any lateral export by underground seepage (see Figure 9.1).

On the large alluvium river basins, before modern barrage-controlled irrigation, there was very little problem from waterlogging, as the water table was generally below 4 m, and annual inflows and outflows were in balance. Even the huge nineteenth and early twentieth century barrage canal commands of the Indian subcontinent did not lead to a general rapid rise in underground water table levels because the design aimed to spread water thinly over a large area. Certainly in some areas, there were local problems when, for example, badly aligned canals cut across natural drainage lines, increasing the risk and duration of periodic flooding (Whitcombe, 1972). For the most part, any rise in groundwater levels was slow because the main objective was to protect as large an area as possible from drought, to minimize famine risk, and to provide the financial benefits of irrigation to as many landowners as possible, which was, in turn, expected to benefit the government exchequer. Typically, these early irrigation projects led to cropping intensities that were less than half of what were theoretically achievable.

The effect of protective irrigation was primarily to encourage farmers to under-irrigate, in an effort to cover as large a part of their land as possible. This was rational for them because water was the scarce factor of production, compared with land and labor, and the highest average return to water came from light irrigations. With simple, traditional agricultural technology and poor infrastructure, the irrigation water response function for a given season is low and very flat; hence, the optimum water application is much less than the potential evapotranspiration which is usually advocated by extension agents. Rational farmers, maximizing the return per unit of scarce water by increasing the area cultivated, helped prevent water table buildup, as seepage losses from field were negligible. Unfortunately, under this irrigation regime, the small quantities of salt present in irrigation water [for example, 300-400 parts per million (ppm) total dissolved solids in Indus water] gradually build up in the profile, and soil leaching is eventually necessary to prevent saline soils and salt damage to crop growth.

Researchers who focus at the watercourse or farm level have found, under extensive and intensive systems, losses ranging from 25-30 percent (Punjab National Bank, 1982) to 40 percent (Lowdermilk et al., 1977). Lattimore (1979) reports team findings that indicated most losses

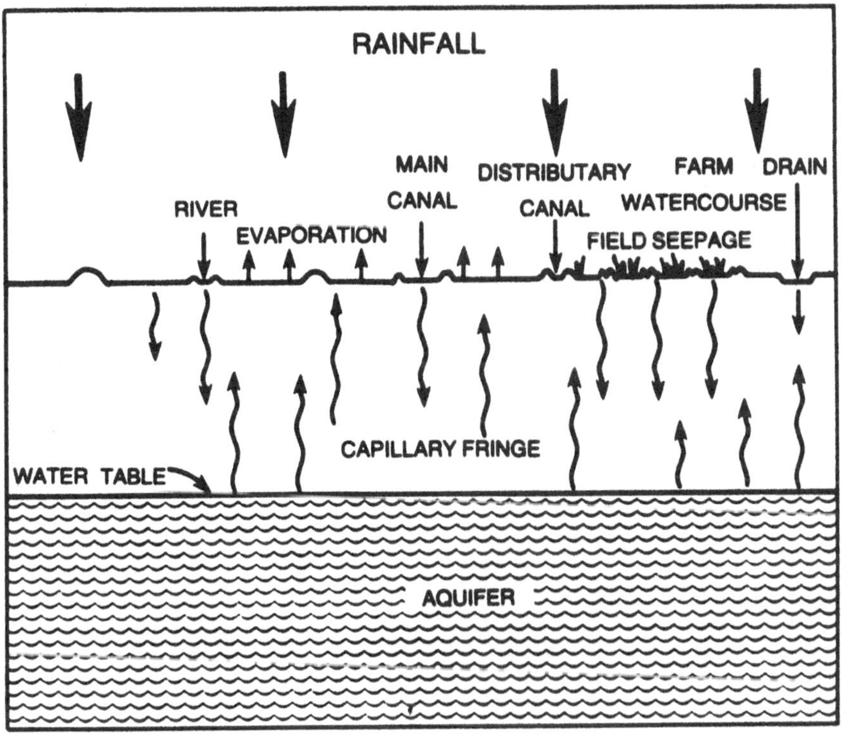

Figure 9.1 Cross section of an irrigated river valley showing sources of seepage

occur through the banks and at junctions. They recommended realignment and consolidation of the banks and concrete structure at junctions, saving up to half the water while achieving more than double the crop production.

Leaching of salts by heavy irrigation, plus seepage from rivers, canals, and watercourses, caused a slow rise in water table in many irrigation projects. Poor field application of efficiency and unlevel fields add to the problem. For example, in Khairpur, Pakistan, the water table rose by 10 cm (0.34 ft) per year from the early 1930s, when Sukkur Barrage was opened, up until 1965. By this time, the position of farmers was serious because the average water table depth was now less than 2 m. This leads to evaporation from the water table resulting in a rapid increase in surface soil salinity (see Figure 9.2). Furthermore, as water tables rise, there are serious negative effects, first upon the rooting patterns of deep-rooting plants (such as tree crops and cotton), then eventually shallow-rooting plants (such as wheat) (see Figure 9.3). (See also Nijland and Guindi, 1984, for Egyptian data.)

Unfortunately, farm-level data availability on the relationship between crop yield and waterlogging or salinity is not available in sufficient quantity to service the large-scale and diverse public investment programs. Drain designers and economists both are working with limited and crude information. Furthermore, where field trials exist, the evidence is obscured by other factors affecting yields, so a clear relationship to aid detailed design is seldom found (see Nijland and Guindi, 1984).

RECENT ACCELERATION IN THE PROBLEM

Over the last 25 years, the rate of salt buildup and the insidious rise in the water table in irrigated lands has substantially increased. This has arisen because of a switch from "protective irrigation" to a drive for increased intensity of irrigation. There are technical engineering, agricultural, and economic reasons for a switch in approach towards intensive irrigation.

Attractive projects to supply more water to agriculture came as a result of advances in the engineering field in water storage dam design and earth moving and other construction technology; an increase in the demand for hydropower and improved ways of creating it; new approaches to groundwater exploitation; and appreciation of

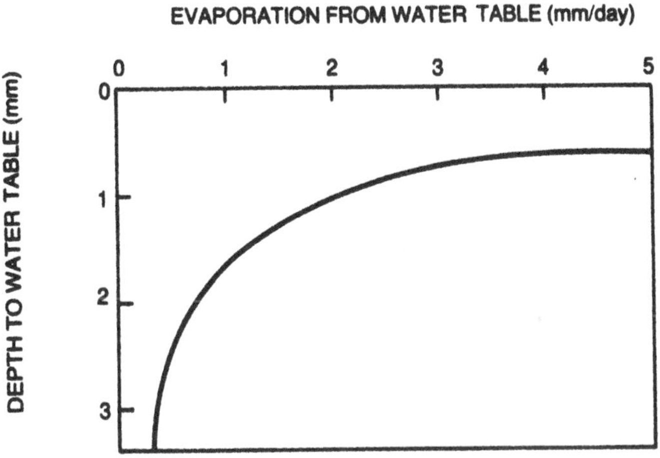

Figure 9.2 Increase in surface soil salinity (MacDonald, 1983)

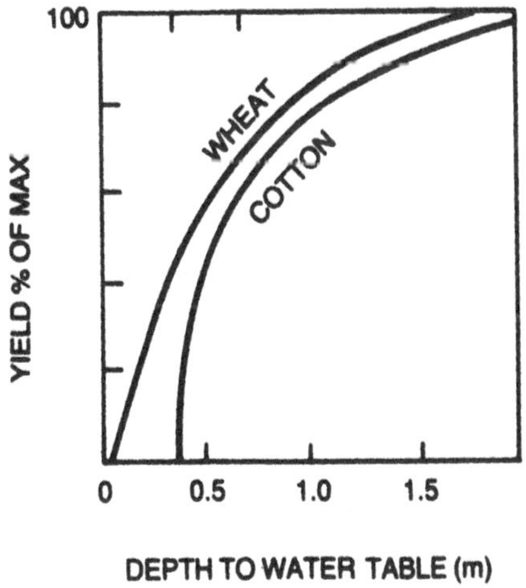

Figure 9.3 Rising water table effects on wheat and cotton (MacDonald, 1983)

opportunities for water saving at the field level. All these developments contributed to the creation of attractive projects to supply more water to agriculture. However, these opportunities were seldom low-cost projects, and it appeared most economic to put additional water into the existing network of the under-used irrigation facilities. In Egypt, India, and Pakistan, the best lands were already irrigated, so intensification began by switching seasonal canals to perennial operation. The next stage was to remodel the existing canals and watercourses to take additional surface water. Subsequently, since the late 1960s, groundwater development has been undertaken on a grand scale. At first, this was not efficiently handled by the farmers, and the seepage increased. This wastage added to the growing drainage problem.

Whenever there is fallow land laid out for irrigation, the cultivators and landowners will pressure the engineers (who typically manage schemes) for additional water. Many water managers have succumbed to those pressures, and many canals have been run bank full, much above design, with increased seepage and much waste when canal bank breaches occurred.

At the time that engineering developments gave an incentive to irrigation investment, there were advances in agronomy characterized as the "Green Revolution" technology, which added further impetus. New varieties of crops emerged from research institutes (particularly wheat, rice, maize, and sugarcane), which responded to fertilizer and could more than repay the costs of crop protection and additional attention to soil cultivation techniques. This shifted the optimum irrigation strategy of the farmer from extensive cultivation of a large area to intensive cultivation. In economic terminology, there was a complementary, or more than additive, response to simultaneous application of the package of modern agricultural inputs, including irrigation water. This implied an upward shift in the response curve for water, giving higher yields for any level of water supply. In short, it paid to apply more water per hectare, and the drainable surplus was again increased.

Figure 9.4 shows the effect of these changes in irrigation intensity in part of the Lower Indus in Pakistan. Before 1923, when perennial irrigation was introduced, the water tables were below 4 m; 50 years later, 75 percent of these areas had water tables less than 2.4 m.

There are limits to this process. Water would not rise to the surface throughout the irrigation project

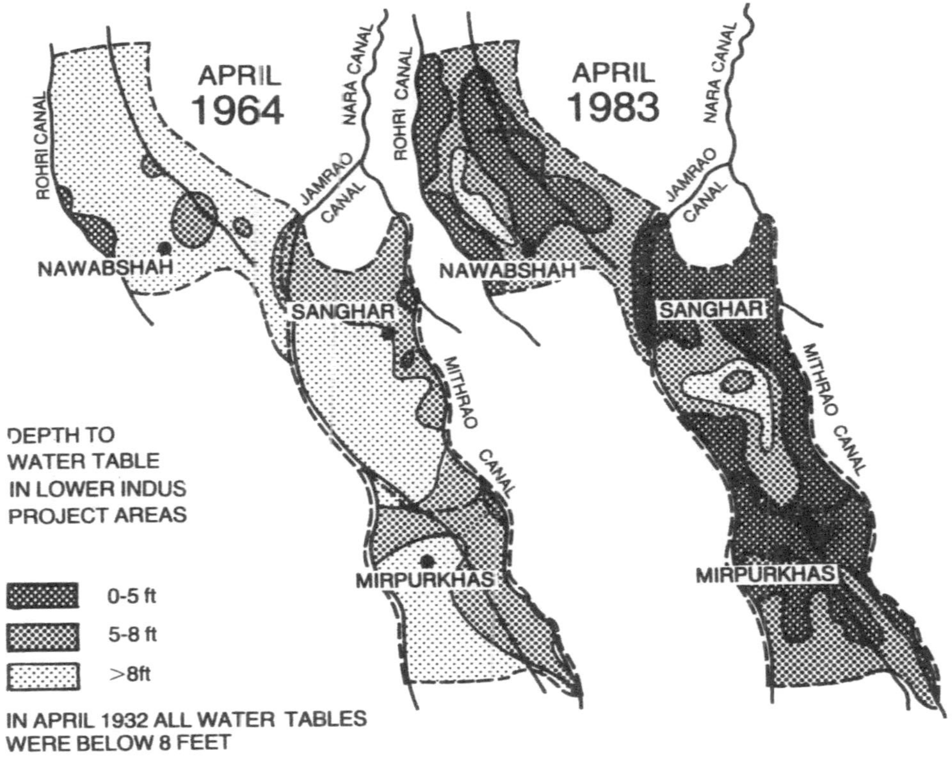

Figure 9.4 Effect of changes in irrigation intensity in part of the lower Indus in Pakistan (adapted from MacDonald, 1983)

areas. Increased salinization of land would reduce irrigated demand, and canal discharges would have to be cut. The high water tables would, in turn, result in greater rates of evaporation from the subsoil water table. In time, reduced inflows and increased evaporation would produce an equilibrium, and the water table would stabilize at perhaps an average of 0.5-1.0 m. To farm this successfully requires high standards of farm management and a regular and reliable irrigation supply. Even with good management, it is likely that severe problems will arise if the groundwater is highly saline (more than 3000 ppm) and after storms (a 10 cm storm may raise water tables by 1 m) for several days. If the millions of hectares of irrigated land that are at risk--but which promise so much for low-income farmers in the arid zones--are to be safeguarded from disastrous deterioration, there appears to be no alternative but to drain them.

Salt export is generally required. It is rare for a satisfactory local solution for disposing of salt to also be a satisfactory long-term solution. Each year, rivers such as the Nile and the Indus, with total dissolved salts between 200-1000 ppm, bring in millions of tons of salts. Whereas, in the past, most salt would be flushed to the sea in floods, now most water is stored and consumed by irrigation. For many months of the year, great rivers such as the Nile and Indus discharge no water to the sea. Therefore, the irrigated lands have virtually become huge evaporating pans and salt stores (Pillsbury, 1981).

ALTERNATIVE TECHNICAL SOLUTIONS

Drainage engineers have to determine the best technical means of obtaining water table control and the optimum depth to the water table. We can see from Figure 9.3 that if we reduce the water table below 2 m, there is no drainage constraint to typical crops. Figure 9.2 also shows that in Indus alluvium, to reduce evaporation from the water table to one-fifth of the maximum level, the water table should be below 2 m.

The economic problem is that the greater the depth that the water is drained, the higher the costs. Pumping costs are directly proportional to height lifted, and capital costs (especially for open drains) also increase substantially. The main technical alternatives for irrigated land drainage being considered in Pakistan are horizontal drains (open or tile drains) and vertical drainage with tubewells. To appraise projects for drainage investment, these options have to be compared with each other and with the "do-nothing" alternative. This will result in the abandonment of certain areas where salt will accumulate as it is leached from cultivated areas. Abandoned land becomes a salt sump and, in effect, provides "dry drainage."

In areas that are already irrigated, open drains are difficult and costly to install, with up to 15 percent of the land area lost, depending upon the soils and their depth. And once the open drains are installed, the government would face enormous costs for bridges and other structures over disposal channels, in addition to land compensation costs. Economists may argue that land compensation costs are merely transfer payments and therefore don't affect the economic assessment. While this is technically correct, we should note that economics is but one test of feasibility, and planners know that raising revenue to pay compensation costs is neither simple nor costless. In addition to presenting funding problems relating to compensation, open drains present management problems; they are the drainage technology most disruptive to the existing pattern of agriculture and will meet with the most problems politically. General maintenance and weed control problems are likely, and open drains are often a source of health hazards; poorly maintained drains are breeding grounds for mosquitoes, bilharzia-infected snails, and other harmful vectors of disease. In silty alluvium soils, slumping of sides of drains will cause difficulties. In short, open drains, while requiring a technology that is simply executed, present severe financial problems in construction, plus management problems in operation.

Tile drains have received a big boost in recent years by development of new tile-laying machines and long-length, perforated plastic pipes. Relatively static oil prices have made plastic pipe a relatively cheap material, compared to brick or earthenware alternatives. Nevertheless, tile drains are extremely expensive--about four times the capital cost of tubewells--with similar running costs, unless the topography allows a gravity outfall. In order to obtain a minimum of 1.5 m depth between collectors, the tiles must be about 2 m deep and spaced at 60 to 150 m, depending upon the soil permeability (ranging from about 0.5 m/day on heavy soils to 1.0 m/day on light soils).

One of the most remarkably successful features of the last two decades in the Indian subcontinent has been the rapid expansion of groundwater development using public and privately installed tubewells. In Pakistan, there are 186,000 private wells installed and 12,500 bigger-capacity, public wells. According to one World Bank estimate, the private wells in 1983 accounted for about 80 percent of the pumpage and approximately 30 percent of irrigation water reaching crops.

Where aquifers are suitable, tubewell drainage is, in principle, more efficient than any alternative. Tubewells are potentially cheap, easy and quick to install, a proven technology, and they can control the water table at any depth. In practice, in Pakistan, public wells have proved difficult to install, maintain, and manage in saline and fresh groundwater areas. Private wells in fresh groundwater areas often have poor designs and suffer from interrupted power and fuel supplies (Johnson, 1982). Tile drains are more expensive to install ($850 per 1250 ha) than tubewells ($100 per 400 ha) and have slightly higher operating costs. Open drains are vastly more expensive and present unacceptable levels of maintenance problems.

Despite engineering confidence that there are effective technological solutions to the admitted growth in salinization problems, not all analysts agree. One recent "ecological" critique concluded (Goldsmith and Hildyard, 1984):

> . . . we have become trapped on a technological treadmill, which can only result in long-term ecological destruction. In that respect, the experience of the U.S. Southwest is, as we have seen, particularly eloquent. Thus, in their thirst for water, the inhabitants of the Southwest have sunk tubewells and built huge reservoirs. In their fight against salinization, America has spent a fortune on technological measures of a type which less prosperous countries can ill afford. Thus, they have lined irrigation canals, dug horizontal drains, and built evaporation basins. Now that those measures have failed to solve the Southwest's water and salination crisis, the search for new 'technical' fixes has become increasingly desperate: river basin transfers and the development of genetically engineered salt-tolerant crops have

become the order of the day, but at what financial--let alone ecological--cost? Sooner or later, the technical fixes will run out: even now, as we have seen, many are proving too costly to implement--witness the massive water transfer schemes which have been proposed for the area. The future is thus bleak for the U.S. Southwest--as, indeed, it is for Sind, Iraq, and South Australia. How long will it be before vast areas of those regions are abandoned, their best farmlands being transformed into uninhabited, salt-encrusted deserts?

COSTS AND BENEFITS OF DRAINAGE

Not all experts agree that drainage is among the highest priority for the irrigation investment. A 1981 U.S. Aid for International Development report on irrigation development options and investment strategies for Pakistan in the 1980s, written by three leading U.S. experts, failed to mention drainage, waterlogging, or salinity (Keller et al., 1981). Young and Haveman in a forthcoming review of the economics of water resources, make only passing reference to drainage (Young and Haveman, 1985).

It is difficult to assess the drainage component of irrigation improvement because the drainage makes feasible and, in turn, depends upon rehabilitation of the irrigation supply system and other complementary investments in water and agricultural improvement. The inadequately named Left Bank Outfall Drain in the Lower Indus includes additional surface water supplies, surface water storage, canal remodeling, intercept or drains for canal seepage, and on-farm water management projects. It is also dependent on a host of ongoing agricultural projects, including credit, seed supply, and improved extension. Indeed, there is a danger that, if each separate component of a development program is forced to justify its inclusion, then essentially the same benefit may be claimed by drainage engineers, agricultural extension workers, and so forth, as the fruit of their own endeavors. Where the overhead costs of development are large and incompletely provided, the attribution of all marginal increases in production to one known additional investment is fraught with problems. The complexity of some investment projects is shown by the following abstract from an Asia Development Bank news release when they approved a $122 million loan to Pakistan as part of a $657 million project to drain 577,000 ha in the Lower Indus Basin.

The project comprises the following major components:

(i) completion of the remaining sections (about 250 km) of the spinal drain, including the construction of a bifurcation structure at the junction of the Kadhan Pateji Outfall Drain (KPOD) and the Dhoro Puran Outfall Drain (DPOD) and the remodelling of both KPOD and DPOD to 57 cumsecs and 85 cumsecs, respectively; and construction of a 38 km long, 65 m wide and 3.6 to 4.9 m deep tidal outfall channel from Pateji to Shah Samando Creek

(ii) construction of a surface drainage network for a gross command area (GCA) of about 578,000 ha, with main, branch and subdrains totaling about 1,280 km in length and having a capacity ranging from 6 cumsecs to 35 cumsecs

(iii) installation of about 1,490 drainage tubewells for a subsurface drainage of about 286,000 ha of CCA

(iv) tile drainage using corrugated polyvinyl chloride pipe laterals for about 21,000 ha, totaling 1,860 km in length

(v) installation of about 550 km of horizontal interceptor drains

(vi) construction of an 11 km distribution system to power pumps for drainage tubewells and other drainage facilities

(vii) remodeling of about 175 km of the Nara Canal and about 88 km of the Jamaro Canal to increase their effectivity capacity

(viii) construction of the Chotiari Reservoir by installing about 56 km of embankment of the existing lake

(ix) improvement of about 920 watercourses and precision land-levelling of about 26,000 ha of CCA

(x) provision of necessary equipment and machinery for operation and maintenance of the Project works

(xi) provision of consultant services to assist Water and Power Development Authority (WAPDA) and the Department of Irrigation and Power of Sind (SDIP) in project implementation, including planning and design, preparation of tender documents and tender evaluation, construction supervision, operation and maintenance, and training of WAPDA and SDIP staff in drainage design, operational planning for drainage systems, operation and maintenance and system management, and staff of the Department of Forestry and Wildlife of Sind who are associated with the Rann of Kutch Wildlife Sanctuary in environmental monitoring.

The interlinking of these components creates an enormous design problem. Selecting the appropriate scale of any particular part of this investment obviously requires some form of partial budgeting, but the possibility of so doing is limited by correlation among the components. In practice, a core model of the final plan is developed and refined by marginal adjustments, then tested using a variety of criteria including technical, economic, financial, political, administrative, legal, and environmental criteria (see Sir M. MacDonald & Partners, 1983).

The impossible task of attempting to estimate the returns to components of an interdependent system is further complicated by the failure of experts to find any agreement on the economics of schemes.

Two recently approved projects in similar areas of an Asian country had the following sets of assumptions in the feasibility studies:

	Project A		Project B	
	Yield	Area	Yield	Area
Cotton and wheat prior to project	+18%	0	-33%	-20%

Clearly, the rates of return will appear much higher for Project B than Project A. Although the same funding agency was involved and therefore the two projects passed through similar monitoring procedures, this anomaly was not spotted. A cynic would note that, if a project needs to pass rate-of-return hurdles, it clearly pays to be pessimistic about the prior-to-project situation.

IS DRAINAGE A COLLECTIVE GOOD?

Conventionally, drainage of irrigated land is considered to be a collective good that cannot be economically undertaken by individuals. This assumption must be questioned because farmers in various parts of the world have produced individual, privately financed, micro-drainage projects. Indeed, in those areas where extensive irrigation has been developed with average intensities over the gross area at, for instance, 50 percent or less of potential, it is dubious whether regional drainage by groundwater pumping will ever be economically feasible. Schemes will have to be localized public or private tile drains or open drainage schemes.

Private investments in drainage are likely to be most critically evaluated by farmers, and, as a result, the form and extent of drainage will be more in line with market-signal benefits than will public schemes. However, in many circumstances, public sector analysts will find that markets are providing distorted price signals, such as high indirect taxes on crops or over-valued exchange rates, that prevent farmers from receiving appropriate economic indicators. This will distort their private investment and consumption patterns, but, in principle, subsidy or tax policies can be devised to correct these distortions.

For example, in Egypt, some private farmers are digging deep, open drains through their farms. If this fails to drain the farm effectively (as revealed by differences in condition of crops close and distant from the drain), they dig two parallel drains on either side of the first drain. The spacing can be halved again until the whole farm has the desired fall in water table. The water may be pumped from a sump back onto their fields, into a canal or to low-lying abandoned areas. Gotsch and Dyer (1982) make an appeal for study of such "homesteader" endeavors before large-scale public schemes are undertaken.

In this way, private farmers are coping with the twin problems of waterlogging and salinity. Private farmers are doing this in a country where farm product prices are depressed far below world prices by government actions designed to maintain low urban food prices and tap agricultural exports for revenue. The opportunity cost of labor is low at certain periods of the agricultural calendar in Egypt, which makes the digging of drains feasible. However, it is unlikely that the drainage water would be pumped if the subsidized energy prices were raised to world levels while output prices stay under the present price regime. In such economies, economists can play an important part in devising tax and subsidy policies or in modifying existing policies that will encourage farmers to make an optimal level of investment from the public sector viewpoint.

Once a public sector drainage scheme is installed, the problems for the public sector are far from over. For example, the prospects for revenue generation from farmers served by new drainage are not very promising. In addition to the normal problems of taxing low-income farmers that are encountered with irrigation charges, drainage faces additional problems, including:

- psychological and political attitudes of the farming community that regard drainage (like roads) as not directly productive and an overhead and, therefore, a government responsibility
- on-farm drainage that may not be completed and maintained if charges are levied. Indeed, many farmers will look for financial compensation for lost land rather than face paying charges
- downstream farmers may argue, often correctly, that it is upstream salt-disposal problems which, in part, create the need for downstream reclamation and drainage. Hence, it follows that downstream costs should be shared by upstream users. This is an argument that is not likely to have much appeal to upstream farmers, whether they are on the upper reaches of the Colorado or the Indus
- current drainage problems created by past mistakes. Farmers in years past have reaped an external economy by farming without drainage, thereby raising the water table and adding salt. Should current farmers pay for these historical unpaid costs or should the government pick up the bill for their previously shortsighted regulatory policies?

CONCLUSION

In view of the growing importance of drainage to sustain the irrigation areas that are now the "food machine" of the arid zone, and in view of the level of ignorance of the technical, financial, econometric, social, political, legal, and administrative aspects, there is clearly a fertile field awaiting basic and applied research.

REFERENCES

Carruthers, Ian, ed. Aid for the Development of Irrigation. Paris, France: OECD, 1983.

Goldsmith, E., and N. Hildyard. The Social and Environmental Effects of Large Dams. Camelford, Cornwall, England: Wadebridge Ecological Centre, 1984.

Gotsch, C. H., and W. M. Dyer. "Rhetoric and Reason in the 'Newlands' Debate," Food Research Institute Studies, Vol. 18, No. 2, 1982.

Hotes, F. "The Experience of the World Bank," Aid for the Development of Irrigation, Ian Carruthers, ed. Paris, France: OECD, 1983.

Johnson, S. H., III. "Large-Scale Irrigation and Drainage Schemes in Pakistan - A Study of Rigidities in Decision-Making," Food Research Institute Studies, Vol. 18, No. 2, 1982, pp. 149-180.

Keller, J., et al. "Irrigation Development Options and Investment Strategies for the 1980s," Water Management Synthesis Project Report No. 4, Pakistan/USAID, 1981.

Kovda, V. A. "Arid Land Irrigation and Soil Fertility: Problems of Salinity, Alkalinity, and Compaction," Arid Land Irrigation in Developing Countries, E. Barton, ed. Oxford: Pergamon Press, 1977.

Lattimore, D., ed. "Improving Agricultural Production Through On-Farm Water Management," Water Management Research Project, Colorado State University, Fort Collins, Colorado, 1979.

Lowdermilk, M. K., A. C. Earley, and M. Freemont. Publication No. 2, Vol. 1. Pakistan: Water and Power Development Authority, 1977.

MacDonald, Sir M., and Partners. Left Bank Outfall Drain, Report for Water and Power Development Authority. Pakistan: WAPDA, 1983.

Nijland, H. J., and S. E. Guindi. "Crop Yields, Water Table Depth, and Soil Salinity in the Nile Delta, Egypt," International Institute for Land Reclamation and Improvement, Annual Report 1983, Wageningen, 1984.

Pillsbury, A. "The Salinity of Rivers," Scientific American, 245(14):54-65, 1981.

Punjab National Bank. "Impact of Lining of Watercourses in Haryana: An Evaluation Study," New Delhi, India, 1982.

Whitcombe, E. Agrarian Record in Uttar Pradesh in the Nineteenth Century. New York: Cambridge University Press, 1972.

White, G. "The Main Effects and Problems with Irrigation," Arid Land Irrigation in Developing Countries, E. Barton Worthington, ed. Oxford: Pergamon Press, 1977.

Young, R. A., and R. A. Haveman. "Economics of Water Resources: A Survey," Handbook of Natural Resource Economics, A. V. Kneese, ed. Amsterdam, North Holland, 1985 (forthcoming).

10
Developing Farm-Level Information for Improved Irrigation Water Management in Developing Countries

Melvin D. Skold and Donald W. Lybecker

INTRODUCTION

Many development assistance projects in recent years have given increased emphasis to activities which are targeted to be of direct benefit to farmers. Further, most agricultural development measures seeking to change resource use and/or agricultural output require implementing change at the farm level. Whether considering efficiency or equity goals, farm and interfarm comparisons are required. Understanding farmers and their decision-making environment is essential to securing technical and institutional change. Examples of technical assistance efforts which recognize the need for understanding farmers are water management projects which focus upon on-farm dimensions for increasing water-use efficiency and farming systems approaches.

Technical assistance activities which benefit farmers require knowledge about farmers, their motivations, and the technical, economic, and institutional constraints which they face. Farms in developing countries, especially those targeted by recent development assistance efforts, tend to be small. Small farming systems often involve a greater degree of complexity than that encountered among commercial farmers in more technically advanced agricultural systems. Increased complexities arise due to more direct farm-household interrelationships, more extensive use of multiple cropping and intercropping systems, lack of knowledge about appropriate measures of performance, and limited secondary data to facilitate even rudimentary analyses (Hardaker, 1979).

It often happens that farmers in small farming systems are highly regulated by government intervention with respect to input supplies and prices, produce markets and prices, and land-leasing and tenure arrangements. There is a high degree of interdependence between crop and livestock enterprises: livestock provide food products for household consumption and sale, contribute manure, and provide draft power for transportation and lifting water for irrigation and the household. Finally, farmers in small farming systems are more interdependent upon each other than are those in larger systems, but generally they lack the social organizations (institutions) to coordinate this interdependence. Small farmers may depend on a common water source for household and irrigation water supplies, and they may pool their efforts and capital to undertake certain activities (e.g., to maintain roads, clean canals, thresh grain, and purchase tractors and other equipment). Further, farmers with fewer resources are subject to greater externalities (positive and negative) because of their interdependence upon their neighbors.

Collection of farm-level data and compiling information to improve understanding of small farmer behavior is further complicated by the fact that farmers are often illiterate and may not deal in weights and measures commonly used for analysis. Lack of literacy limits data collection alternatives and makes it necessary to use enumerators. Measures such as "donkey loads" or "camel loads" are common but lack uniform definition.

The amount of time (labor) associated with a certain activity is not recognized as the least amount of time required to complete the task; time taken for completion of the task depends on social interactions and a number of other factors, all of which may be more important than labor-use efficiency. Thus, problems arise in labor requirement specification for a given task or enterprise.

It is also characteristic that farmers in small farming systems are more isolated (independent) from other sectors of the economy and from other social groups. Because of this, they may tend to inject an important amount of emotional bias in their answers to queries (Zarkovich, 1966). Their emotional background causes them to be wary of outsiders asking questions, as these queries may be related to taxation or regulation, or they may touch on matters important to the farmers integrity, authority, or dignity (Zarkovich, 1966).

Purpose

This paper draws on the experience of the authors in applying farm-level data collection methods to analyze alternative techniques for improving farm-level irrigation systems in Egypt. It recognizes that data must be collected with a specific end in view. Data collection procedures and the kind of data collected must be guided by the anticipated use or purpose of the data for the conduct of analyses of technical assistance alternatives. General or comprehensive data collection schemes are costly and seldom provide the detail necessary to make specific evaluations of technical assistance problems. Recognizing that specific analytical needs cannot usually be anticipated, the data must be capable of providing basic farm economic information which is timely, reliable, and flexible, for a variety of analytical needs.

Consequently, data collection procedures must recognize the characteristics of the farmer population and the farming system in which they operate. Even though the observations or illustrations included here are based on the Egypt Water Use and Management Project (EWUP), which is directed toward improving farm-level irrigation practices, it is expected that the problems and procedures discussed have applicability to the broad spectrum of farmer-oriented technical assistance activities.

It is recognized that unique farm-level data must be collected by each of the disciplines involved with improving the irrigation system; some data will be unique to disciplinary analyses, and other data will be of use to more than one discipline. Other papers in this seminar series relate strategies to obtain the data necessary to assess the agronomic, engineering, institutional, and sociological parameters of the workings of an irrigation system. This paper focuses on the data useful for economists to evaluate and understand the farm economic situation and to perform the financial and economic evaluations of alternatives to improve the irrigation system.

Data Versus Information

At the outset, it is important to distinguish between data and information. Data, or a data system, is an attempt to represent reality empirically. Generally, categorization and classification are associated with data compilation, which is necessary because of the complexities of most real-life phenomena. If data collection is properly guided, the data set often becomes a useful descriptor of reality. Facts, principles, and numbers can be related to each other in such a way as to provide an understanding of the real-life situation under scrutiny (Riemenschneider and Bonnen, 1979). But the data system does not produce information. Data requires analysis and interpretation to become information (Riemenschneider and Bonne, 1979). Processing or analysis of data makes it useful to decision makers; the analysis and interpretation of data converts data into information (Barnard, 1979; Blackie and Dent, 1979; Riemenschneider and Bonnen, 1979). Following this distinction, it is necessary to link data collection to a specific problem or problem set. General or comprehensive data systems may or may not be capable of providing information necessary to aid decision makers. Data collected with a particular end in view, such as to provide analysis of farm-level irrigation system improvement alternatives or other technical assistance technologies, is targeted to a particular use. Appropriate analysis can render the data into useful information.

Thus, as development of farm-level information systems is considered, it is important to keep in mind (a) the users of the data, (b) the use of the data, (c) the most appropriate means by which to collect the data, and (d) the kinds of analyses which will be required of the data. In the sections which follow, the users of farm-level data for technical assistance analyses will first be considered. Then, consideration will be given to the uses of farm-level data, followed by a discussion of the advantages and disadvantages of various approaches to data collection. Finally, selected analysis formats using the data will be discussed and inferences drawn as to the appropriate data collection strategies.

Basic Farm Economic Information

Information basic to the management of any economic activity are records and budgets. Farms are no exception. Farmers base their decisions on past experiences. These experiences may be recalled, or they may be recorded in some sort of record-keeping system. On the basis of past experience and other received information, farmers make decisions and/or plans for the crop year, a crop rotation cycle, or, in the case of capital investments, for years into the future. The process of planning for future actions is called budgeting. Thus, farm decision-making utilizes experience and other acquired knowledge to plan or budget for the future (Calkins and DiPietre, 1983; Osburn and Schneeberger, 1978; Brown, 1979; Dillon and Hardaker, 1980).

The budgets which guide decision-making may involve the entire farm, as in whole-farm budgeting. If only a part of the farming operation is affected, partial budgeting would be applied, or a budget may represent a single enterprise. In each case, the budgeting process may be completed "in the farmers head," "on the back of an envelope," or in a more formal and systematized way. Regardless of the method, the weighing of past events and other information is important to understanding farm economic decisions. The past events are captured in farm record systems, baseline surveys, and other approaches to data collection. Data are also provided from the transfer of experiences from other farms. Further, results from research at experiment stations, weather and other natural phenomenon records, and knowledge of government policies and other institutional arrangements affecting the farmer's decisions become important to understanding his use of the resources under his control. Plans for the future and decisions to select from among alternatives are based on formal and/or informal budgeting procedures. Purposeful data collection via record keeping, sample surveys, and the compilation of available secondary statistics are essential to understanding farmer behavior and the complex environment in which farmers make decisions. Also, the financial and economic feasibility of proposed physical, biological, or institutional changes to be provided to their proponents requires an appropriate, systematic, and timely data collection system.

USERS OF FARM-LEVEL DATA

Data such as that provided by farm records, surveys, and budgeting are commonly used by farmers operating in commercial agricultural systems to facilitate their decision-making. Similar farm-management data are used by a variety of other users to guide decisions from different perspectives. Financial institutions apply farm records and budgets in their appraisal of agricultural loans; researchers use the data to analyze the efficacy of agricultural technology and institutional change; and policymakers use budgets to evaluate the production cost structure to guide price support and supply control programs (Tinnermeier, 1983; Miller and Skold, 1980). The same basic data set, when subjected to different analyses for different users, becomes valuable information to several groups of decision makers.

The primary users of farm-level data in developed countries are farmers. These data systems are designed to be maintained and used by farmers for tax management, financial planning, and evaluating farm resource use and investment decisions. Consequently, the data systems and budgeting procedures have been designed to serve the needs of the farmer-user. In applying these data systems to developing countries, however, two important differences are apparent: (a) the farmer is not the primary user; rather, analysts in some bureau, department, or ministry are the primary users, in their attempt to bring about agricultural or irrigation system development and (b) the farm records and budgets developed to serve tax and financial planning needs of farmers in developed countries are not designed to meet the needs of analysts in developing countries. Rather, the records and budgets must be redesigned to provide the user with a thorough understanding of factors which underlie farmers' decisions, the relationships between the farm and the household and between enterprises on the farm, and the constraints (physical, biological, institutional, economic, and social) which affect the farm operation.

It is also important to remember that the government agency and project analysts who are the primary users of farm record and budget data are not always agricultural economists. The increasing emphasis on systems analysis and the systems approach places farm-management analysis in the context of other disciplines (Spedding, 1979; Collinson, 1972). Further, while farm management economists have generally been the practitioners of farm record keeping and budgeting, the field of farm management has always been recognized as an interdisciplinary area (Jensen, 1977).

Agricultural economists working on agricultural or irrigation systems in development projects are part of an interdisciplinary team of physical and biological scientists and other social scientists. Thus the data set generated by the agricultural economists must not only be sufficient for economic analysis; it must also have relevance for analyses of a wide range of problems, and these analyses must provide useful information to decision makers (Barnard, 1979; Blackie and Dent, 1979; Perrin et al., 1976; CIMMYT, 1980). The analyses generally relate to the economic design and evaluation of new technologies (Hardaker, 1979; Candler and Slade, 1981). Subsequent discussions will explain how farm-level data and information can meet the needs for economic analysis, as well as serve the analytical needs of noneconomists.

USES OF FARM-LEVEL DATA

Data related to understanding the operations of a farm may be of direct use to any of the disciplines functioning in the interdisciplinary or systems mode. When focusing on water management, knowledge of the resource base, enterprise production practices, enterprise interrelationships, relationships between the farm and the household, and constraints affecting the use of resources by the farm irrigator is useful to the engineer, the agronomist, and the social scientist associated with the efforts to develop improved water-management practices (Tinsley, 1984; Horsey, 1984; Abdel Al, 1984). Most likely, however, it is the agricultural economist who performs the analysis and interpretation to convert farm-level economic data into information useful to his counterparts in other disciplines. Farm-level economic data are directly associated with the physical system and the biological and social relationships; comprehensive farm-level data are of direct use to several other disciplines associated with initiating change at the farm level (Brown, 1979).

Analysis of farm-level data can result in several types of information about the potentials for technical and operational changes in the irrigation system. The analyses may involve the following:

- Comparative analyses can be made (a) between enterprises on the farm or (b) with other farms producing the same enterprise. Such comparisons are useful in the problem identification phase of a project, providing information to farmers about the profitability of alternative production techniques or considering the effects of a change in the irrigation system on various enterprises.
- Other analysis with farm-level data may identify physical resource constraints, whether water,

land, labor, or some form of capital is the most limiting resource. Further, examination of the resource requirements of the various enterprises reflects much about the relationship between those enterprises. Often, enterprises are directly competitive with respect to a certain resource at a particular time. For example, two crops may place heavy demands for water at the same time.

- For other analyses, examination of cash flows from input expenditures and from revenues when commodities are sold is important. Improvements in the irrigation system may involve both financial and economic costs to the farmers (the differences in these two types of costs will be explained in subsequent discussion). The farmer's ability to adopt new irrigation technologies or the reasons for following existing practices may be based on cash-flow requirements. Several reports have indicated that farmers' inability or unwillingness to apply new technologies is related to their risk-bearing ability; small, subsistence-oriented farmers often cannot bear the risk associated with new irrigation practices which increase their cash expenditures for inputs.
- The analyses may also be directed toward evaluating the profitability of new investments in items such as irrigation pumps, the design of field distribution systems, or improvements in a canal serving a number of farmers. Again, these analyses require both financial and economic comparisons. Farm-level data are the basis for the micro-level feasibility or cost-benefit analyses needed in the evaluation of the irrigation system changes being tested.
- If the farm-level data are collected over time, both between-year and through-time comparative analyses can also be made (Lybecker et al., 1984). Evaluations can be made of the progress of farmers operating under new irrigation practices relative to those following traditional practices. The pay-back of investments in improvements to the irrigation system can be analyzed, and/or sources of year-to-year variability in the farming environment can be examined.
- Measurement of the quantities of inputs used will allow for the estimation of relevant production functions at the field, farm, and more aggregate

levels. Such production functions will allow for the imputation of input (e.g., water or fertilizer) values and optimum use levels.

Most analyses of the potentials of changes in the irrigation system (or any aspect of the farming system) at the farm level fit within one or more of these types of analyses. An important requisite on the farm-level data system is that it be comprehensive enough to make any of these types of analyses possible.

METHODS OF DATA COLLECTION

Primary data to describe and form the basis for analysis and understanding of farms can be obtained by three different approaches. Often, the approaches are used in combination, as each method has its advantages and disadvantages. The major approaches to farm-level data collection are (a) controlled experiments, (b) farm surveys, and (c) farm records.

Controlled Experiments

Controlled experiments have a distinct advantage for generating certain kinds of data. Only by holding most aspects of the environment constant and by varying others, while observing their effects, can one isolate the relationships between certain variables. Such experiments, generally conducted on experiment stations or experimental farms, have been particularly valuable in physical and biological research advances. Agricultural economists often unite their efforts with those in other disciplines in the analysis of experimental data (Heady and Dillon, 1961; Hexem and Heady, 1978). Through controlled experiments, agricultural scientists have gained understanding of plant selection and varietal improvement for irrigated agricultural situations, the complex soil-plant-water relations, and the optimal design of water applications systems on farmers' fields.

One of the problems of conducting irrigation-related experiments under controlled or "laboratory" conditions is the lack of direct transferability of the results to farmers' fields. Farmers cannot control the environment to the extent it is controlled in the experiment, or they may lack the managerial ability or incentive necessary to provide the required controls. Consequently, many promising advances in the soil-plant-water system have not been adopted because of their failure to be tested in the financial-institutional-social context in which the farmer operates.

This lack of transferability of experimental research knowledge to farm application has been an important stimulus to the irrigation systems and farming systems approach to technology transfer (Shaner et al., 1982; EWUP, 1984). Under these approaches, technical evaluations and experiments involving the manipulation of physical and biological variables are conducted on farmers' fields. Successes achieved with these approaches indicate that many of the limitations of experiments conducted in more laboratory-type conditions are overcome by focusing the analyses on the farmers' fields.

It remains true, however, that even on-farm experiments are limited in their contribution to knowledge of farm operations and farmer behavior. Experiments are limited to the observation of a few controlled variables. Farming involves the complex interaction of many physical, biological, and human variables, some controllable and others uncontrollable. To include human elements in experiments complicates the experimental design beyond the possibility of most research budgets. Consequently, social science research tends to be based on observations of human behavior rather than experimental approaches.

While experimental research, both on special experimental institutions and on farms, will continue to provide important data for on-farm water-management research, it must be complemented by other data based more on observations of farmer behavior. The focus of the remainder of this paper will be on data systems which reflect farmer behavior. Farmers' reactions to information, resource and other constraints, improved irrigation techniques, and changing financial incentives--as observed within an agricultural season and through time--will be shown to be essential components of the data set necessary to gauge the financial, economic, and social acceptability of improved irrigation practices designed for implementation by farmers.

Farm Surveys and Farm Records

Two approaches to collecting data about the behavior of farmers are surveys and record keeping. If a survey includes all members of the population, it is a census. Because the populations are generally large, most surveys are based on samples; the samples are generally taken following some strategy to meet statistical reliability objectives. If a survey is conducted many times (i.e., "repeated sampling"), the researcher can overcome some of the problems (discussed below) associated with survey sampling. The sample size must be sufficiently large so that statistical inferences can be made with reasonable reliability.

Farm records analyses are case studies including a number of observations; records involve a much more detailed examination of farm production activities. Data are collected and recorded on a frequent and regular basis.

Collinson describes the major difference between the two collection techniques as a "...compromise between sampling error and observational error" (1972, p. 116). Sampling error is the random error inherent in sampling, due to the large number of uncontrolled factors which may, by chance, affect the value of the parameter estimated by the survey. Nonsampling errors (which include observational and measurement errors) are those systematic biases, both response and methodological, which do not tend to cancel out (Mansfield, 1980). Sampling error can be reduced by increasing the size of the sample; this is less costly and more easily accomplished with the sample survey method than with record systems. Within a given budget for collecting data, increased sample size means fewer visits to each farm, thus increasing the probability of observational and measurement error. Observational error can be controlled by taking more time to ascertain the true value of the variable in question. The frequent-visit, case-study technique accomplishes this goal. However, for a given budget constraint, the greater number of visits required by the case-study technique means fewer farms in the sample, thus increasing the probability of sampling error (Jakus, 1984).

It was mentioned earlier that, because illiteracy is widespread among farmers in developing countries, data solicited from farmers requires enumeration. Further, even those farmers who are literate are generally not accustomed to recording, weighing, and measuring. Thus, the record systems discussed here are compiled by junior-level professionals and find their primary use for the analysis of the potentials for improved irrigation practices by agricultural economists working as a part of an interdisciplinary team. This is in contrast to the record systems, as they have evolved in developed countries, which tend to be kept by farmers for tax purposes and management decisions.

Farm Surveys

A farm survey is an examination of a number of representative farms chosen randomly according to some known probability distribution. It is desirable to attain a sufficient number of observations within the sample to generalize reliably on the results of the sample survey for the population. With a limited budget, an increase in the size of the sample will require that less detail be collected on a given farm, perhaps forcing the survey team to make only one visit per farm. Given the limits of data collection budgets, surveys have been found to be almost the only practical means of collecting data about a large number of farmers (Upton, 1973).

Sample surveys are generally preferable to a census of the entire population, for obvious reasons of economy. Except for certain types of data, single-visit surveys are desirable because they allow a larger sample size (Casley and Lury, 1981; Collinson, 1972).

It has been noted that with an increase in the size of the sample, on-farm visits become less frequent. With decreased frequency of visits, there is a corresponding increase in dependence on the memory of the respondent. As has been shown, certain types of data, such as labor, are subject to severe memory bias (Coleman, 1983). To eliminate this bias, such data should be recorded as soon as possible after the event. Therein lies the advantage of the farm-record approach to data collection (Jakus, 1984).

Farm Records (Abdel Al, Martella, and Ayad, 1984)

The general approach of farm-records survey is described by some authors as that of case study (Casley and Lury, 1981). Rather than for a single case, records are maintained for a few or a number of farmers. Often, analyses of records data involves between-farm comparative analyses. Farmers and/or analysts can observe similarities and differences in the operation of successful versus unsuccessful farmers and, by induction, make prescriptive statements for managerial recommendations (Johnson, 1969).

Analytical needs of projects to improve the irrigation system require the ability to budget a number of different alternatives into the future. Budgets provide the basis for evaluating and comparing the relative profitability of alternative investments, which may involve changing the irrigation practices on a single farm or a group of farms. Records are considered by some to be an absolute prerequisite to effective budgeting (Calkins and DiPietre, 1983). Others acknowledge the advantage of records for budgeting but recognize that budgets to analyze alternatives for the future can be based on other data sets as well (Brown, 1979).

Farm records, then, provide a baseline of data and a format for budgeting. They provide information on current levels of efficiency and a comprehensive view of available resources (Calkins and DiPietre, 1983). Records also provide information for understanding the relationships between enterprises on a farm, between the farm and the household, and between enterprises and the set of resources available (Abdel Al and Skold, 1982; Lybecker et al., 1984).

The more detailed data provided by records are less subject to certain kinds of statistical error. The increased reliability does not occur without cost, however. Records can seldom be applied to a sufficient number of farms or to a randomly selected set of farms necessary for valid statistical inferences. The intensity of data requirements for the farm-record system prohibits a large number of cases and requires the full cooperation of record keepers. Thus, the selection of a statistically valid sample is difficult. The former results in sampling error, and the latter ends up with bias.

Data Errors (Jakus, 1984)

Before proceeding to a discussion of sampling and nonsampling errors, it is necessary that "error" and "bias" be defined. Error is simply the difference between the sample value and the corresponding true value. Error is composed of two parts: "sampling error" and "bias." Sampling error is the difference "...between the estimator and the true value of the parameter to be estimated" (Kmenta, 1971, p. 156). Sampling errors vary from sample to sample, and they are expected to cancel themselves out over a large number of observations. Bias is the difference between the expected value of the sample estimator and the corresponding true value of the parameter (Wonnacott and Wonnacott, 1979). In contrast to sampling error, bias is a systematic and consistent type of error which is reflected in nonsampling errors.

Sampling errors can arise if the size of the sample is not sufficient to make reliable statistical inferences. With relatively small samples, a probability exists that the sample is not truly representative of the larger population. If the sample is not representative, then valid statistical inferences are not possible. By increasing the sample size, the probability of sampling error is reduced.

Bias is rooted in nonsampling error. Nonsampling error exists due to the presence of biased sampling procedures, a nonrepresentative sample, biased tools (e.g., a questionnaire), respondent bias, and enumerator bias. Nonsampling errors can be reduced by the attentive researcher. While no objective measure of bias can be calculated, the researcher can formulate a subjective "gut feeling" about the magnitude of the bias problem (Hursh-Cesar, 1976).

Respondent Error (Jakus, 1984)

Respondent error can take many forms. Among the influencing factors are the emotional background of the respondent, prestige errors, the "rounding-off" effect, memory errors, end effect, and conditioning.

A respondent's reply to a question may be altered conclusively due to his emotional feelings regarding the survey. There are many sources for this reluctance to answer truthfully. An obvious example is that, if the respondent feels that the data are to be used for tax collection purposes, he will deliberately bias his response in a manner which he believes will result in less taxation. Further, if the respondent cannot be persuaded that the survey will work to his benefit, he will not see reason to expend the effort necessary to supply accurate replies. Also, the respondent's desire not to offend his visitor may result in the respondent giving answers designed to please the interviewer (Herzog, 1976).

If a respondent wishes to impress the interviewer, he will bias the response to achieve this purpose, introducing prestige error. There are many kinds of this self-lifting bias: women report themselves to be younger than they truly are, the young report themselves to be older, the illiterate say that they are literate, and others claim to have read a nonexistent book if it is introduced as a famous work by a well-known author (Zarkovich, 1966). Within the context of this paper, prestige errors may take the form of biasing yield estimates upward or claiming the use of improved varieties when they are not actually used.

People also tend to round off estimates as a matter of practicality. Bias is introduced when the sample population as a whole has a tendency toward a rounded-off estimate. It has been observed in various samples that a large proportion of farms reported statistics ending in zero.

In developing countries, much of the rural population is illiterate and probably does not keep written records; all methods of data collection are memory dependent. Memory dependence results in memory error, and the bias which results from memory error is known as memory bias. In general, the "memory-fading process" is such that the longer the period of recall (length of time since the event took place), the greater the memory error. But this fading process varies with the characteristics of the item in question. The key characteristics associated with memory fading are frequency of occurrence, regularity of occurrence, and the significance of the event. Distinction is made between "single-point" and "continuous" data and between "registered" and "nonregistered" data. The first classification deals with the length of time needed to complete the activity, while the latter classification deals with the respondent's ability to remember an activity. Labor use can occur at any time during the production process; it is quite routine (nearing continuous) and recurring (nonregistered). Estimation of yield (for crops which mature at one time) is a single-point, registered activity, which is less likely to be subject to severe memory error. For continuous, nonregistered events--those activities which are routine and recurring, measurement error is a serious problem. Frequent interviewing is needed to keep the measurement error "reasonable" when collecting data of the continuous, nonregistered type (Norman, 1976).

The length of the recall period is intractably tied to memory error. In general, the shorter the recall period, the better the quality of the data. But, once again, the appropriate recall period to use depends upon the type of data to be collected. Data on food expenditures may involve an appropriate recall period of one day or one week, while data on expenditures for bicycles or radios may have an appropriate recall period of one year. In an analysis of memory bias in agricultural labor data, Coleman found severe memory bias in records of activity using a seven-day recall period (1983). Daily estimates of labor were smallest for the day immediately preceding the interview and largest on the day furthest removed from the interview. But, the use of short recall period (such as one day) will result in a large number of zero responses and will create greater variance about the sample mean because the data will contain true daily variation in the item. A recall period of one week or one month will tend to dampen these variations.

Before moving into a discussion of the "end effect," it is necessary to establish a clear understanding of the "reference period." The reference period has to do with the period of time to which the data refer. The distinction between the reference period and the recall period is that the former relates to the block of time for which the data is being collected, while the latter refers to the length of time that has elapsed since the event took place.

The two important aspects of the reference period are (a) the length of the period and (b) the location in time of the reference period. With respect to the length of the period, in general, "...the longer the reference period the more important becomes the effect of memory errors" (Zarkovich, 1966, p. 198). When collecting data which is susceptible to the memory-fading process, it is desirable to use a short reference period and a short recall period. Any reference period should be designed to take account of natural cycles which may be present.

The degree of memory error is also influenced by the location in time of the reference period. If the respondent cannot properly identify in his own mind the period to which the data refer, a potential exists for error to be introduced, through the transfer of events into and out of the reference period (particularly those events located near the periphery of the reference period). This transfer is known as the end effect.

The end points of the reference period are the beginning and ending dates of the time period for which the data are being collected. If the reference period has both end points located in the past, the period is said to be "open"; neither end point is clearly defined in the mind of the respondent. Transferences of events can occur at both ends of the reference period. A closed reference period has its end points clearly distinguished, and transferences are less likely to occur. The half-open (half-closed) reference period is one in which one end point is clearly defined while the other is not. Transferences will likely occur at only the poorly defined end point.

Once again, the severity of the end effect will depend upon the type of data being collected. Events which occur rarely are not as subject to confusion as are events which are routine and recurring. Data with which farm management surveys are concerned are of a continuous, nonregistered nature. Data on number and frequency of irrigations, labor, input application, and home consumption of products are subject to severe end effect. The choice of the reference period should reflect the type of data being collected.

The end effect is more severe for this type of data when a short, artificial, open reference period is used. If a longer reference period is adopted when collecting continuous, nonregistered data, respondents often answer with some kind of average (Casley and Lury, 1981). Thus, when asked the number of times a particular crop is irrigated, farmers may think of an "average" year or may think "once a week," whether such is actually the case or not.

Respondent error may also be due to conditioning. Conditioning results when the respondent reacts to previous queries or requests for similar data. Conditioning may result in any of several different effects. Because the respondent is being asked the same or similar questions, he may pay more attention to his activities, thus his responses are correspondingly more accurate. On the other hand, the respondent might grow a bit tired of repeatedly answering the same questions, gradually losing interest in the activity altogether. The responses in this case are of poorer quality. Another possible turn of events is that, in the course of the first interview, the respondent will make the effort to come up with an accurate response to the question, but in subsequent interviews, his replies will be based upon the answer formulated during the first interview. The respondent is conditioned by the initial response. Obviously, conditioning will be a greater concern to those types of data collection techniques which involve more frequent visits, such as record keeping, than those which involve one or few visits.

Other Sources of Error

The enumerator's presence during the interview is another potential source of bias. Whether collecting survey data or farm records, enumerators must be trained to be sensitized to ways by which they can bias the answers to questions (Hershfield et al., 1976). This bias may even result from lack of technical knowledge and information about agriculture.

All work which is directed at eliminating respondent error and enumerator error will go for naught if a biased tool is the basis of the interview. That is, if the questionnaire itself is poorly worded and ill designed, error will result. The length of the interview itself will have an effect on the quality of response; accuracy of response generally declines with the length of the interview.

With this review of types and sources of error possible in the collection of farm-level data, discussion can now focus on implications of these errors to specific data used to generate information about the merits of alternative irrigation practices. As established at the outset, the primary concern is on sampling error versus observation error. Farm records are more prone to sampling error; observation error is a problem more associated with surveys.

RELIABLE FARM-LEVEL INFORMATION

Since budgets are the primary tool for evaluating future alternatives, examination of the components of budgets can reveal the reliability of data which underlie the budget estimates. Even if evaluations focus on the project level rather than the farm level, it is the same basic data set on which evaluations are based (Brown, 1979).

Budgets may refer to an entire farm, to only a portion of the farm that might be affected by changes in the irrigation system, or to a given enterprise. Thus, distinction can be made between whole-farm, partial, and enterprise budgeting. Enterprise budgets can be considered the most basic data sets, and compilation of all enterprise budgets for a given farm should provide a rather complete reflection of the whole farm. Examination of the components included in an enterprise budget can be generalized to whole-farm and partial budgeting as well. Further, evaluations at the project level will be subject to the same base-data error problems as evaluations of on-farm irrigation improvements. After discussion of the components of an enterprise budget and the errors inherent in deriving estimates of each component, discussion will turn to estimating other parameters important to assessing the potentials for improving the irrigation system.

Components of an Enterprise Budget

Table 10.1 is an enterprise budget for the cotton crop at one of the project sites of the Egypt Water Use and Management Project (1984). The budget reflects the costs and returns from one feddan of cotton. (One feddan is equivalent to 1.038 acres or 0.42 ha.) The first section of the budget shows the income from one feddan of cotton, including returns from the lint and stalks (the latter are used for fuel on the farm). The next section includes an itemization of variable cost items; variable costs vary with the level of output or require annual (within the crop-growing season) decisions by the farmer as to their level of use. The final section includes fixed costs. Fixed costs occur regardless of the choice of crop or level of variable input use. Fixed costs are prorated to each crop on a monthly basis as crops vary in their growing season length and the amount of time they occupy the land.

The section at the bottom of the enterprise budget shows the monthly distribution of labor, by age and gender, required for cotton and the estimated schedule of irrigation water demands, also by month. While this information is not always included as part of an enterprise budget, the schedule of requirements for these two critical inputs is valuable for analyses to examine the potentials for improved irrigation practices using the enterprise budgets.

To contrast the advantages and disadvantages of the surveys versus records to develop farm-level data, the merits of each technique relative to each component of the budget included in Table 10.1 are be discussed below.

Income

Gross income per feddan consists of two parts: the yield of lint and stalks and the price or value associated with those crop output categories. Crop yields are

Table 10.1 Crop enterprise budget for cotton grown at the Abu Raya Project site; 1980-81 crop year.*

Item	Units	Number of Units	Per Unit Value Income or Cost L.E.	L.E.
Income				
Cotton[a,b]	Kentar	6.0	55.832	335.0
Cotton stalks	Camel load	5.0	5.000	25.0
Total income				360.0
Variable Costs				
Land Preparation				
Organic Fertilizer	Donkey Load	175.0	0.050	8.8
Transportation				
Donkey Rental	Donkey Hour	25.9	0.100	2.6
Labor to Drive Animal	Boy/Girl Hour	25.9	0.136	3.5
Labor for Spreading	Man/Hour	24.2	0.299	7.2
Plowing	Tractor Hour	4.6	2.233	10.3
Smoothing	Tractor Hour	2.8	2.174	6.1
Furrowing	Tractor Hour	0.5	2.540	1.3
Labor for Furrowing	Man Hour	3.2	0.250	0.8
Planting				
Seed	Kela	6.4	0.333	2.1
Labor for Planting	Boy/Girl Hour	15.6	0.162	2.5
Weeding				
Labor for Weeding	Man Hour	11.2	0.255	2.9
Labor for Weeding	Boy/Girl Hour	27.3	0.151	4.1
Hoeing				
Labor for Hoeing	Man Hour	22.1	0.328	7.2
Labor for Hoeing	Boy/Girl Hour	29.5	0.160	4.7
Donkey Plow[c]	Plow Hour	4.4	0.300	1.3
Thinning	Boy/Girl Hour	15.0	0.114	1.7
Insect Control				
Remove Insect Eggs	Boy/Girl Hour	182.5	0.080	14.6
Insecticides	Feddan	1.0	16.675	16.7
Labor to Apply Insecticide	Man Hour	2.6	0.357	0.9
Chemical Fertilizer[d]				
Super Phosphate (0-15, 5-0)	Kilogram	80.8	0.032	2.6
Ammonium Nitrate (31-0-0)	Kilogram	80.0	0.066	5.3
Urea (46-0-0)	Kilogram	60.0	0.096	5.8
Labor to Spread Fertilizer	Man Hour	9.0	0.250	2.3
Irrigation				
Saqia Rental	Saqia Hour	32.8	0.050	1.6
Cow or Buffalo Rental	Cow/Buff. Hour	32.8	0.350	11.5
Labor to Drive Animal	Boy/Girl Hour	32.8	0.075	2.5
Labor to Spread Water	Man Hour	32.8	0.250	8.2
Harvesting				
Picking	Boy/Girl Hour	374.4	0.205	76.8
Transport Lint	Kentar	6.0	0.300	1.8
Cutting Stalks	Man Hour	20.3	0.405	8.2
Transportation[e]				
Car Rental	Hour	0.8	1.200	1.0
Labor to Lead	Boy/Girl Hour	1.6	0.200	0.3
Total Variable Costs				227.1
Return Above Variable Costs				132.9

Table 10.1 (cont'd)

Fixed Costs				
Land Rent[f]	Month	8.0	4.680	37.4
Management Charge	Month	8.0	1.500	12.0
Total Fixed Costs				49.4
Grand Total Costs				276.6
Return Above All Costs				83.4

FOOTNOTES:

* This study for an area of one feddan.
EWUP Farm Record Data for Kafr el Sheikh, 1980-1981.

a Cotton planted in April and harvested in October. Land preparation begins in mid-February.

b One kentar of unginned cotton weighs approximately 157.5 kilograms.

c The cost of the donkey plow includes the cost of the plow rental, animal rental, and labor.

d The fertilizer price is the average price for fertilizer purchased from the cooperative and the free market.

e Transportation for stalks only.

f The rental rate for land is computed as seven times taxes (legal rental rate).

	Labor Distribution			Water Distribution (cu meters)			
Month	Man Hours	Woman Hours	Boy/Girl Hours	First Irrig.	Second Irrig.	Third Irrig.	Fourth Irrig.
November	0	0	0	0	0	0	0
December	0	0	0	0	0	0	0
January	0	0	0	0	0	0	0
February	0	0	0	0	0	0	0
March	41	0	35	853	0	0	0
April	19	0	34	432	0	0	0
May	17	0	34	270	0	0	0
June	17	0	170	306	315	0	0
July	8	0	36	360	360	0	0
August	3	0	21	270	0	0	0
September	0	0	251	0	0	0	0
October	20	0	125	0	0	0	0
Total	125	0	705				

Total Water Applied = 3166 Cu Meters

Ratio of Return over Variable Costs to Water Applied = 0.0420

Ratio of Return over All Costs to Water Applied = 0.0264

single-point, registered types of data. Sample surveys which rely on recall are known to be capable of providing good estimates of such data. The sample surveys can be conducted to include a sufficient number of farmers and may be randomized either by area or by farmer. In the case of Egypt, cotton yields are registered very strongly in the mind of the farmer; the farmer is obliged by government policy to deliver all cotton production to the local cooperative for sale at a fixed (and very low) price. Thus, estimation of cotton yields with sample surveys would seem to result in estimates which are relatively free of both sampling error and measurement error.

Deriving yield estimates from the averages of farm record keepers may result in sampling errors, however. Record keepers are not likely to be randomly selected by either an area or a farm criterion. The number of record keepers is likely to be too small to adequately account for the variation between observations, thus leading to increased possibilities of sampling error. Of course, if yields vary only a small amount between observations, the likelihood of obtaining a reliable estimate of crop yields from records data increases.

It should be noted that yields for some crops may not be as "registered" as is the case for cotton. For example, berseem (Egyptian clover), which is used almost exclusively on the farm and whose yield is measured in kerat-cuts may be a nonregistered item. (A kerat is one twenty-fourth of a feddan; several cuttings of berseem are made during the growing season.) Further, the amount of berseem actually produced under each cutting is only crudely measured. The measurement problem will apply to yield estimates derived from records as well as surveys, but records may provide a more accurate estimate of the number of cuttings than surveys which require recall over some lapsed period of time.

Crop yields are an item of dignity and prestige among farmers; thus, respondent bias may be an important element in enumerator-received crop yield estimates. Again, because all cotton must be delivered to the cooperative, it is possible to check for respondent bias in cotton yields. Government policies require that a fixed amount, approximately 50 percent, of the output of rice and wheat be marketed to the cooperative. Farmers may purposefully bias yield estimates of rice and wheat downward so that the government does not increase its claimed share of output. Or, in the case of wheat, the government's claim on yield is less for "native" wheat than for improved wheat varieties. The advantage to the farmer to bias estimates of seed used toward "native" varieties is obvious (Haider, 1982).

Respondent bias can be more easily detected by record keeping than by sample survey. Crop yields which are non-registered data or for which prestige or advantage may accrue to the farmers from intentionally biased answers may be more accurately estimated by records than by surveys.

Prices associated with quantities are generally registered, single-point data. Prices can be verified by the purchaser of the commodity. However, it is important to recognize that prices for some items may be affected by black market operations and may have important local and seasonal variation.

Variable Costs

Variable costs also include quantities and prices or values. Some quantities are for purchased inputs, and others are estimates of amounts of input contributed by the operator and the operator's household.

Seed, fertilizer, and the services of a tractor are typical purchased items. The quantity and cost of these inputs would tend to be registered and single point. On the surface, it would seem that these items would tend to be among those items where surveys would provide the most reliable cost estimates. As long as all seed is purchased through the local cooperative, respondent bias or error in seed estimates can be detected. If, however, improved seeds are available through private as well as governmental cooperative suppliers, survey estimates of seed may be less reliable. As mentioned above, farmers have a clear advantage in claiming more "native" wheat than improved varieties.

Egyptian fertilizer allocation policies hold the potential for inducing respondent bias in fertilizer use per feddan. Farmers receive a specific allocation of cotton fertilizer in proportion to their required allotment of cotton. But, because all cotton must be delivered to the cooperative at a fixed, low price, farmers may choose to apply their "cotton" fertilizer to another crop for which their share of output is superior. In responding to a survey, farmers would tend to given the "correct" answer, that is, the official government allocation of fertilizer for cotton. Record keepers who are in almost day-to-day contact with farmers are much more likely to detect the actual amounts of fertilizer applied to each crop. Thus, policies of the government with respect to input allocation and crop marketing can induce significant respondent biases in seemingly single-point, registered data. These respondent biases are more likely to be detected by record keepers than by enumerative surveys.

Hiring the services of tractors for plowing or other machine operations holds some potential for prestige bias. Farmers may feel it is prestigious to claim to be using tractors, even if they do not. Such bias would be easily detected by a record keeper, but it would be less likely observed by an enumerator in a survey.

The quantities for other input items, hours of animal use, hours of farmer and farm-family-member labor use, and quantities of farm-supplied inputs (such as animal manure applied) are both nonpoint and nonregistered data. Their use is intermittent and routine. Recall bias for such data is known to be a problem. The cost of these items makes up about 70 percent of the variable input costs associated with cotton production. Severe recall bias in the estimation of these elements in an enterprise or a whole-farm budget makes estimation of such budgets by techniques for which recall bias is a problem a questionable process.

Table 10.2 illustrates the differences in estimates of hours of labor per feddan for important crops at one project site in Egypt. There is a general consistency between the two estimating procedures in the relative amount of labor required per crop. Data from records tend to reflect smaller amounts of adult labor per crop than the sample survey, but estimates of child labor are larger than those given on the sample survey. Thus, the upward, end-effect bias expected from responses to routine events appears to be present in the adult labor estimates but not in the estimate of child labor. It could very well be that farmer respondents remember themselves as having worked much more, and their children much less, than is actually the case (perhaps a universal trait among fathers trying to get effort from their children).

The potential for statistical analyses with the 50 observations included in the survey is clearly superior to the 15 observations available from farm records. But, given the differences in labor utilization estimates, one has to wonder if the potential measurement error in survey data would result in reliable statistical parameters. The estimates of functions relating output to input level and

Table 10.2 Estimates of amount of labor per feddan as derived from sample survey and farm record data, Kafr el Sheikh Governorate, 1980-1981 crop year

	Hours of Labor per Feddan as Estimated by:			
	Sample Survey		Farm Records	
Crop	Adult	Child	Adult	Child
Cotton	136	352	125	705
Rice	166	82	122	199
Wheat	87	32	126	26
Maize	181	46	109	74
Flax	109	116	60	155
Berseem	130	57	232	32

the associated marginal-value productivity estimates could be given only tentative interpretation.

Fixed Costs and Other Components

The fixed costs section of the enterprise budget includes land rent and management charge. Land rent is the estimated cost for using the land resource. Land use costs are generally estimated as an opportunity cost on the value of land or the cash rental value for land. If the land is owned, either approach to estimating land costs considers land costs as economic rather than financial costs. For rented land, however, either cash or share (in-kind) rents are financial or cash costs.

The management charge is the opportunity cost of the organization and coordination of the production process. This charge may be based on the production period of the enterprise or on a percentage of gross returns. "Return above all costs" is the payment for risk taken during the production of the enterprise and is a residual factor; thus, it may be either positive or negative.

Other useful information is also sometimes included. The enterprise budget in Table 10.1 reports the labor use distribution of labor by type and month and irrigation water distribution by month. This latter information is important in irrigation projects. Depending upon the enterprise, monthly distribution of other scarce resources (animal power or tractor power, for example) may be shown.

Consideration of an enterprise budget will serve to identify the problems associated with farm-level data collection. Certain items included in the budget are single-point data and are registered. However, under certain circumstances, respondent bias may be present in these data. For other items, oftentimes accounting for the major portion of variable costs, the data are nonpoint and nonregistered. High recall error is to be expected, and respondent bias may also be prevalent.

The farm records approach offers a clear potential to reduce both respondent bias and recall error, thereby reducing measurement error. But the number of farms in the record keeping approach leaves open the potential for serious sampling error. In the subsequent discussion of information from farm-level data, ways to insure against sampling error are discussed.

INFORMATION FROM FARM-LEVEL DATA

Reliable data are required for the analyses which lead to useful information for decision makers. From the agricultural economist's viewpoint, the information produced from analysis of data is important for (a) evaluating the extent of a problem (Is it confined to a few farms, or is it common to most farms? Is the observation statistically valid?); (b) evaluating the efficiency of resource use of other irrigation management alternatives; and (c) evaluating the financial and the economic costs and benefits of present versus improved irrigation water management practices.

Evaluating the Extent of the Problem

When the irrigation improvement activity is confined to a small area or a group of farms and there are no plans to extend the results of efforts to improve the system to other areas, whether the fields or farms being studied are representative is not an issue. Generally, however, investigations--even on one field or one farm--are part of a process which intends to extend the results to other fields, farms, and even areas. It is important, then, to know if the field- and farm-level data collected are typical or characteristic of a larger geographic space, a set of farmers, or only a particular physical and institutional environment.

Evaluating the extent of the problem is especially important when equity goals are to be explicitly addressed. Improved irrigation practices can have equity implications for farmers, depending on their location relative to the water source, the size of their operation, their tenancy status, and the nature of the farm resource endowments.

To secure data which are representative of a defined population, the procedure of random sampling is employed. If properly drawn, the sample can accurately reflect the population, and the analyst can make inferences about the population. Survey data collection procedures enable the analyst to collect certain kinds of data which are sufficiently free of sampling error. Farm records can seldom be collected from a sufficient number of randomly selected respondents; sampling error tends to be an inescapable problem for farm record data.

Farm-level data collected by sample surveys are necessary to assess the extent or scope of a problem associated with the irrigation system. It is important for analysts of the irrigation system to know if the problem discovered is associated with the physical environment (e.g., soil type, location along a canal), institutional factors (e.g., tenancy status, local water delivery scheme, availability of inputs and services), farm size factors, enterprise mix and enterprise choices, management level applied, or farm-household interrelationship. Most of these items of data are included in what agricultural economists refer to as the structure of the farming sector.

It is fortunate that the data required to specify the structural characteristics of a farming population are usually single-point, registered data. It does not require extensive recall for a farmer to relate the size of his holdings, whether land is owned or rented, the number of farm workers included in his household, the location of the farm along a watercourse, the timing and reliability of water deliveries, and the amount of each crop and livestock enterprise produced.

These structural data are necessary for evaluation of the equity issues, which are often an objective of efforts to improve the irrigation system (Skold, 1984). While some insights into structural characteristics can be gained by extensively applied farm record systems, sample surveys are clearly a more efficient tool for compiling these single-point, registered data.

Evaluating the Efficiency of Resource Use

Perhaps no term has greater acceptance among all disciplines included in interdisciplinary water-management investigations than "efficiency." All disciplines can relate to it. Engineers concern themselves with efficiency of the water-delivery system and field-application efficiency. Agronomists are prone to apply efficiency measures in recommending optimal planting rates, fertilizer-use levels, and in evaluating the plants' use of stored soil water. Agricultural economists often use economic efficiency measures to assess the use of resources on a farm, along a watercourse, or over a larger area. While not so likely to use the term "efficiency," sociologists are concerned about the performance (efficiency) of local infrastructure, the performance--or lack thereof--of farmer organizations, and the extent to which economic and noneconomic values influence farmer behavior (departures from efficiency).

Even though all disciplines salute the efficiency flag, there is little agreement about measures of efficiency. Consider the diagram presented in Figure 10.1. Efficiency can be measured in physical terms or in economic terms. Physical efficiency for the variable input (X) is maximized when about 15 units of X are used. When comparative analyses reveal which crops provide the greatest return per unit of water or per hour of labor, this efficiency concept is applied. Engineering efficiency concepts tend to reflect this measure of efficiency as well. The ratio of water at the end of a structure (output) relative to water entering a structure (input) applies the same concept. Similarly, selection among alternatives on the basis of the benefit-cost ratio can be viewed as arraying irrigation-improvement investments and choosing the one with the greatest ratio of benefits (outputs) to costs (inputs).

Alternatively, efficiency of the fixed resource is maximized at 20 units of X. The variable input is used at the level to maximize yields per hectare or gain per animal. Agronomists often use the term "optimum yield" as the level of water or fertilizer use which maximizes yield per unit of land. Agricultural economists expect input use levels which maximize efficiency of the fixed input to occur when the price of the variable input is inexpensive. Thus, if labor is redundant or water is provided without cost to the farmer, use levels which maximize efficiency of the fixed input are expected.

Economic efficiency, however, includes the relationship between the combination of fixed and variable inputs and the relative values or prices of the input and the output. By the tenets of economic theory, economic efficiency is maximized when the cost of gaining an increment to output is equal to the value of the increment of output. These conditions are satisfied when the slope of the response curve (total product) is equal to the input-output price ratio. That is, the marginal product, dy/dx, is equal to the price ratio, Px/Py, where Px and Py are the prices of the variable input and the output, respectively. Economic efficiency will always be between the level which maximizes efficiency of the variable input and the level which results in maximum efficiency for the fixed input. If X is expensive, the economically efficient point will be closer to 15 (Figure 10.1). If X is free, the economically efficient level of input use will be closer to 20 (Figure 10.1).

Enterprise budgets, such as the one presented in Table 10.1, can reflect each of these concepts of efficiency. From enterprise budgets, choosing between cotton and maize on the basis of the greatest return above variable costs per feddan involves selecting the crop which gives the greatest return for the fixed resource (in this case, a feddan of land). Choice between enterprises can also be based on the ratio of return above variable costs to the amount of water applied, perhaps return per 1,000 m^3 of water. Then, we consider water as a variable input to be applied at different intensities to land, depending on the crop or other factors.

Only when we can observe varied amounts of water applied per unit of land and the associated crop output can we begin to apply the economic efficiency concept. An enterprise budget could be developed for a number of levels of application of a variable input, such as water, to a fixed input, a unit of land. Given a value for water and for the crop output, the economically efficient level of an input use can be estimated.

Farm-level data, as represented in Table 10.1, is sufficient to budget the relative efficiencies of the

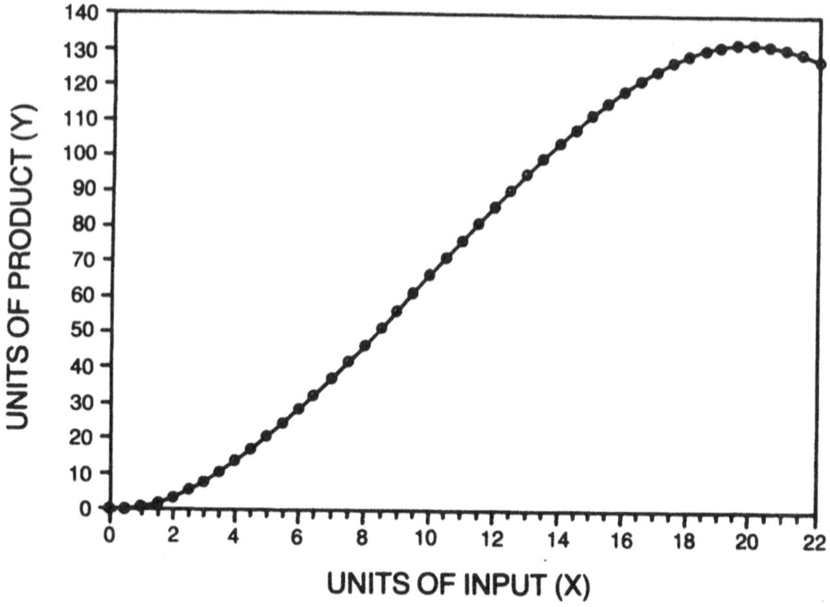

Figure 10.1a Total product curve

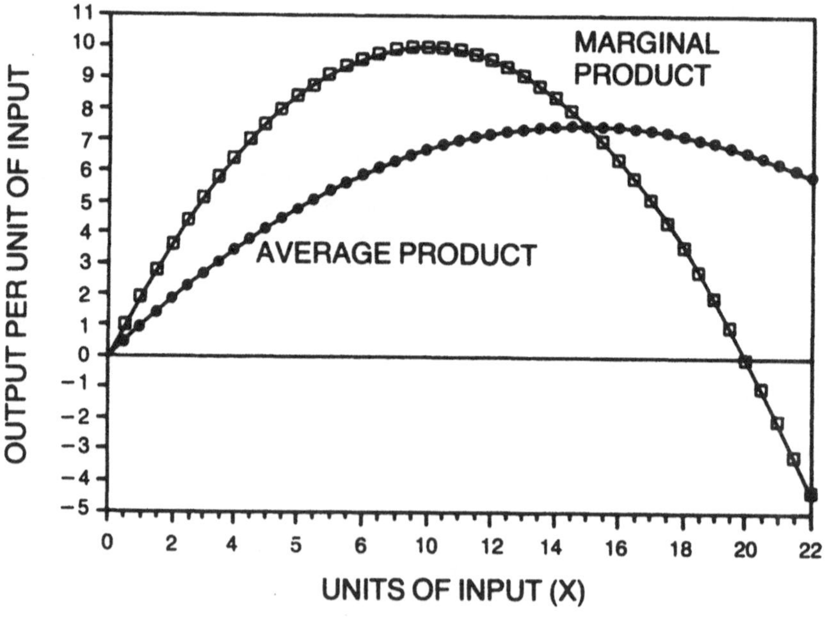

Figure 10.1b Marginal and average product curves

existing and proposed water-management technologies. Furthermore, such data are flexible enough to accommodate efficiency concepts as applied by the various disciplines. The reliability of the data elements included in an enterprise budget, then, is important to all disciplines concerned with improving irrigation systems.

Evaluating Financial and Economic Conditions

Both experience and logic have shown that, to be adopted by a farmer, improved irrigation technologies must be viewed by farmers as advantageous. Often these objectives can be assessed in financial or economic terms. Sometimes the objectives are related to status and the mores of the community; in such cases, the budgeting techniques of economists provide only partial insight.

It should also be remembered that some financial and economic considerations are not immediately reflected in current-year or between-year analyses. Some improved irrigation practices may not result in a measurable change in financial or economic conditions, but changes in wealth or net worth might occur. Farmers may persist with activities which defy explanation on financial and economic grounds but which can be understood when aspects of wealth and status are considered.

To evaluate the potentials for alternatives to improve the irrigation system, it is important to distinguish between financial and economic evaluations. The likelihood of adoption of an improved irrigation practice at either the farm or the project level depends on both financial and economic conditions. Both financial and economic analyses can be conducted from the viewpoint of the farmer, the agency involved with operating the irrigation system, and society as a whole (Gittinger, 1982).

A premise which underlies financial analysis is that prices reflect value. While this may appear to be a statement of the obvious to the noneconomist, there are some often-experienced circumstances for which this premise may not hold. Some of these cases will be discussed when economic analyses are considered.

The enterprise budget in Table 10.1 includes quantities and values for the output of the cotton crop and the inputs used in the production of cotton. Some of the values or prices are easily observed because market exchange occurs and the price is so determined. For other items, however, prices are not established in a day-to-day or annual market exchange. Economists use a variety of techniques to assign or impute value to such items; sometimes accounting procedures are used, at other times values are assigned on an alternative use concept, or values are also imputed as estimates of the value of an item in the production process. Attributing value by any other approach is referred to as opportunity cost or economic cost valuation.

The differences between financial and economic analyses and their implications for evaluation of improved irrigation practices can best be seen by examination of Table 10.1. From the financial viewpoint of the farmer, the cotton crop produces L.E. 335.0 of gross income per feddan; this amount is received from the cooperative as cash or value in kind for the delivered crop. A value of L.E. 25 is also assigned to the cotton stalks. If, in fact, the stalks were sold to a neighbor, the farmer may receive L.E. 25. But the stalks are generally used on the farm; no cash is exchanged. The L.E. 25 is an opportunity return, not a realized financial return; thus, it is not considered in a financial analysis.

Among the variable-cost items, cash or financial outlays occur for hiring the services of a tractor and for purchasing seed, insecticides, and chemical fertilizers. If labor is hired, it is a cash outlay, but, in most cases, the labor is farmer or farmer-family supplied and is assigned an opportunity cost value. A financial analysis of the cotton enterprise would appear as:

Income	L.E. 335.0
Variable costs	
Plowing	10.3
Smoothing	6.1
Furrowing	1.3
Seed	2.1
Insecticides	16.7
Chemical fertilizers	13.7
Total variable costs	50.2
Return above variable (cash) costs	L.E. 284.8

For comparative analysis, the financial returns above variable (cash) costs for cotton would be contrasted to returns from other enterprises.

More important, consider a change in an irrigation practice. Suppose, for example, the change in the irrigation system is to place a pump at the head of an elevated canal to supply water to the farmer by gravity, rather than requiring the farmer to lift water by animal power. To pay for the improved system, farmers would be assessed L.E. 15 per feddan of cotton and similar amounts for other crops.

A cursory analysis would indicate that the farmer may view this improved irrigation practice as attractive. Presently it "costs" the farmer L.E. 23.8 to irrigate a feddan of cotton, but none of the presently incurred costs are financial or cash costs. The farmer owns the saqia (water wheel) and the cow; human labor is supplied by the farmer and farm-family members. The value of the contribution of these items is in economic terms, not financial terms. To be assessed L.E. 15 per feddan for water for the cotton crop increases the financial costs and reduces the return above variable (cash) costs. The choice between the improved practice and the existing one depends on how an individual farmer values the opportunity cost of saqia and cow ownership and the rate of return for labor expended by the farmer and farm family.

The same investment may be evaluated from the government agency's position. Perhaps the agency plans to pay for the construction cost of the pump and elevated canal, and the L.E. 15 per feddan of cotton charge to the farmer is their estimate of operation and maintenance cost. The cost to the agency is the construction cost, which may include financial and economic costs. Costs of materials, special equipment, etc., are likely to be financial costs. But if construction is completed by salaried employees under permanent employment by the agency, the labor costs are economic rather than financial costs.

A further distinction between financial and economic analysis occurs when one makes these evaluations from the position of the state or society. The price per unit of cotton is a price which is set by the government at about 55 percent of the estimated world market "farm gate" price (Haider, 1982). Farmers are "taxed" by receiving less than a fair market value for their cotton. Alternatively, farmers are subsidized with free irrigation water, reduced fertilizer and insecticide prices, reduced petroleum fuel prices, and reduced interest rates on operating loans. The government of Egypt must consider the real costs of maintaining the existing irrigation systems versus the real costs of an improved system. Again, both financial and economic analyses are relevant. To reflect the economic costs for the government of Egypt, the cotton enterprise budget must reflect the "actual" values for the taxed and subsidized items, not the values established by government policies.

Economic analysis of the improved irrigation system from the societal point of view must also reflect the appropriate prices. If the pump to lift water into the elevated canal uses petroleum or electricity, the real economic costs of those energy sources must be reflected. If the elevated canal is to be lined with concrete, and concrete is a subsidized item, the full cost of concrete should be used. The improved irrigation system may lead to increased crop yields. It is the full value of the increased production which should be counted as a benefit, not the government-set price.

Information for Diagnostic Analysis

Clyma (1986) showed the basic similarities among the various approaches used to conduct agricultural development activities. Farm-level data collected to evaluate (a) the extent of the problem, (b) the efficiency of alternatives, and (c) financial and economic implications also serve well for diagnostic and other analytical approaches. The data, with analysis and interpretation, provide information for the identification of problems and insights into alternative solutions and are sufficient to evaluate practices implemented to improve irrigation practices.

Crop enterprise budgets developed from early reconnaissance surveys, refined by follow-up surveys and records, assisted with the identification of problems in that phase of the Egypt Water Use and Management Project (EWUP). The enterprise budgets provided comparisons between enterprises of the returns to farmers above production costs, amounts of labor (by age and gender) required by each crop, and the number of irrigations applied to each crop. The process of gathering data necessary to develop a budget revealed important relationships between the government cooperative and the farmer, the procurement policies of the Ministries of Agriculture as they affect each crop, and the irrigation policies of the Irrigation Ministry. Further, the relationship between crops, between crops and livestock, and between the farm and household are reflected by an enterprise budget (EWUP, 1979, 1980).

The same data are applied to evaluations of proposed improvements in the irrigation system (McConnen et al., 1982). Examination of the financial and economic feasibility of the technical improvements advanced provides a means to narrow the set of possible improvements and to gain some idea about the efficacy of each proposal.

In the same way, the enterprise budgets serve well for evaluations of the success of implemented practices (McConnen, 1984; Lybecker, 1984). Thus, the data generated have been demonstrated to be sufficiently flexible to serve a variety of uses and users. And, if collected following strategies which recognize the errors and biases potentially associated with particular pieces of data, the resulting information will be reliable for the evaluation of alternative irrigation practices.

PRESENTING INFORMATION

As stated earlier, information about the financial and economic aspects of the farm must be related to a variety of users with background and training in other disciplines as well as economics. Further, the interdisciplinary systems approach requires that the analyses performed by economists be related to the contributions of other disciplines. Excellent analysis and information are useless if they are not adequately presented in a form useful to a decision maker.

Information developed must be both problem- and user-specific. It must be free of disciplinary jargon, or, if disciplinary terms are applied, they must be defined, and the concepts to which terms are applied must be explained. Often, the format of presentation can serve to reduce language barriers between disciplines. Tables and graphics serve well for communication; microcomputers have greatly enhanced the ease and quality of presentation.

Economics deals with choices among alternative courses of action. At minimum, evaluations are made between the existing situation and a proposed improvement in the irrigation system. Often, improvements can be achieved along a continuum; an improved distribution canal can be accomplished by a variety of means, requiring different combinations of capital, labor, and operating costs. Presentation of information which considers alternatives and the sensitivity of alternatives to technical performance and prices best serves the needs of decision makers.

CONCLUSIONS

The demands on farm-level data are great. Because most farmers in developing nations are illiterate or are not adept at recording, weighing, and measuring, crucial farm-level data must be enumerated. The data collected must be comprehensive, so that the needs of a number of disciplines involved with improving water management systems are met, but it must also be directed towards the needs of a variety of uses and users. Strategies to collect the data must be cognizant of the needs for statistical accuracy and the possibilities of error and bias which cannot be overcome by statistically designed collection schemes. Finally, the analyses of the data must consider both the financial and economic situations as viewed by the farmer, the agency initiating change, and the society in which both the farmer and the government agency operate.

A comprehensive data system will utilize laboratory and farm-field experiments, statistically valid sample surveys, and record keeping. Interpretation and analysis of these data will result in the information necessary to test the feasibility of and to evaluate proposed changes in the irrigation system (McConnen, 1984).

Experiments which focus on a few key variables in the soil-plant-water system are necessary to understand and predict the relationships between those variables in question. When the feasibility of improved irrigation technologies requires consideration of farmer behavior, data necessary to understand that behavior are also required. Sample surveys and record keeping are means by which the behavioral characteristics can be examined.

Short of time-and-motion studies, data reporting the observed behavior of farmers are based on recall. If the data solicited are from single-point events which are registered in the mind of the farmer, sample surveys are an efficient tool for solicitation. Examples of such data are observations about the size and structure of farming operations, size of household, amounts of land given to various crops, cropping patterns and sequences, seeding and commercial fertilizer applications, and the yields of marketed crops.

Data which are of a routine and recurring nature and which are continuous tend not to be registered in the mind of the farmer. Other data sought may involve intentional respondent bias. For such data, recall errors and respondent bias, even in statistically valid samples, are so severe as to make their validity questionable. Record-keeping systems can greatly reduce the recall period and enable enumerators to establish rapport sufficient to reduce the error and bias. Data for which records offer the greatest advantage are those which relate to farmer-supplied and farm-utilized items. Examples are the amount of farmer and farm household labor used, the timing and frequency of irrigations, the use of organic fertilizers produced on the farm, the amount and timing of animal labor used, the use of farm products consumed by the household or livestock on the farm, and the yield of those crops destined for on-farm consumption.

Farm records are subject to statistical error. When used in combination with sample surveys, analysts can evaluate the extent to which records reflect the characteristics of the population. Budgets that examine the feasibility of proposed changes in the irrigation system or evaluate changes which have been implemented result in information necessary to understand the potentials for improved irrigation practices. Information about both the financial and economic performance of the improved irrigation practice is sufficient to understand farmer response to monetary incentives, estimate the impact on the agencies initiating change, and explore the implications to the government.

REFERENCES

Abdel Al, Farouk, David Martella, and Gamal Ayad. "EWUP Farm Record System," Egypt Water Use and Management Project Manual No. 10, Colorado State University, Fort Collins, Colorado, 1984.

Abdel Al, Farouk, and Melvin D. Skold. "Farm Records Summary and Analysis for Study Cases at Abyuha, Mansuriya, and Abu Raya Sites, 1979/80," Egypt Water Use and Management Project Technical Report No. 23, Colorado State Univerity, Fort Collins, Colorado, 1982.

Abdel Al, Farouk, M. D. Skold, and David Martella. "Irrigation Practices Reported by EWUP Farm Record Keepers (Abyuha and Abu Raya Sites, 1979/80 and 1980/81)," Egypt Water Use and Management Project Technical Report No. 29, Colorado State University, Fort Collins, Colorado, 1982.

Barnard, C. S. "Concepts and Structures," Information Systems for Agriculture, M. J. Blackie and J. B.

Dent, eds. London, England: Applied Science Publishers, 1979.

Blackie, M. J., and J. B. Dent. "Structure and Operation of Existing Farm Information Systems," Information Systems for Agriculture, M. J. Blackie and J. B. Dent, eds. London, England: Applied Science Publishers, 1979.

Brown, Maxwell L. "Farm Budgets from Farm Income Analysis to Agricultural Project Analysis," World Bank Staff Occasional Paper No. 29. Baltimore, Maryland: Johns Hopkins University Press, 1979.

Calkins, Peter H., and Dennis D. DiPietre. Farm Business Management: Successful Decisions in a Changing Environment. New York: Macmillan, 1983.

Candler, Wilfred, and Roger Slade. "Collection of Reliable Farm-Level Data in LDCs," Journal of Agricultural Economics, 32:65-70, 1981.

Casley, D. J., and D. A. Lury. Data Collection in Developing Countries. Oxford, England: Clarendon Press, 1981.

Coleman, Gilroy. "An Analysis of Memory Bias in Agricultural Labour Data Collection: A Case Study of Small Farmers in Nigeria," Journal of Agricultural Economics, 34:79-86, 1983.

Collinson, M. P. Farm Management in Peasant Agriculture: A Handbook for Rural Development Planning in Africa. New York: Praeger, 1972.

CIMMYT (International Maize and Wheat Information Center). Planning Technologies Appropriate to Farmers: Concepts and Procedures, Economics Program, International Maize and Wheat Improvement Center, Apartado Postal, Mexico, 1980.

Clyma, Wayne. "Irrigated Agriculture: A Comparative Analysis of Development Concepts," Chapter 1, Irrigation Management in Developing Countries, K. C. Nobe and R. K. Sampath, eds. Boulder, Colorado: Westview Press, 1986.

Dillon, J. L., and J. Brian Hardaker. Farm Management Research for Small Farmer Development. Rome, Italy: Food and Agricultural Organization of the United Nations, 1980.

Egypt Water Use and Management Project (EWUP). "Problem Identification Report for Mansuriya Study Area," EWUP Technical Report No. 1, Colorado State University, Fort Collins, Colorado, 1979.

Egypt Water Use and Management Project (EWUP). "Problem Identification Report for Kafr El-Sheikh Study Area,"

EWUP Technical Report No. 6, Colorado State University, Fort Collins, Colorado, 1980.

Egypt Water Use and Management Project (EWUP). "Improving Egypt's Irrigation System in the Old Lands," final report, Egypt Water Use and Management Project, Cairo, Egypt, 1984.

Gittinger, J. Price. Economic Analysis of Agricultural Projects. 2nd ed. Baltimore, Maryland: Johns Hopkins University Press, 1982.

Haider, Mohamed I. "The Impact of Egyptian Agricultural Policies on Farm Income and Resource Use," Ph.D. dissertation, Department of Economics, Colorado State University, Fort Collins, Colorado, 1982.

Hardaker, J. B. "A Review of Some Farm Management Research Methods for Small-Farm Development in LDCs," Journal of Agricultural Economics, 30:315-327, 1979.

Heady, Earl O., and John L. Dillon. Agricultural Production Functions. Ames, Iowa: Iowa State University Press, 1961.

Hershfield, Allen F., Niels G. Roling, Graham B. Kerr, Gerald Hursch-Cesar. "Problems in Interviewing," Third World Surveys: Survey Research in Developing Nations, Gerald Hursh-Cesar and Prodipto Roy, eds. New Delhi, India: Macmillan, 1976.

Herzog, William, David Stanfield, and Gerald Hursch-Cesar. "Problems in Measurement," Third World Surveys: Survey Research in Developing Nations, Gerald Hursh-Cesar and Prodipto Roy, eds. New Delhi, India: Macmillan, 1976.

Hexem, Roger, and E. O. Heady. Water Production Functions for Irrigated Agriculture. Ames, Iowa: Iowa State University Press, 1978.

Horsey, Henry R., E. V. Richardson, M. D. Skold, and D. K. Sunada. "Analysis of Low-Lift Irrigation Pumping," presented to Irrigation and Drainage Division, American Society of Civil Engineers, Flagstaff, Arizona, 1984.

Hursh-Cesar, Gerald, and Prodipto Roy. "Problems in Sampling," Third World Surveys: Survey Research in Developing Nations, Gerald Hursh-Cesar and Prodipto Roy, eds. New Delhi, India: Macmillan, 1976.

Jakus, Paul M. "Evaluation of Alternative Data Collection Techniques," M.S. thesis, Department of Agricultural and Natural Resource Economics, Colorado State University, Fort Collins, Colorado, 1984.

Jensen, Harold R. "Farm Management and Production Economics, 1946-70," A Survey of Agricultural Economics

Literature, Vol. 1, Traditional Fields of Agricultural Economics, Lee R. Martin, ed. Minneapolis, Minnesota: University of Minnesota Press, 1977.

Johnson, Glenn L. "Stress on Production Economics," AEA Readings in the Economics of Agriculture, K. A. Fox and D. Gale Johnson, eds. Homewood, Illinois: Richard D. Irwin, 1969.

Kmenta, Jan. Elements of Econometrics. New York: Macmillan, 1971.

Lybecker, D., Elia Sorial, F. Abdel Al, and Nabil Farag. "Abyuha Farm Record Summary and Analysis Over Years (1979-83)," Egypt Water Use and Management Project Technical Report No. 75, Colorado State University, Fort Collins, Colorado, 1984.

Mansfield, Edwin. Statistics for Business and Economics. New York: W. W. Norton, 1980.

McConnen, R. J., F. Abdel Al, M. Skold, G. Ayad, and E. Sorial. "Feasibility Studies and Evaluation of Irrigation Projects: Procedures for Analyzing Alternative Water Distribution Systems in Egypt," Egypt Water Use and Management Project Technical Report No. 12, Colorado State University, Fort Collins, Colorado, 1982.

McConnen, R. J., E. Sorial, and G. Fawzy. "Watercourse Improvement Evaluation," Egypt Water Use and Management Project Technical Report No. 63, Colorado State University, Fort Collins, Colorado, 1984.

Miller, Thomas A., and Melvin D. Skold. "Uses and Users of Costs and Returns Data: A Needs Analysis," Developing and Using Farm and Ranch Cost of Production and Return Data, Great Plains Agricultural Council Publication No. 95, University of Nebraska, Lincoln, Nebraska, 1980.

Norman, David. Field Data Collection in Social Sciences. Bryant Kearl, ed. New York: Agricultural Development Council, 1976.

Osburn, Donald D., and Kenneth C. Schneeberger. Modern Agricultural Management. Reston, Virginia: Reston Publishing Company, 1978.

Perrin, R., D. L. Winkelmann, E. R. Moscardi, and J. R. Anderson. "From Agronomic Data to Farmer Recommendations: An Economics Training Manual," International Maize and Wheat Improvement Center Information Bulletin 27, Apartado Postal, Mexico, 1976.

Riemenschneider, C. H., and J. T. Bonnen. "National Agricultural Information Systems," Information Systems for Agriculture, M. J. Blackie and J. B. Dent, eds. London, England: Applied Science Publishers, 1979.

Shaner, W. W., P. F. Philipp, and W. R. Schmehl. Farming Systems Research and Development: Guidelines for Developing Countries. Boulder, Colorado: Westview Press, 1982.

Skold, Melvin D., Shinnawi Abdel, Atty El Shinnawi, and M. Lofty Nasr. "Irrigation Water Distribution Along Branch Canals in Egypt: Economic Effects," Economic Development and Cultural Change, 32:547-67, 1984.

Spedding, C. R. W. An Introduction to Agricultural Systems. London, England: Applied Science Publishers, 1979.

Tinnermeier, Ronald L. "Enterprise Budgets for Credit Programs: Guidelines for Developing Countries," Department of Agricultural and Natural Resource Economics, Colorado State University, Fort Collins, Colorado, c. 1983.

Tinsley, R. L. "Planning and Evaluating Water Delivery with Farm Records and Water Level Recorders," paper presented at the Invited Seminar Series, "Current Issues in and Approaches to Irrigation Water Management in Developing Countries," International School for Agricultural and Resource Development, Colorado State University, Fort Collins, Colorado, 1984.

Upton, Martin. Farm Management in Africa. London, England: Oxford University Press, 1973.

Wonnacott, Ronald J., and Thomas H. Wonnacott. Econometrics. New York: John Wiley and Son, 1979.

Zarkovich, S. S. Quality of Statistical Data. Rome, Italy: Food and Agricultural Organization of the United Nations, 1966.

Part 3

Management and Institutional Aspects

11
Irrigation System Management

Jack Keller

When considering irrigation systems in terms of their actual performance relative to potential or designed performance, we are usually disappointed. Dean Peterson (1984) asks:

> . . . why is it that irrigation systems have fallen so notoriously short in terms of what reasonably could be expected? Authorities generally now believe that the difficulty lies in the failure of irrigation systems to perform as systems. This is especially true if the irrigated farms are viewed as part of this irrigation system. Water deliveries at fields do not match crop needs; other production needs--seeds, pesticides, fertilizer, labor, and technical know-how--are not available to farmers in timely fashion. The real managers--the farmers themselves--are not involved in the planning and management. The system with its human parts really is multidisciplinary. Systems can only be studied in vivo, not on experiment stations or by controlled experiments on farms. Diagnosis requires an inductive or clinical approach, rather than the deductive approach of the agricultural and physical sciences. These are the ideas one hears about the new approach--called water management.

This paper presents an overview of concepts related to irrigation system management which I have reached as a result of extensive interdisciplinary field study and consulting activities involving irrigated agricultural development throughout the developed and developing world. Table 11.1 provides a perspective of worldwide trends in irrigation development. Data for the mid-1980s reflect the long-range national plans for development of the different regions. However, the target figures may be unrealistically ambitious and may more nearly represent what can be done, rather than reasonable targets. In both developed and developing countries, improved irrigation system management has the potential of increasing water and energy use efficiency by 10 to 15 percent. In addition, by improving irrigation system and crop management in the developing countries, both the area irrigated and the production from it can be at least doubled in many cases.

The discussions which follow will deal with a number of rather specific points which are ultimately relative to and reflect on the management and/or manageability of irrigation systems. The specific points which will be covered include: definition of the objective; structure of irrigation systems; irrigation as a happening; water and social tension; evolution of irrigation systems; control and management levels; and water pricing policy and law.

DEVELOPMENT OBJECTIVES

As a beginning point, it appears that many irrigation projects are developed and designed without having a clear concept of the objective. I can see three somewhat different objectives for developing an irrigation project. These are: for commerical production, for sociopolitical reasons, and/or for geopolitical reasons. A commercial production objective refers to a project where the principal purpose is to produce food and fiber for markets. A social benefit objective refers to a project which is principally directed to improving the well-being of a rather large number of farmers with small land holdings. By geopolitical, I refer to projects which are initiated for security or impressionistic reasons, rather than for either of the above reasons.

Obviously, most projects contain elements of all three objectives. However, it appears useful to delineate the main thrust of the objective at the onset so that it can be optimized, rather than making all projects appear to be commercially oriented using standard benefit-cost analysis techniques. Clearly, for commerical projects, a

Table 11.1 Areas of irrigated land in the world, in millions of acres

Location	Mid-1970s[a]		Mid-1980s[b]		Targets[b]	
Developed Countries		78.3		91		140
Developing Countries		224.4		264		498
Africa	5.4		7		28	
Latin America	30.1		35		70	
Near East	41.0		45		78	
S. and S.E. Asia	147.9		175		322	
Centrally Planned		257.6		278		582
Asia	213.1					
Europe, USSR	44.5					
World Total		560.3		633	1220	

[a]Taken from A. Aboukhaled, A. Felleke, D. Hillel, and A. A. Moursi, Opportunities for Increase of World Food Production (Report to the Technical Advisory Committee of the Consultative Group on International Agricultural Research, I.D.R.C., Ottawa, Canada, April 1979), p. 161.

[b]Taken from J. Doorenbos, "The Role of Irrigation in Food Production," Agriculture and Environment 2, (1975):39-54.

relatively high discount rate is appropriate. However, for sociopolitical projects directed essentially to providing social benefits to poor peasant farmers, a much lower discount rate may be in order. On the other hand, a geopolitical project which is developed for security reasons is an undertaking similar to a military operation, in which the economic benefit-cost analysis is inappropriate.

The above does not imply that a project with the principal objective being social benefits should not be productive, and in a sense, economically feasible. However, at the beginning of the project, the principal interests might be institution building and improving the well-being of the local population, with the commercial economic benefits being delayed until this has taken place. Thus, the project might be conceived in such a way as to optimize the social benefits at the onset in a manner in which they can be sustained by the productivity of the project as the building of needed institutions is put in place. For such projects, a maximum amount of community self-help and involvement is essential. An alternative to using reduced discount rates might be to allocate a portion of the project cost to a subjective evaluation of social benefits. This could be done in a similar manner to allocating environmental benefits within projects in developed nations.

Shifting to the social or geopolitical objectives should not be used as an excuse for careless planning, without rigorous technical and economic analyses, or a lack of attention to broader social, environmental, and economic consequences. If there are social or political values, these ought to be accounted for by surrogate values in the decision process and the best commercial project under these circumstances built. Otherwise, irrigation projects justified for sociopolitical or geopolitical reasons are very apt to run counter to the objectives and have severe consequences which are not rigorously assessed.

The principal argument I wish to put forth in this section is that all projects should not be designed, implemented, and managed as though they are commercially viable. When this is done, systems are often designed and implemented in such a way as to be practically unmanageable. Thus, ultimate project performance is apt to be considerably worse than what might have been achieved had more realistic appraisal techniques been utilized.

STRUCTURE OF IRRIGATION SYSTEMS

Irrigation systems involve both physical works and human activity. There is little need to discuss the physical works or the fact that human activity, such as management and the knowledge and effort to irrigate, are essential. I believe a point that is often missed is the need for communication between the users and suppliers so that the physical delivery and application of water can take place in a meaningful way. The management efforts of the various people involved and either an automatic anticipatory system or a communication system are needed to tie the decision-making processes together.

As a matter of explanation, consider a typical municipal water supply system; such a system is totally anticipatory and requires little effort at communication to achieve effective and efficient delivery of water. Each homeowner is provided with a supply which can be tapped upon demand, and their collective desires (demands) are communicated to the municipal supply through the pipe network itself. This is ideal, from the standpoint of the simplicity of the communication network; however, it is expensive. Typical irrigation systems are much more open-ended and do not have sufficient water or a supply network which can deliver water indiscriminately according to the demands of any and all users. To cut costs and spread benefits, water supplies are stretched, and deliveries require communication from the farmers and/or estimates based on predicted crop water demands to function. Furthermore, the operation of systems involving small farmers requires that the farmers coordinate their efforts to allocate the water amongst themselves.

To picture the social stress within a typical irrigation system, think of being supplied by a municipal water supply where you would need to request water a considerable time before you wanted to use it in order to flush the toilet or wash your clothes or take a bath; or, worse yet, you were placed on some sort of rotational mode, where you could only have water periodically if it happened to be available in the system.

In systems with many small farmers, it is virtually impossible to conceive of a bureaucratic delivery system which can deal with each and every farmer. Thus, it appears essential that the farmers organize in groups so that they can deal collectively with the main system management; and the main system management, in turn, only needs to deal with a relatively few headgates serving collective groups below the outlets along the main canals (see Figure 11.1).

IRRIGATION AS A HAPPENING

Except for fully automated mechanical systems (such as trickle and center pivot sprinkle), irrigation is essentially a "happening." I use this term in reference to the fact that a traditional irrigation system does not irrigate--it is merely a network of channels feeding prepared fields. Human enterprise does the irrigating. Furthermore, the control and allocation of the water to the fields also requires continuous and direct human action. In other words, irrigation involves people and their tools. People provide the labor and management. The tools or hardware require capital to obtain and energy to operate. In addition, for successful irrigated agricultural production, other physical inputs such as seed, fertilizer, and pesticides are required. These, in turn, take additional management, labor, capital, and energy.

The delivery of all of the above must come together in a more or less optimum mix in order to achieve high production. This production, to be meaningful and sustainable, involves harvesting, transportation, storage, and marketing. Obviously, this postproduction phase also requires people and tools, and the overall situation must take place in a hospitable physical, political, economic, and social environment.

With the above viewpoint, it is impossible for me to visualize irrigated agricultural development in anything less than an integrated, interdisciplinary framework. Unfortunately, field experience has shown that irrigation projects have been designed with insufficient regard for the very interdisciplinary nature of both the management of the deliveries and the efficient on-farm use of water. The quality of management, quantity and timing of labor, microeconomics, other needed agronomic requirements, and marketing have often received insufficient study. This has caused "technically feasible" systems to fail to come anywhere close to meeting expected production goals.

WATER AND SOCIAL TENSION

I like to think that the function of an irrigation system is to supply the water in the fashion of a more or

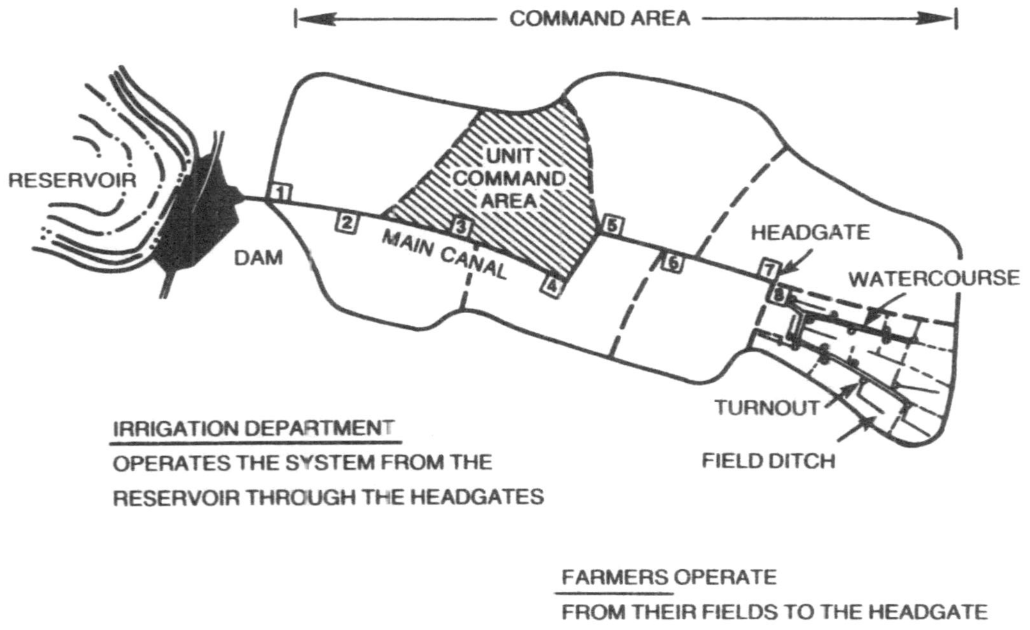

Figure 11.1 Typical surface irrigation components

less uniform membrane covering the design command area (see Figure 11.2). With this view in mind, I think in terms of a relative water supply (RWS) or water density, which is the ratio of the amount of water the crops preferred by the farmers can beneficially use to the average amount of water which would be available at each farm turnout if the total supply were uniformly delivered throughout the system. Using this conceptual model, one can think of the RWS in terms of both the social and water tension created. The higher the RWS, the less the tension. For example, if there were twice as much water available as required by the most opportune crops, such as is often the case in the United States, the water tension might be relatively low, providing the delivery system was reasonably efficient. On the other hand, typical irrigation systems in India are designed with RWS values as low as one-third, giving rise to high tensions. This is because Indian systems are extensive; that is, they are designed to supply less than optimum benefits to a maximum number of beneficiaries. Figure 11.3 depicts the intensity of management required as a function of the RWS. The lower the RWS, the greater the tension and, thus, the greater the relative management intensity required to achieve an equitable distribution of the limited supply of water.

High tension systems require extensive physical works and diligent management to stretch and hold the water membrane in place. Without the necessary capital and management inputs, the membrane merely relaxes, and the limited water supply is captured by the proverbial head enders (see Figure 11.4). This is not only inequitable in terms of the original project goals, but also uneconomic and counterproductive politically. Even if the overall productivity remained the same, an extensive irrigation system is much more costly (perhaps as much as 50 percent more on an actual per-hectare of irrigation basis) than a system designed to serve a limited number of beneficiaries with a relatively high RWS and, consequently, lower tension. This is because extensive systems necessitate construction of longer canals, and if the water membrane is not held in place, much of the system is essentially unused. One might also add that extensive systems which are not managed to hold the membrane in place are politically very undesirable because of the unfulfilled expectations of all the potential beneficiaries who do not receive a reliable and equitable share of water.

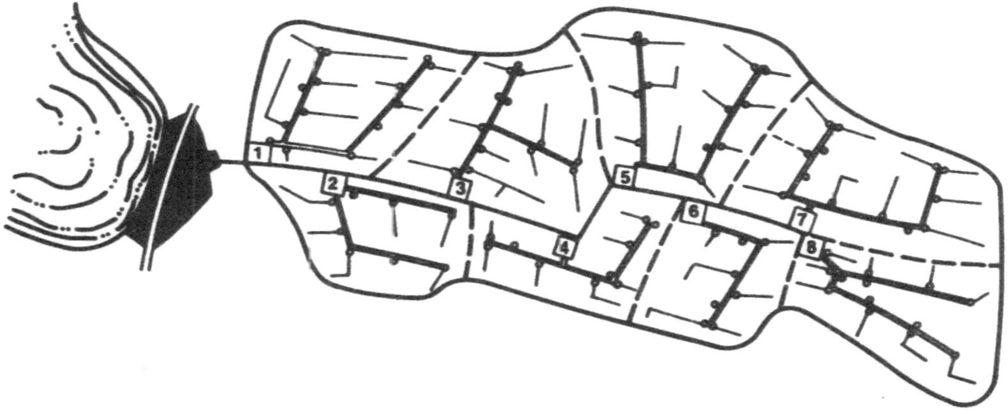

Figure 11.2 Adequate irrigation system with effective management "stretching" the water like a uniform membrane over the entire command area

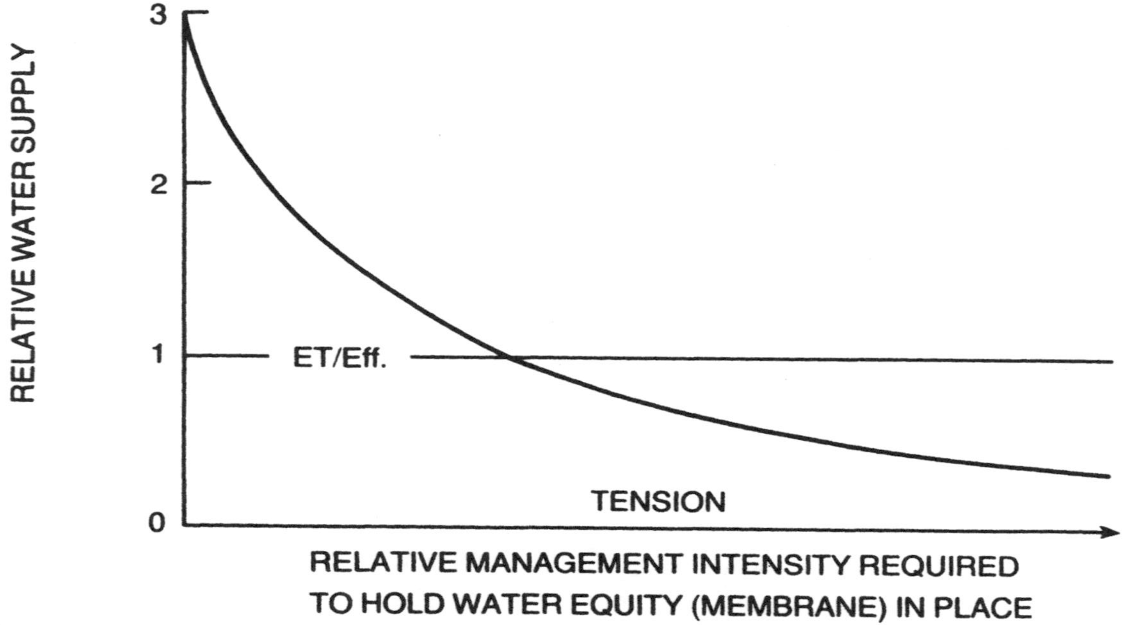

Figure 11.3 Relationship between relative water supply and management intensity required for equitable distribution

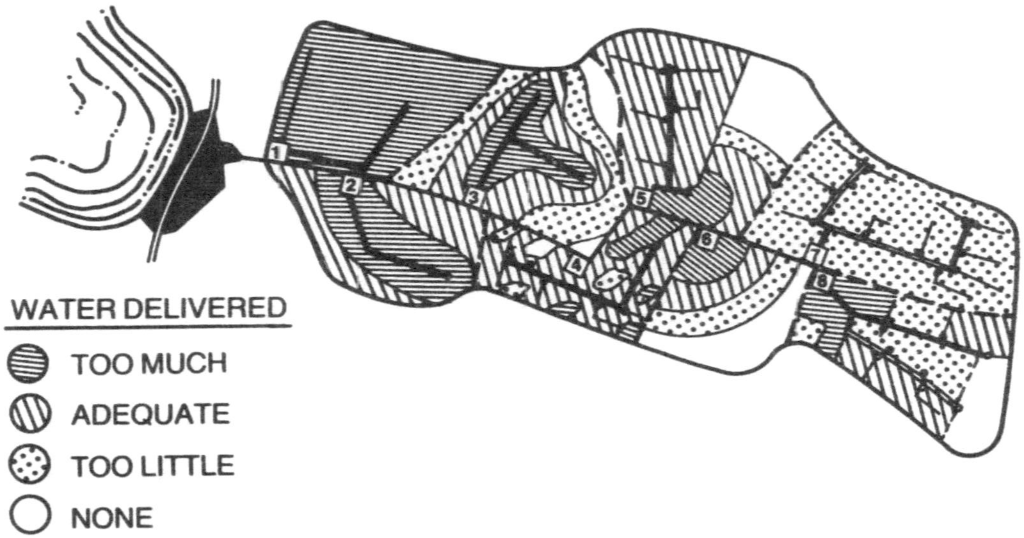

Figure 11.4 Command area with various types ~f water distribution problems

I wish to make two more points concerning water and the social tension. First of all, one of the principal concerns of farmers is the reliability of the water supply. In fact, on many farmer surveys, this is the paramount benefit of irrigation water, in the farmers' eyes. Without a reliable system, farmers view irrigation more or less the same as rainfall and act accordingly. They can hardly be expected to level their land or add the other high-cost inputs which are necessary to obtain the benefits needed to justify irrigation development. So the water which is delivered has limited value to the farmers, and they spend little effort in utilizing it efficiently.

A final statement on the RWS (or water density) and resulting social and water tension is due concerning tubewells. One element of large extensive irrigation projects is that, no matter what else, they often tend to serve as large groundwater recharging systems. This, in turn, enhances the availability of groundwater and the development of pump irrigation, which is an important component of many large irrigation schemes. Water pumped from wells is essentially available on demand. It gives another possibility for stretching the membrane, using private entrepreneurship; however, it may not reduce the tension element because conjunctive use also increases the marginal value of whatever canal water is available. By augmenting undependable surface supplies with more costly pumped water, the overall cost of the irrigation water is not elevated too high, but its availability can be more or less optimized. Moreover, the necessary inputs for higher production are attracted and the hoped-for economic benefits of the project may become available, in spite of the relatively poor operational control and management of the overall surface delivery system.

EVOLUTION OF IRRIGATION SYSTEMS

When I think of an irrigation system or project, I visualize a system which involves both social and physical aspects and is organic in nature and, thus, evolving. Essentially, the direction of the evolution is opportunistic. That is, the system evolves in accordance with its environment, moving toward what attracts it. For example, think of a tree: the branches grow toward the sunshine, not toward the shade, and the roots concentrate where there are water and nutrients, not in dry or sterile soil. Even if laid on its side, the tree grows toward the sun again. One might say irrigation is an organic happening, with capital as its nutrient and income as its attraction, and human in its nature.

From this evolutionary viewpoint, one expects institutional frameworks, as well as the physical systems, to change as they mature. The directions of change are affected by such things as main- and middle-system management, water changes, the type of distribution (continuous flow or rotational), the degree of emphasis on farmer participation, the reliability and scarcity of water, the general topography, soils, crops, the climate, markets, settlement patterns, land tenure, etc. For example, even a low-tension system which is poorly managed may provide excessive amounts of water to head enders, at the expense of providing water to the tail enders. If viewed early during the development of the project, there would appear to be a great deal of waste. However, if the topography were such that the wasted water could be rediverted throughout the drainways, in time one might find whole new irrigation settlements outside of the original design command area where (either individually or collectively) farmers have developed the "newfound streams and springs" which resulted from the water wasted at higher elevations. Thus, what might appear to be a wasteful system at first could evolve into an efficient one, with the return-flow irrigation farmers even more productive than the primary or initial users. This can happen because return-flow waters may be more reliable, as they are buffered from the individual short duration of main-system flow events.

The unfortunate problems with return-flow development situations, i.e., where the return-flow water is eventually opportunistically used, are that they are politically embarrassing, they often result in an uneconomic approach to development (as discussed earlier), and they are often associated with drainage problems. If planners could have visualized the outcome in the first place, they might have developed a smaller initial system and planned for the community action and uptake of the return-flow waters. This would have saved money and reached the same end point more quickly, without being politically embarrassing.

The important message in the organic concept is that if irrigation institutions and systems are conceived of as being opportunistic in their evolution process, then one can expect that it will be very difficult to make them conform to some set of desired goals through regimentation. On the other hand, it should be relatively easy to achieve the desired project goals if we can only understand how to create an environment which attracts the evolutionary process to the desired ends. This is obviously a challenge which is quite site specific and requires our most astute interdisciplinary capabilities; for not only is each system evolving, but the entire institutional, economic, and social environment in which it exists is also evolving.

CONTROL AND MANAGEMENT LEVELS

There are three potential management levels: the main system, the middle system, and the farmer system. Typically, in the United States, the farm systems are so large that they encompass what might be called the "middle system." That is, water is delivered directly from the main system to the individual farmer's holding. In fact, the main system usually delivers water to more than one outlet serving a given farmer's contiguous holdings. Thus, the main system actually forms part of the on-farm irrigation infrastructure, and there is only one management interface--that is, between the farmer and the main system. Furthermore, there are relatively few farm systems, and, thus, the main system only needs to communicate with a few users.

In projects involving main systems in developing countries where farm sizes are small, there is need for managing a middle system. This is because the bureaucracy operating the main system can hardly be expected to communicate and deal with each farmer. The best opportunity for eliminating this problem is to induce the farmers within each unit command area (UCA) to organize a water user association to maintain the watercourses and distribute the water within the UCA (see Figure 11.1). This gives the farmers within each UCA access to local management, and the main-system bureaucracy only needs to control water deliveries to a relatively few headgates.

Farmer participation in the management of the middle system is not only important from the standpoint of communication and reducing the points of interface between the bureaucracy and the farmers themselves, but also from the standpoint of maintenance. For example, if the bureaucracy operating the main system endeavors to deal directly with each farmer, it also overtly assumes the responsibility of maintaining the entire canal network down to each farm holding. Thus, an inordinate operation and maintenance burden is placed on the bureaucracy. The bureaucracy usually fails in this area, and the middle system becomes a no-man's-land with deteriorated, inefficient, tertiary watercourses. There is increasing evidence that if the farmers are involved in the middle system at the onset of new projects, they can be expected to maintain and manage it, as well as help construct it.

PHYSICAL INTERFACE BETWEEN MANAGEMENT LEVELS

As referred to earlier, an irrigation system is composed of a water supply and a main distribution system which provides water to a number of UCAs. Each UCA is comprised of the group of farmers residing within it and the middle system serving the farmers. The dynamic physical relationship at the interface between each UCA and the main system can be described by a combination of the actual water supplied as a function of time (supply hydrograph, SH) and the water supply required by the irrigated crops as a function of the time throughout the growing season (demand hydrograph, DH). The ratio of the demand to the supply hydrograph might be called the relative water supply hydrograph, RWSH, as depicted in Figure 11.5.

The demand and supply hydrograph lines are really probabilistic bands which vary according to weather and crop conditions. Seasonal and annual variance is often high, so farmers operate in a sort of actuarial environment. The summation of all the UCA demand or supply hydrographs make up the system demand and supply hydrographs. Each UCA hydrograph is the summation of the respective individual farm hydrographs. The overall objective of system management might be to strive for some sort of optimum fit between the demand and supply hydrographs at all three management levels. This requires coordinated effort between the farmers and their middle-system managers and between these middle managers and project personnel.

Figure 11.5 shows a plot of a hypothetical supply hydrograph, SH (dashed line), with a continuous delivery of 50 percent of the peak demand volume (dashed line). A hypothetical demand hydrograph assuming 100 percent of the command area is irrigated, DH_{100}, is also plotted as a function of the relative volume compared to the peak demand volume per day. The relative water supply hydrograph, $RWSH_{100}$, is a plot of the ratio of SH to the DH_{100},

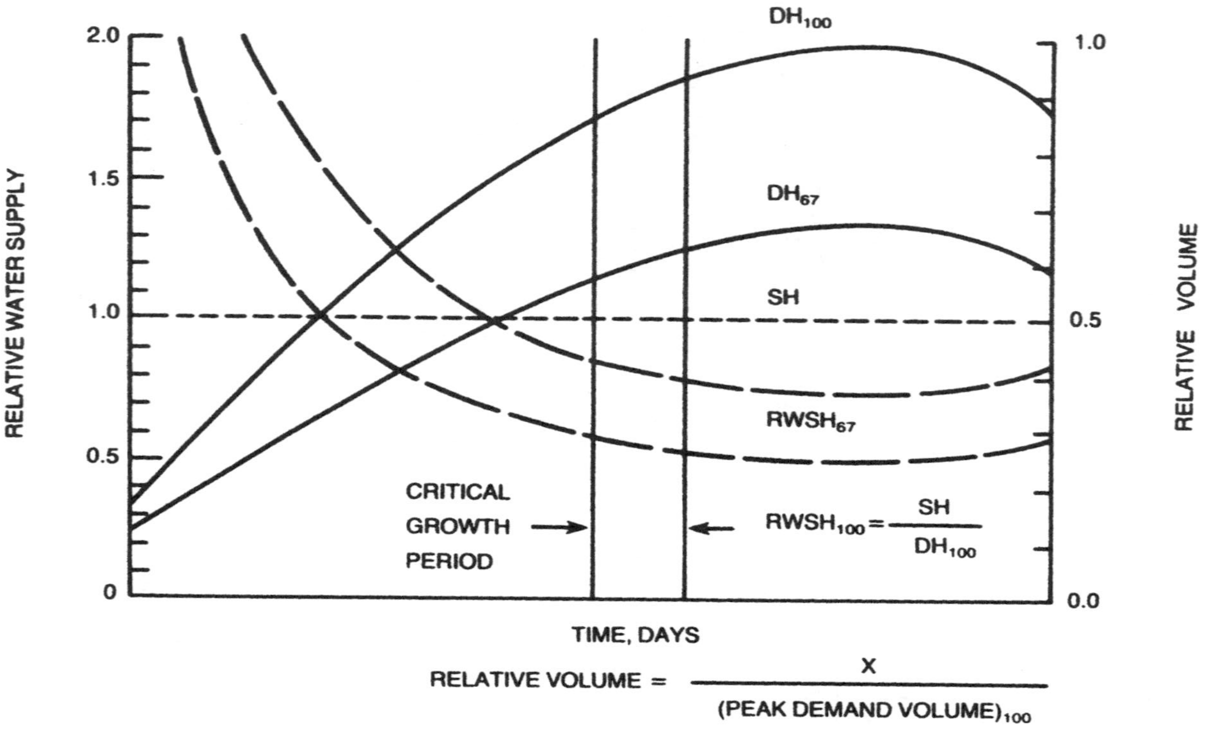

Figure 11.5 Demand and relative water supply hydrographs for 100 percent and for 67 percent of area planted, assuming a supply hydrograph (SH) of 50 percent of peak demand

which dips to a low relative water supply of RWS = 0.5. From Figure 11.3, it is apparent that such a low RWS will undoubtedly result in a very "high-tension" system requiring very strong system management to distribute water equitably to all the land area within and between UCAs.

On grain crops, it is usually not economically efficient to irrigate with an RWS less than about 0.75. Therefore, if the water is equitably distributed within and between UCAs, individual farmers might opt to irrigate only part of their land. As demonstrated in Figure 11.5, by irrigating only 67 percent of their land, the RWSH would bottom out at RWS = 0.75. For high-value fruit, vegetable, and root crops, it is usually not economically efficient to irrigate with an RWS of less than 1.0 to 1.25. Therefore, prudent farmers growing such crops would probably opt to irrigate only 40 to 50 percent of their land. The point here is that farmers can rationally control the effective RWS for the land they elect to irrigate, providing they know in advance what this individual water allocation will be. Consequently, we should consider the effective RWS in socioeconomic and biophysical terms.

In most irrigation projects or systems, infrastructure and/or management are such that the water is not equitably distributed to all the land. Depending on the degree of inequity, this may or may not offset the overall productivity of the project. It is still conceivable that 67 percent of the land might be irrigated with a minimum RWS = 0.75. However, rather than each farmer sharing in the shortage of water, one-third of the farms may not receive any water during the peak-use period. The lack of equity can occur at the main system level between UCAs or within the UCAs.

The intriguing management question is: How can each level of management have an attractive incentive system to induce an optimal fit between the demand and supply hydrographs at the interfaces between the main system and UCAs and the UCAs and farm units? To achieve this optimum, farmers need to make their planting decisions based on the limitations of the water supply, and the main-system management must provide reliable deliveries which meet, as well as possible, the demand hydrographs of the UCAs. Finally, the middle-system (WUA) managers must distribute the water received from the main system in an efficient, reliable, and equitable fashion to the various farm units.

WATER PRICING POLICY

I view water pricing and water law as the means for creating an environment to induce the system to evolve in a desirable direction. Water law provides a legal framework for the development to take place. It assures farmers of their right to receive the benefits from their capital and labor inputs, which are needed to develop their holdings for efficient irrigation.

I do not view increasing the high price of water as an effective means for inducing optimum irrigation water-use efficiency. In fact, on the contrary, high-priced water may actually hinder on-farm development because it captures resources needed for other activities. Consequently, high prices may actually create a disincentive for optimum water use in the long run. However, if the farmers were like residential users on a demand system such as a municipal water supply, water pricing would be effective because there is little cost associated with saving water and using it more effectively. This is also true for sprinkle- and trickle-irrigated farms supplied from demand systems, providing the cost of water does not price them out of business. Other methods of creating incentives for residential users to be efficient, which I believe are less desirable because they leave too little discretion to the users, are rationing or merely limiting the allowable use and shutting off the supply accordingly (or charging a penalty for overuse).

When we think of farmers, however, we must realize that for them to increase their water-use efficiency, they must usually increase the labor, capital, and management inputs to their own farm irrigation practices. Thus, it costs them considerably more to use water efficiently than to misuse it. This extra cost is not necessarily offset by additional benefits from higher crop yields unless the water is in short supply. Increasing the price of water actually allows them less leeway for providing the additional on-farm cost of using it well.

There is ample experience with the use of subsidies to enhance better on-farm irrigation development to improve water-use efficiency. For example, through the SCS, ACP, FHA, and tax-incentive programs in the United States, farmers can apply for and receive grants to offset much of the cost associated with lining canals, leveling fields, and improving irrigation practices. One way the use of high-tech efficient irrigation systems has been stimulated has been through our tax-credit program. These are all, in effect, negative water pricing policies.

In actuality, the attraction for using irrigation water efficiently is the degree of water scarcity and the productive value of the water itself. Reducing the price of water to farmers (even to zero) but making it scarce by rationing and reliable or on demand, provides the farmers with maximum incentive and potential for effective on-farm water management and use. For the marginal value of water to be greater than zero, it must be allocated so that farmers can expand or contract their irrigated area or select better cropping patterns. Unfortunately, most allocation and water charge systems are linked to irrigated crop areas and, thus, inhibit (or even eliminate) the opportunity necessary for a nonzero marginal value.

The argument presented above challenges the idea that charging (more) for irrigation water will necessarily stimulate farmers to use it (more) efficiently. However, I do not mean to imply that water pricing does not have a place. For one thing, revenues based on the quantity delivered provide a defensible means for meeting investment and/or recurring costs. Also, high water charges may provide a means for allocating a scarce resource, reducing easily controlled overuse and waste, and/or drawing it away from crops which give a low economic return per unit of irrigation water required.

A final point relative to the above discussions is that in order to manage most effectively, distribute equitably, and charge fairly for the quantity of water delivered, the water must be measured volumetrically. This requires metering of some type, which is usually done by measuring the rate of flow and multiplying the rate by the delivery time. Unfortunately, volumetric measurement and the recording necessary are fairly expensive in terms of management, labor, and the hardware required--especially for projects serving numerous small farmers under demand systems. However, with rotational water deliveries, timing is usually done, and the additional cost of measuring flow rates and computing volumes delivered should be relatively inexpensive.

ADDITIONAL THOUGHTS

A few additional comments or lessons appear in order. One is that, no matter how carefully an irrigation development is planned, there will undoubtedly be reasons to make corrective changes as the development and implementation of the system takes place. Thus, it appears wise to leave room for flexibility and change in the very design of the system so that advantage can be taken of new opportunities for system improvements and more promising objectives, as new insights are gained and the holistic view of the environment evolves. Two good questions to ask relative to project development or rehabilitation are: What are the project's real chances for success relative to the objective functions? and How can the objective functions best be optimized?

Figure 11.6 gives a breakdown of the existing training given to civil and agricultural engineers, as well as agricultural science majors. The dotted curve in Figure 11.6 indicates what might be a more ideal mix of engineering and agricultural science course content for these three professions. In addition to the curriculum content showing engineering and agriculture, some acquaintance with agricultural economics, sociology, and management skills should also be included as indicated. The main purpose of Figure 11.6 is merely to show a breakdown of the present and a more ideal mixture of engineering and agricultural science courses, plus the need to place more emphasis on the social science courses.

Unlike industrial systems, irrigation systems are managed not only from the top toward the bottom, but also from the bottom up. This is because the very act of irrigation requires farmers to be entrepreneurs and take risks. Farmers are not on the "payroll" but must gamble their time, capital, and talents in the real present, in hopes of accruing future benefits. Because of this, a top-down management system is unworkable, since the bureaucracy and its employees do not take the risks.

To elaborate on the above, perhaps an irrigation system is more like a free school--merely putting the building in place and providing staff does not produce education. Education occurs when students take the risks, the time, and the effort to attend the school and learn, in hopes that what they have learned will be of benefit to them in the future. Thus, to get the students into a program, they must be attracted to it. At least one major attraction, even in view of extremely difficult study programs, is a lucrative and/or interesting job opportunity upon completion.

Irrigated agriculture requires more labor per unit of land than rainfed agriculture. The amount of capital, management, and/or labor increases as the quantity per

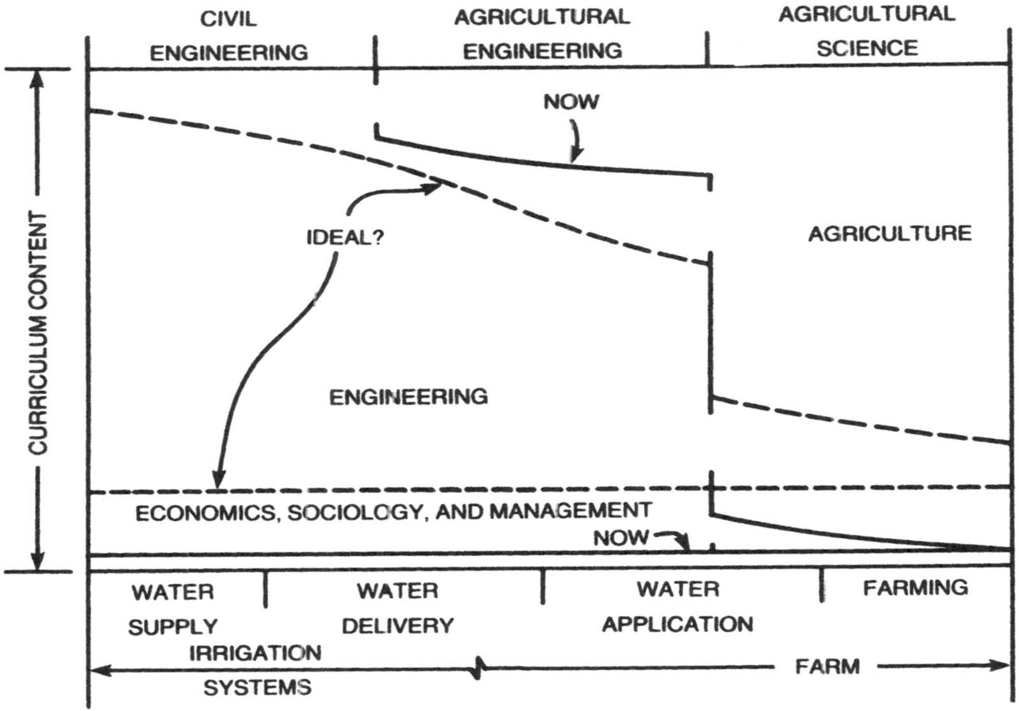

Figure 11.6 Existing and potential curriculum content for training physical scientists dealing with agricultural water management

unit of land is decreased (tension increased). In developing irrigated agriculture, the restraints on efficient and full development as affected by labor shortfalls are often ignored. The typical result is partial development of the plots allocated to each farmer and inefficient irrigation. By only partially developing the allocated land, there is less land to plant, cultivate, irrigate, weed, and harvest; consequently, less labor is required. Furthermore, by allocating all the water to only part of the land, the water tension is reduced; thus, less labor is required per unit of land irrigated. A few typical labor-requirement intensities for farming with only hand implements (no animal or machine power) are: transplanted paddy (rice) requires approximately 3,000 person hrs/ha, potatoes require approximately 2,000 hrs/ha, and corn requires 1,000 hrs/ha.

CONCLUSION

We began by noting that the vast majority of irrigation systems fall short of expectations as a result of poor system management. With only about half of the targeted basic irrigation development now in place, there is ample scope for rehabilitating old systems and constructing new systems to make them more manageable. First, however, the public planning objectives of each system--be it for commercial production, sociopolitical, and/or geopolitical reasons--must be more clearly defined and system analysis and design pursued accordingly.

The objective of the farmers who are the beneficiaries of public irrigation systems is to maximize their net benefits from irrigation by maximizing the productivity per unit of land, which is usually their scarce resource. For an individual farmer, water may not be his scarce resource unless it is rationed, allocated, and/or distributed inequitably. Therefore, he is not usually concerned about water-use efficiency or fair and equitable distribution of water to other farmers, although these are the typical operational objectives for public irigation systems. This dichotomy of the operational objectives of the public irrigation systems and the private beneficiaries is the root cause of many problems in managing them.

Areas that are often overlooked in planning for better water management are:

- o the need for communication between the users and suppliers of water so that the physical delivery and application of water can take place in a meaningful way
- o the need for visualizing irrigated agricultural development in an integrated interdisciplinary framework
- o the fact that systems with low relative water supplies require extensive physical works and diligent management to achieve equitable distribution and high water-use efficiencies
- o the evolutionary nature of irrigation institutions and systems and the need to create an environment which attracts the evolutionary process to the desired ends
- o an understanding of the three potential management levels where small farms are involved: the main system, the middle system, and the farmer system; and the nature of the physical interface between these three management levels
- o the potential countereffectiveness of increasing the price of irrigation water as a means for inducing optimum water-use efficiency
- o the need for leaving room for flexibility and change when designing systems, a more integrated study curriculum for the technicians involved in system design and management, and a better understanding of entrepreneurship and labor requirements of irrigated agriculture.

All of the above leads us full circle to whence we started, which is improving irrigation system management to enhance human well-being by having more successful irrigated agricultural development. To achieve the hoped-for results requires some alteration of the viewpoints and knowledge levels of everyone concerned. This includes the national politicians, the financiers, the planners, the designers, the contractors, the managers, the local politicians, the research and extension services, the agro-business suppliers, the farmers, the marketing people, and, perhaps most important of all, we consultants who are involved in technology transfer and the entire development program. Without new insights on our part, we may be the only beneficiaries of the development process as it reduces to welfare for the politicians and technocrats.

However, I feel more comfortable thinking that my major incentive is more successful irrigated agricultural development and the evolution toward a better world.

NOTES

1. This paper was prepared for presentation for the International School for Agricultural and Resource Development's 1984 Invited Seminar Series, "Current Issues in and Approaches to Irrigation Water Management in Developing Countries," Colorado State University, Fort Collins, Colorado. It was adapted from the paper "Consulting on Overseas Projects: An Overview Worldview of Irrigated Agricultural Development," presented at the Irrigation Association Annual Conference, Denver, Colorado, December 1983.

2. Taken from A. Aboukhaled, A. Felleke, D. Hillel, and A. A. Moursi, Opportunities for Increase of World Food Production, report to the Technical Advisory Committee of the Consultative Group on International Agricultural Research, I.D.R.C., Ottawa, Canada, April 1979, p. 161.

3. Taken from J. Doorenbos, "The Role of Irrigation in Food Production," Agriculture and Environment 2, 1975, pp. 39-54.

REFERENCES

Aboukhaled, A., A. Felleke, D. Hillel, and A. A. Moursi. Opportunities for Increase of World Food Production. Ottawa, Canada: Report to the Technical Advisory Committee of the Consultative Group on International Agricultural Research, I.D.R.C., April 1979.

Doorenbos, J. "The Role of Irrigation in Food Production," Agriculture and Environment 2, 1975.

Peterson, Dean F. "Systems and Technology for Improved Irrigation Water Management," paper presented at Agricultural and Rural Development Training Workshop, Washington, D.C., June 18-22, 1984.

12
Improving Management of Irrigation Projects in Developing Countries: Translating Theory into Practice

Warren Fairchild and Kenneth C. Nobe

INTRODUCTION

In recent years, increasing attention has been given to improving the management of irrigation projects in developing countries. We have recently had the opportunity to help design and implement such a project effort in Pakistan--the Command Water Management Project--in which design of a management-oriented organizational structure received considerable emphasis. This project was designed for joint World Bank/U.S. Agency for International Development (USAID) involvement. The background and nature of this project will be the "centerpiece" of our presentation. Before we get into our basic subject matter, however, a short explanation of the International Bank for Reconstruction and Development, better known as the "World Bank," is in order.

World Bank Organization and Operation

The name "World Bank" conjures up many unusual visages. Many believe the World Bank is a U.S. agency; others see it as a large commercial bank with branches all over the world; most have few ideas about the Bank but, because of the name, believe it must be large. However, it is neither a U.S. agency nor a commercial bank. Large it is, the largest development agency in the world, which in FY84 made loans of about U.S.$15.5 billion (B). Of this amount, U.S.$11.9 B was in the form of loans, and about U.S.$36 B was credit. The Bank's lending program has grown significantly in recent years. Even so, the loans in FY84 were somewhat below the projected target because borrowing countries have been forced to reduce the level of their involvement due to the requirements of prudent financial management in light of the world's depressed economic situation.

World Bank is an international investment and lending institution. Its genesis came from World War II, when the 44 allied nations saw need for such an institution to assist in developing the economies of poorer nations. The Bank's Articles of Agreement was signed by all Allies except the Soviet Union. It became operational on June 25, 1946. Today, the Bank is owned by its 146 member nations that have subscribed over U.S.$35 B in capital. The Soviet Union and most other communist bloc nations still are not members.

Officially, the World Bank is a specialized agency of the United Nations (UN); however, it operates independently of the UN, with its own board of directors. Whereas voting in the UN is based on one vote for each member nation, voting in World Bank is similar to a corporation and is based upon subscribed capital, with the U.S. share now being about 21 percent.

The Bank has two affiliates: the International Finance Corporation (IFC), which became operational in 1956, and the International Development Association (IDA), which became operational in 1960. IFC encourages the growth of the private sector in developing countries. World Bank and IFC finance their loans through the sale of bonds in the world money market. These loans now carry a conventional rate of a little over 9 percent and are repayable in about 40 years. IDA credits are made available to the 50 poorest of the developing nations, each with an annual per capita gross national product of less than U.S.$520. These credits are for a period of 50 years without interest; however, there is a small carrying charge (about 0.75 percent). IDA credits make up about one-third of the Bank's lending program. Since it is funded by replenishment from the member nations, one hears a lot about IDA when its funding is considered by the U.S. Congress.

Bank loans and credits cover a wide range of sectors, including agriculture, education, energy, industry, population, telecommunication, transportation, urban and rural development, and water supply and sewage. Lending for agriculture and rural development is the largest program and generally makes up 30 to 40 percent of total investments. A broad range of activities is financed in the agricultural sector, such as: irrigation and water management, extension, research, seeds, fertilizer, forestry, watershed management, soil erosion control, and credit. The thrust of the agricultural sectoral program is, naturally, to increase production, but more specifically, it is focused on the rural areas where poverty is heavily concentrated in the developing countries.

The general procedure for World Bank lending is for a recipient member nation to identify a project and prepare a project report. This "feasibility" report is appraised by a World Bank mission. If the appraisal mission finds the project is technically and economically feasible (with greater than 15 percent economic rate of return) the project goes to the board of directors for approval. Following approval and signing of an agreement, the project is implemented by the member country, with minimal Bank supervision. As a condition for a Bank loan, however, the Bank often stipulates certain institutional improvements that must be met, such as strengthening the implementing agency and requiring revisions in government policies relating to subsidies, agriculture pricing, water charges, the role of the private sector, etc. It is these conditions that place World Bank in the position of a developmental agency with tremendous influence in bringing about change. To manage its lending program, World Bank has about 2,500 professionals on its multidisciplined staff, assembled from most of the member countries.

Since our paper will focus on the manangement factor in a new Bank-funded project in Pakistan, some data on the Bank's role and its investment strategy will help place this project into perspective. In general, the strategy for water and agricultural sectors has been to assist the government of Pakistan (GOP) in implementing priority projects and programs by mutually reinforcing and integrating financial and nonfinancial assistance. This approach is evidenced by the Bank serving as executing agency for the UN Development Programme-financed Indus Basin Planning Study that resulted in the Revised Action Program for agriculture (RAP).

The GOP took the preliminary findings of RAP as a basis for many of the policy and project decisions made during the Fifth Five-Year Plan period (FY78-83), which emphasized improved distribution of agricultural inputs and services and granted top priority to water sector drainage, irrigation water management, and rehabilitation projects. Bank Group-financed projects during this period were consistent with and supportive of the objectives of GOP's Fifth Plan. The Bank Group-financed projects included, among others: Reservoir Maintenance (Ln. 2166-PAK and Cr. 1255-PAK); On-Farm Water Management (OFWM) (Cr. 1163-PAK); Agricultural Extension and Adaptive Research (Cr. 922-PAK); and Punjab Extension and Agricultural Development (Cr. 813-PAK). For the Sixth Five-Year Plan period (FY84-88), the Bank lending program in the water sector will continue to support these kinds of successful efforts, including the newly approved Command Water Management Project, the Left Bank Outfall Drain now being appraised in Sind Province, and such other proposed projects as SCARP Transition, Private Tubewells, and Rural Electrification, as well as continuing to support the successful OFWM and Irrigation Systems Rehabilitation Projects initiated during the Fifth Plan period.

Importance of Irrigation to World Food Production

During the last 10 years, roughly 40 percent of all increases in food production in developing countries have come from expanded irrigation. In the last 50 years, the land under irrigation has increased threefold, with the cost of development far exceeding the rate of inflation. Despite this, water has been generally treated as a free good, with water charges seldom covering the operation and maintenance (O&M) costs. This situation results in waste of water and a loss of food-production potential, which is directly translated to a drain on the meager financial resources of these countries. Current studies indicate that water for irrigation, rather than the land area available for cultivation, will become the critical natural resource in agricultural development in the future.

It is very risky to make near-term estimates of food production. Climatic extremes, as evidenced by the 1972 harvest that resulted in a 32 million metric ton (mmt) shortfall over the previous year, markedly influenced near-term and year-to-year production. This was further evidenced by the 1979 grain harvest, which had a 90 mmt shortfall, and its relationship to the ensuing U.S. grain embargo. The mid- and long-term agricultural production picture is framed by several very definite conditions. We do know that (a) world population is now doubling about every 35 years; (b) there is less new land available for increased production; and (c) higher energy costs have resulted in substantial increases of fertilizer and other production costs. Therefore, it would seem logical that a substantial increase (3 to 4 percent annually) in food production is required simply to meet current nutritional standards[1] and the ability to meet such increased targets will be more difficult and costly to achieve than past production levels.

Nonetheless, annual worldwide food production has grown substantially during the past 20 to 25 years.[2] This increase has been generally greater than the population increase and reflects emphasis placed on food production by developing countries and development agencies, as well as technological advances such as those which led to production of "miracle rice and wheat"--the Green Revolution of the late 1960s and early 1970s.

Through the 1960s, most growth in worldwide food production was associated with expanding the cultivated area. Since 1955, about 375 million acres (M ac) of land have been brought into cultivation in the developing countries, which is a larger total area than that devoted to cereals in the U.S., Canada, Western Europe, and Japan combined. The increase in food production through expanded acreage will be relatively small in the future, however, given the limited remaining unused area suitable for crop production, particularly under irrigation.

Irrigation and fertilizer were crucial to the success of the Green Revolution, which began with the introduction of high-yielding varieties (HYV) of rice and wheat in 1965-66. Only 12 years later, over 135 M ac of these HYVs (which constituted more than one-third of the area sown to wheat and rice in the developing countries) were sown to these two cereals, thus making this the most rapid period of technology adoption in agricultural history. Almost all of these adjustments occurred on irrigated lands. It is estimated that fertilizer alone has been responsible for only 30 percent of recent increases in food production in the developing countries. In these countries, fertilizer usage has been increasing at about 8 percent annually; however, application rates are still only about 15 percent of those in the U.S.

Irrigated agriculture will continue to be a major factor in meeting future world food grain production targets. Moreover, because of the scarcity of good land still available for irrigation development, the capital cost of new irrigation development has increased significantly--endemic of recent worldwide cost increases for civil works. Therefore, the need for better management of existing systems, including on-farm water management, becomes paramount. Hence, the World Bank and such bilateral aid agencies as USAID are now emphasizing irrigation water management in their agricultural lending programs. The Pakistan Command Water Management Project is a prime example of this new emphasis. Significantly, this project will involve both the management of water and nonwater inputs, in an effort to efficiently improve agricultural production in selected project areas.

THE COMMAND WATER MANAGEMENT PROJECT IN PAKISTAN

The Command Water Management (CWM) Project is a five-year project (FY85-89) which is highly divisible.[3] It comprises seven subprojects which are located in all four provinces and total about 510,000 acres. The project consists of four components: canal rehabilitation and remodeling, drainage, on-farm water management, and project management.

The core planning concept for this project is the strengthened management of the existing institutions, infrastructure, agricultural inputs, and services, so as to efficiently remove major constraints to irrigated agricultural production in selected subproject areas. In irrigated areas, the adequacy and reliability of the water supply are the keys to agricultural production. Small farmers cannot afford the risk of applying expensive inputs unless they are assured of a reliable supply, which has not generally been the case in Pakistan. Hence, in this project, improvements to and strengthened management of the irrigation infrastructure are emphasized in association with improved deliveries of nonwater inputs and services. In the subproject areas, a wide range of institutional and physical improvements were considered. Components, and portions thereof, that would most efficiently remove major agricultural production constraints have been included in this project. Unprogrammed but feasible components would be installed in the postproject period as funding becomes available.

The project objectives are to:

- o increase agricultural production through improved water management, along with efficient supplying of agricultural services and nonwater inputs
- o develop water management techniques and programs replicable over a wide range of agroclimatic zones
- o build within the provincial agencies a continuing capability for planning, implementing, operating, and maintaining integrated and efficient programs of irrigated agriculture
- o strengthen farmer participation in formal water user associations to improve their overall water and nonwater input management and provide them an opportunity to have a stronger voice in public decision-making.

Background

To better understand the urgent need for this project, it is necessary to give a background explanation of the current situation in Pakistan on which this project is based.

Demographic and Physical Features. Pakistan has a population estimated at 85 M and is growing at a rate of about 3 percent per annum. Per capita annual income is only about U.S.$385 (Bank's estimate for 1983), with approximately 30 percent of the population having an income below U.S.$150, which was the estimated absolute poverty level for FY80. The adult literacy rate of 24 percent is well below the 50 percent level recorded in comparable developing countries.

Pakistan has a total land area of 197 M ac, of which only 40 percent, or about 78 M ac, is suitable for crop, range, and forest production. About 40 M ac of this is commanded for irrigation and generates about 90 percent of the nation's total agricultural production. Most of the cultivable land is located in the Indus Plain, where the soils are well suited for irrigated agriculture.

The climate in the Indus Plain is arid to semi-arid and subtropical. The temperature in most cultivable areas, reaching 120 degrees F in May and June, allows year-round cultivation. Annual precipitation is less than 8 inches over much of the Indus Plain, whereas pan evaporation may exceed 5 feet, making irrigation a necessity for agricultural production. Even so, heavy rainfall does occur during monsoon seasons, causing substantial periodic flood damage. (Flood losses were estimated to be in excess of U.S.$1,800 M in 1973 and again in 1976.)

Past Agricultural Performance and Policies. Despite the gradually declining relative importance of agriculture in Pakistan's economy, it remains the largest commodity-producing sector; the principal source of income for a majority of the country's population (almost 55 percent of the labor force); the most important source of exports (about 42 percent in FY82); the principal supplier of raw materials for industry; and the primary market for output of nonagricultural sectors.

During the period FY77-83, Pakistan's annual agricultural growth rate exceeded 4 percent, a marked increase over the 1.7 percent for the FY71-76 period. This doubling of the agricultural sector growth rate during the late 1970s and early 1980s can be attributed partly to favorable weather. However, a major factor was farmer response to changes in GOP's policies during the Fifth Five-Year Plan period (FY78-83) for improving farm-level incentives through adjusting prices, reducing subsidies, encouraging private sector investment in areas such as input distribution, concentrating on funding more efficient agricultural extension and research services, improving water management at the farm level, and rehabilitating some of the irrigation systems.

Even with the recent significant increase in agricultural production, the average yields for major field crops in Pakistan remain well below that of similar developing countries. For example, in FY81 the average yield of wheat was about 665 kg/ac, rice 650, cotton lint 140, sugar cane 15.850, and maize 515. Hence, there remains ample opportunity and potential to further improve Pakistan's agricultural production. Based on various empirical and field level surveys and policy studies in Pakistan,[4] it has been determined that the major factors which would best contribute to increased growth in the crop sector include (a) providing a timely and assured irrigation water supply; (b) improving the delivery and availability of quality seeds, fertilizer, insecticides, extension, agricultural credit, and machinery; (c) furthering farmer education; and (d) improving the rural infrastructure--particularly roads, rural electrification, and communications. The interaction and the combination of these factors and their effect on agricultural production were the basis for agricultural production targets developed by the GOP for the Sixth Five-Year Plan. Optimizing benefits from the interaction of these factors, however, will require strengthened farmer involvement and improved institutional management.

The Indus Irrigation System. The single most significant infrastructural work relating to agricultural production in Pakistan is the Indus Irrigation System (34.5 M ac), which is the largest contiguous irrigation system in the world. The Indus system encompasses the Indus River and its tributaries, three major storage reservoirs, 19 barrages/headworks, 12 link canals, and 43 canal commands covering about 90,000 chaks.[5] The total length of the canal system is about 39,000 miles, with watercourses,[6] field channels, and field ditches running another 1.0 M miles. Approximately 100 M ac/ft (maf) of surface irrigation supplies are diverted annually into this canal system. In addition, approximately 12,500 public tubewells and 100,000 private tubewells pump annually an additional 25 maf of groundwater for irrigation.

Conveyance of water supplies in excess of design capacity has placed great stress on the system in recent years. This situation, coupled with deferred maintenance, has resulted in frequent canal bank breaches and interruptions in service. Overall, system irrigation delivery efficiencies are extremely low. Various studies by the Water and Power Development Authority (WAPDA) indicate that about 25 percent of the diverted surface supplies are lost through deep percolation and evaporation prior to reaching the chaks. The irrigation efficiencies in the chaks average between 40 and 50 percent. Inadequate or nonexistent surface and subsurface drainage has resulted in serious surface flooding, a high groundwater table, and associated problems of waterlogging and salinity.[7] In addition, there have been continuous technical, financial, and operational problems associated with the public tubewell program, which, in turn, led to a recent divestiture policy decision by the GOP.

Institutional Arrangements. Improving crop production in irrigated agriculture requires a high degree of coordination and/or integration of agencies and firms supplying the water and nonwater inputs and services. Currently, the major responsibility for supplying these inputs and services at the provincial and field levels is split mainly among the provincial irrigation departments (PIDs), agricultural departments, and the private sector. There is little evidence to indicate that adequate coordination exists to ensure the timely availability of these agricultural inputs and services. Also, a deficiency exists in coordinating the supply of surface water and groundwater within the PIDs where responsibilities for operation of SCARP tubewells and surface irrigation systems are split at the field level. Until recently, the farmers did not have an organization through which their views could be effectively aired. With time, however, formal water user associations (WUAs) could serve as the vehicle for improved farmer participation in the planning, construction, and the operation and maintenance (O&M) of irrigation facilities, as well as facilitating distribution of required nonwater inputs and services. And prior to the CWM Project, an institutional arrangement for coordinating the activities of public and private input supplies in project areas did not exist; hence, the heavy emphasis on institutional and organizational considerations in this project.

Project Location. The Command Water Management Project comprises seven subprojects, four in Punjab and one each in the Northwest Frontier, Sind, and Baluchistan provinces. Except for Las Bela in Baluchistan Province, these subprojects are located in the Indus River Basin and are served by the Indus Irrigation System. Las Bela is located near Karachi and received its first water delivery in 1983 from the Hub River Dam. In total, these subprojects encompass some 610,000 ac, with about 510,000 being commanded by irrigation canals (Table 12.1).

Water Resources in the Subproject Areas

Water Resources Availability. The seven subprojects receive irrigation supplies mainly from surface deliveries, which are generally of excellent quality. In addition, Sehra-Naulakhi, Niazbeg, Pakpattan, and the upper reaches of Shakhot receive some public and/or private tubewell water. Groundwater is generally not available in the Las Bela and Warsak Lift subproject areas, and the groundwater in 6-R and in lower Shahkot is brackish and unsuitable for crop production. Table 12.2 gives a summary, by subprojects, of the ac/ft of surface water delivered to the head of the watercourses, the estimated groundwater pumpage, and the crop water requirements (based on the present cropping pattern and intensities), and irrigation efficiencies. WAPDA estimates that, on an annual basis, these supplies meet a percentage of the crop requirement ranging from a low of 52 percent in Shahkot to

Table 12.1 Subproject area by unit (000 ac)

Subproject/Command	Gross Area (ga)	Command Area (cca)
Punjab		
Pakpattan Canal	119	97
Shahkot Disty	63	49
6-R Disty	133	104
Niazbeg Disty	45	41
	(360)	(291)
Sind		
Sehra-Naulakhi Branches	165	164
NWFP		
Warsak Lift Canal	55	43
Baluchistan		
Las Bela Branch	34	12[a]
Total	614	510

[a]Limited to upper five minors.

a high of 85 percent in Sehra-Naulakhi. During periods of peak crop water requirements, however, the water supply may meet only 25-35 percent of such requirements.

The Warsak and Las Bela canals and the Niazbeg distributary have physical constraints that limit surface deliveries. The Warsak Lift Canal pumping station is currently pumping only 160 cusecs, rather than the 200 cusecs planned. Negotiations are under way with USAID to install new pumps capable of pumping 200 cusecs, which would increase the canal deliveries to watercourse head to about 91,000 ac/ft annually. WAPDA wants to rehabilitate the Las Bela Canal so that it can deliver the full 160 cusecs as planned, rather than the 50 cusecs now being

Table 12.2 Irrigation water supplies and requirements at watercourse head

Subproject	Canal	Public TW	Private TW	Total	Crop Water Required	Canal	Public TW	Private TW	Total
	[ac/ft annually (000)]					[cusec/1000 ac]			
Pakpattan	217	--	54	271	420	3.1	--	0.8	3.9
Shahkot	80	25	--	105	202	2.3	0.7	--	3.0
6-R	250	--	--	250	439	3.3	--	--	3.3
Niazbeg	62	27	39	128	168	2.1	0.9	1.3	4.3
Sehra-Naulakhi	319	200	31	551	650	2.7	1.7	0.3	4.7
Warsak Lift	73	--	--	73	89	2.3	--	--	2.3
Las Bela	31	--	N/A	31	--	2.3	--	--	2.2

delivered. This would increase the surface water diverted for irrigation to the Las Bela Canal to over 100,000 ac/ft annually. In the lower reaches of the Niazbeg Canal, much of the commanded area cannot be adequately served by gravity irrigation because of insufficient canal head. Generally, the Indus Irrigation System was designed to modestly deliver about 3 cusecs/1000 ac to the chak (head of a watercourse). To meet this planned figure in all subprojects, improvements indicated above would need to be made, and additional supplies would need to be provided to the Niazbeg, Shahkot, and Sehra-Naulakhi subproject areas.

Irrigation supplies for subprojects with fresh groundwater (FGW)--Pakpattan, Niazbeg, and Sehra-Naulakhi --are significantly higher than those in the saline groundwater (SGW) areas of 6-R and a portion of Shahkot and in areas where groundwater is unavailable (Warsak Lift and Las Bela). The data in Table 12.2 would appear to indicate that, in SCARP FGW areas (Upper Shahkot and Sehra-Naulakhi), farmers do not exploit the groundwater potential if subsidized public tubewell water is available, whereas in non-SCARP tubewell areas (Pakpattan and Niazbeg), farmers do have the incentive to install their own private tubewells.

The Surface Distribution System. Through the years, the distribution systems have deteriorated because of deferred maintenance. This deterioration is being slowly rectified through increased donor funding to PIDs for required O&M and for such projects as the Irrigation Systems Rehabilitation Project (Cr. 1239-PAK). Not only are water losses in the system excessive (only about 50 percent of the diverted surface and pumped groundwater supplies reach the crop) but the supply is unreliable because of frequent breaches in canal banks. Because of sediment deposition in canals and problems with outlets to watercourses (moghas), PIDs have difficulty in delivering planned supplies to many chaks located at the lower end of distributaries and minors. This situation was verified by WAPDA's measurement of discharges from 186 moghas, which indicated that about 30 percent in the tail reaches were drawing less water than the authorized full supply (afs), compared to 10 percent in the head and middle reaches (Ministry of Water and Power, 1983). Also in this survey, it was found that 67 percent of the moghas were drawing more than the allotted afs. The only control structures installed in the system are at the heads of canals, branches, distributaries, and minors.

The head regulator on the Naulakhi branch canal is in need of repair and is being replaced by Sind's PID, with funding provided in the provincial Agricultural Development Program (ADP). Warsak Lift and Las Bela subprojects have problems in the distribution system that are unique to them. The upper reaches of the Warsak Canal traverse along the side of the unstable mountain. Frequent rock slides block the channel, causing reduced flows and/or breaches in the canal banks. The lower 20 to 25 percent of the Warsak Lift command is currently without water service because of limited pumping capacity. Because of the low water deliveries in the Las Bela Canal, only the upper four minors are now receiving water. These minors are constructed of rubble masonry and are in extremely poor condition. Water losses from the main canal, minors, and watercourses are excessive because of poor layout and construction and the coarse-textured soils of the area. Some of the minors do not have proper head to command their service areas. Watercourses in the Warsak Lift and Sehra-Naulakhi subprojects are extremely long for the areas served (e.g., over 60 ft/ac, compared with about 30 ft/ac in Punjab's subprojects).

Subsurface Drainage. Drainage problems in the subproject areas are currently limited to the saline groundwater (SGW) areas in 6-R and lower Shahkot, where recently the groundwater table has been rising by about 0.9 ft and 0.4 ft, respectively, each year. The Shahkot subproject is located in the original SCARP I project area. In about the lower 20 percent of this subproject area, the groundwater has turned brackish, and public tubewell pumping of water for irrigation service has been discontinued; hence, the recent rise in the groundwater table. WAPDA now estimates that about 36,000 ac in 6-R and about 10,000 ac in lower Shahkot have a groundwater table within 5 ft of the surface throughout the year (MWP, 1983). Thus, these areas are eligible for GOP's designation as disaster areas. It is anticipated that, without treatment, high groundwater areas will continue to expand in the future in these two subprojects.

Las Bela received its first irrigation service in 1983; however, it appears that a serious subsurface drainage problem may quickly develop. This conclusion is predicated on the available but somewhat limited data which indicates that bedrock underlays the command area at a depth of 10 to 35 ft. In addition, the soils are generally coarse textured, with very high infiltration rates (e.g., 0.5 to 1.5 in/hr, over much of the command). It is estimated that some subsurface drainage via horizontal tile would probably be required within 10 to 15 years after full project development. WAPDA and the Baluchistan PID will need to monitor this situation closely and take corrective actions if and when the situation becomes critical.

Private and public tubewell pumping generally takes care of the subsurface drainage requirement in the fresh groundwater (FGW) areas. However, groundwater investigations indicate that the FGW levels in Pakpattan and Niazbeg subprojects are rising slowly (less than 0.2 ft annually), which would indicate opportunity for further groundwater exploitation in these areas. The operation of the public tubewells has generally been on a schedule reflecting the warabandi[8] deliveries of the surface system. Opportunity exists for such tubewells to be operated on a schedule that would better reflect crop water requirements. Pumpage on such a revised schedule could still meet the dual objectives of irrigation and drainage. Farmers operate private tubewells during periods that they determine critical for crop production, however, with little or no concern for drainage requirements within the total command area.

Surface Drainage. The topography in all subprojects is generally quite flat and featureless, with much of it having a gradient of less than 0.3 percent. Natural drains have generally been obliterated by farmers as they have reclaimed natural drainageways for farming operations. Also, roads, railways, flood embankments, and irrigation works have further obstructed natural drainage flows. The government has been actively involved in constructing surface drains since the 1940s. As evidenced by the substantial flood damage during some recent monsoons, however, there remains a need for vastly expanding the existing surface drainage system, including drainage outlets for chaks. Also, surface drains will be required for outlets for the subsurface drainage effluent in the SGW areas.

Soil Resources

Soils in the project area are predominantly moderately coarse to medium textured, except for the Sehra-Naulakhi subproject soils (in Sind Province), which are fine textured. The soils are relatively well drained except for those located in the waterlogged area of 6-R and lower

Shahkot. According to available data (WAPDA, 1979), most of the soils are relatively free of salinity/sodicity problems but, as noted earlier, such problems are now emerging in local areas.

WAPDA has divided the soils in the subproject areas into eight land-capability classes, with Classes I-IV being suitable for crop production (Class IV is marginal), Classes V-VII being suitable for forestry and grazing, and Class VIII being unproductive. Most of the land is in Classes I and II, which means that these soils have the inherent characteristics of being very productive under irrigation. Las Bela and 6-R have significant amounts of Class VII and VIII lands, but these areas are located mostly outside the commanded areas.

Farming Characteristics

Farm Size and Tenure. The average size farm unit in Pakistan is about 12.5 ac. In the project area there are about 50,000 units, of which about 66 percent are less than 12.5 ac, 25 percent range from 12.5 to 25 ac, and 9 percent are greater than 25 ac. However, about 40 percent of the canal commanded area (cca) is in farms of less than 12.5 ac, 33 percent between 12.5 to 25 ac, and 27 percent over 25 ac. About 57 percent of the farms are either owner or owner-cum-tenant operated, while 43 percent are tenant operated. Moreover, there are large variations in these characteristics among subproject areas.

Cropping Intensities and Yields. The cropping intensities in the subprojects are quite high, relative to average Pakistan conditions--except in Warsak Lift, where reduced pumping capacity limits delivery of surface water supply; in Las Bela, where irrigation service was first initiated in 1983; and in lower Niazbeg because of poor command resulting from an inadequate head in the distributary.

Crop yields in the subproject areas are generally low, but rice and wheat yields in some subproject areas are slightly higher than the national averages. Maize and sugar cane yields in all subprojects are lower than the national average. As the waterlogging condition expands in 6-R and lower Shahkot, it would be expected that the yields in these subprojects would decrease further. Table 12.3 gives the estimated average FY81 crop yields for the subprojects, compared with national averages.

Table 12.3 Average crop yields (FY81, kg/ac)

Crop	Pakpattan	Shahkot	6-R	Niazbeg	Sehra/ Naulakhi	Warsak Lift	Las Bela	National Average
Kharif								
Rice	780	800	590	560	860	--	--	650
Cotton	340	120	280	100	420	--	--	420
Maize	--	--	360	380	380	330	--	520
Sugarcane	13,280	14,960	14,620	12,010	14,400	9,140	--	15,850
Fodder	5,820	6,680	5,330	4,000	6,580	5,480	5,580	--
Rabi								
Wheat	770	970	830	880	760	530	400	660
Fodder	15,550	22,870	12,940	15,440	17,460	16,880	9,100	--

Agricultural Input Supply and Services. The current annual fertilizer consumption, in terms of nutrient per cropped acre, is estimated at around 30 kg. Government-owned, *tehsil*-based bulk depots and village-based sub-depots originally supplied most fertilizers, pesticides, and seeds. Beginning in FY80, however, the private sector began selling fertilizers. As presently planned, all fertilizer subsidies are to be phased out by FY86 (Walters, 1984). Pesticide usage is sparse. In February 1980, the responsibility for sale and distribution of pesticides was transferred from the Extension Service to the private sector. All pesticide subsidy was withdrawn in 1983. The use of certified seeds in the subproject areas is rare, with less than 10 percent of required wheat, rice, and cotton seed being certified at the present time.

Major sources of agricultural credit in the subproject areas are the Agricultural Development Bank, federal and provincial banks for cooperatives, and commercial banks. Small farmers are granted interest-free production loans up to Rs 6,000 in kind for seed, fertilizer, and pesticides, provided they repay within two months after harvesting (Nobe, 1982).

Mechanization has been strengthened in various districts of Punjab and Sind provinces on a pilot basis to maximize production of specific crops. Cotton maximization schemes cover seven districts in Punjab (including the 6-R subproject). The Training and Visits (T&V) program of agricultural extension covers five districts in Punjab (Pakpattan and Shahkot subprojects) and five in Sind (Sehra-Naulakhi); further expansion is planned.

Surveys conducted by WAPDA during preparation for the CWM Project (MWP, 1983) indicated that farmers receive inadequate assistance from the Agricultural Extension Service, and small farmers have extreme difficulty in availing themselves of agricultural credit. Obtaining the required fertilizer and pesticides is also difficult because of logistical problems, and this becomes particularly acute in the more remote villages. For these reasons, the CWM Project seeks to coordinate timely deliveries of both water and nonwater inputs in the subproject areas.

CRITICAL ENGINEERING DESIGN ELEMENTS FOR SUCCESSFUL IMPLEMENTATION OF THE CWM PROJECT

Earlier, we described the core project concept and objectives for the Pakistan Command Water Management

Project. However, it would be well to review these items briefly as a base for discussion in this section of the critical elements of project implementation. First, we stated the core project concept was to strengthen management of the existing institutions, infrastructure, agricultural inputs and services, so as to efficiently remove major constraints to irrigated agricultural production in the subproject areas. Second, we stated the project objectives as:

- o increase agricultural production
- o develop a wide range of appropriate water-management techniques and programs
- o strengthen provincial institutional capability
- o strengthen farmer participation through creation of formal water user associations.

The project contains four components, i.e., canal rehabilitation and remodeling, drainage, on-farm water management, and project management. We will discuss the first three components under this heading. Because of the importance of the project management component, however, we have reserved a special section following this one for discussion of it.

Canal Rehabilitation and Remodeling

Canal rehabilitation and remodeling will be undertaken in all seven subproject areas. In addition to the required civil works, the project provides funds for a limited amount of equipment, investigations, surveys, and recurring costs for this component.

Canal rehabilitation would be modeled after the ongoing Irrigation Systems Rehabilitation Project (ISRP) (Cr. 1239-PAK). This work would be carried out in all seven subprojects and would be governed by the general specifications and criteria of ISRP. This rehabilitation would restore canals to levels required for the safe and efficient conveyance of irrigation water, based on present operating or design capacities. Except for lining of some canal sections in the Las Bela and Warsak Lift subprojects, rehabilitation would be limited to earth work and repair and replacement of hydraulic structures. The major quantity of earth work would be to (a) raise the height of the canal banks to secure required operating freeboard and (b) widen canal banks to improve safety of operation and to provide for an adequate canal road. The Warsak Lift rehabilitation would consist of improving the masonry, as well as the earthen portion of the canal. The upper reaches of the canal traverse along the steep slope of an unstable mountain. In this section, covered escapes would be provided to bypass debris from rock slides. Rehabilitation in Las Bela would include repair of the existing lined canal.

Canal remodeling to be provided under this project will include (a) lining of selected distributaries and minors, (b) remodeling of moghas, and (c) improvements to canal structures and lift facilities, as well as installation of other water control structures in order to meet water management requirements. In the Warsak Lift subproject area, nonproject improvements (e.g., installing new pumps) are planned for the pumping plant. The only additional water supplies planned under the project in all subprojects are significant water savings expected to be associated with planned project improvements to canals and watercourses. Also, it is expected that planned improvements would lead to a more assured water supply, afford greater equity to water users, and (within physical constraints of the system) improve the timeliness of water deliveries so as to better reflect crop water requirements.

The project would provide for lining certain sections of minors and smaller distributaries to reduce current excessive seepage losses and to improve water control. Generally, reaches of canals selected for lining would be less than 30 cusecs capacity and would be dependent on (a) excessive seepage losses in the canal (generally greater than 4 cusecs/million sq ft surface area); (b) the relationship of canal and water table elevations; (c) the quality of groundwater, with high priority given to SGW areas; and (d) the value of water. Moreover, in areas where a high water table is not a problem and groundwater quality is good, an economic evaluation would be made of a balanced program of canal lining and private tubewell installation to achieve a desirable conjunctive surface/groundwater operating system. The canal lengths proposed for lining are estimated as follows: Punjab, about 93 mi; Sind, 73 mi; NEFP, 10 mi; and Baluchistan, 12 mi. It is estimated that about 50,000 ac/ft of water would be saved annually by these civil works. (See Table 12.4 for lengths of distributaries and minors proposed for lining and estimated water savings.)

Table 12.4 Planned lining and estimated water savings

Subproject	10 cfs	10-20 cfs	20-30 cfs	Total	Savings (ac/ft)
	[length of lining (mi)]				
Pakpattan	13.9	11.3	13.5	38.7	8,700
Shahkot	6.5	3.5	--	10.0	500
6-R	1.2	--	26.7	27.9	8,600
Niazbeg	7.1	9.1	--	16.2	4,400
Sehra/Naulakhi	23.5	9.6	40.1	73.2	16,750
Warsak Lift	1.0	5.0	4.0	10.0	2,500
Las Bela	3.0	3.7	4.9	11.6	8,800
Total	56.2	42.2	89.2	187.6	50,250

Remodeling of moghas would be provided in all the subproject areas to improve water distribution and ensure equitable water deliveries to chaks, especially to those in the tail-end reaches. About 1,325 outlets would be remodeled, including about 665 in Punjab (Pakpattan, 205; Shahkot, 95; 6-R, 280; and Niazbeg, 85), 450 in Sind, 150 in NWFP, and 60 in Baluchistan). Gating of outlets would be considered in cases where continuous watercourse flows caused damage to crops, or where water management considerations dictated.

The project would provide for remodeling of structures at the headworks and silt ejectors near the pump station for the Warsak Lift subproject. The lower reaches of the Niazbeg distributary would be remodeled, and a low-lift pumping station would be installed to provide additional head of about 5 ft, for better command of the service area. Other improvements would involve replacement and repair of regulating structures in various subprojects to facilitate better water control and water management.

Drainage

The project would provide drainage tubewells and disposal channels for SGW areas in the Shahkot and 6-R subprojects, civil works for surface drainage in four subprojects, as well as required transport and surveying equipment, investigations and surveys, and related recurring costs for this component.

Subsurface Drainage. WAPDA determined in recent field examinations (November 1983) that some 36,000 ac in the upper portion of the 6-R subproject and about 10,000 ac in the lower portion of the Shahkot subproject require immediate subsurface drainage. The CWM Project would provide for a grid network of automated drainage tubewells (about 60 in 6-R and 20 in Shahkot), including electrical hookups. In addition, about 55 mi of lined channels would be constructed to convey drainage effluent to the nearest surface drain outlets. Effluent from 6-R would be carried to a desert evaporation pond. Shahkot subproject effluent would outlet into the Maduana drain and thence to the Ravi River.

Recent investigations by WAPDA also indicate that the groundwater tables in 6-R and Shahkot are currently rising at annual rates of about 0.9 and 0.4 ft, respectively. If these increasing rates continue, it would appear that up to another 40,000 ac in each subproject may require drainage within the next 5 to 10 years, under GOP's definition of a disaster area. The Las Bela subproject also appears to have an imminent subsurface drainage problem, the timing of which would depend upon the time and extent of development and the degree to which a water management program is implemented. Expenditures for these extended subsurface drainage programs, however, would likely come during the postproject period. However, the project would provide for monitoring of groundwater levels in these three potentially critical subprojects in the interim.

Surface Drainage. The project would provide funding for improving the existing surface drainage systems in the Shahkot, 6-R, Warsak Lift, and Las Bela subprojects, which, according to WAPDA and PIDs, urgently require such improvement. The surface drainage systems would provide outlets to (a) drain excess storm rainwater in a timely manner and (b) dispose of effluent from the subsurface drainage systems for the 6-R and lower Shahkot subprojects. Priority for surface drain construction would be given to the SGW areas in 6-R and Shahkot. The project would provide about U.S.$5.9 M (base cost) for the surface drainage program, including funds to acquire about 800 ac of required lands. We estimate that this level of funding would provide about 20 percent of that required for completing the entire surface drainage system. The remaining work would be carried out during the postproject period.

On-Farm Water Management

The CWM Project would provide technical staff from the provincial Department of Agriculture, OFWM directorates, and Extension Service to aid farmers in the layout of (a) watercourse renovation, (b) farm irrigation and surface drainage ditches, (c) precision land leveling (about 20,000 ac targeted per project area), and (d) demonstration plots (one ac plot planned per watercourse renovated). In addition, this provincial staff would give guidance to farmers for applying and scheduling irrigation water to their fields. OFWM directorates' land-leveling equipment would be made accessible to the farmers on a rental basis. The project would provide construction materials for watercourse renovation, transport, office and surveying equipment, administration, engineering, and funds for related recurring costs for this component.

The project would provide for renovating about 1,050 watercourses (530 in Punjab, 360 in Sind, 100 in NWF, and 60 in Baluchistan subproject areas). Except for the percentage of cement-brick lining, this activity is modeled after that in the bank-financed, ongoing OFWM Project (Cr. 1163-PAK). It is assumed that about 80 percent of the watercourses in the Punjab and Sind subprojects would be renovated under this project. About 67 percent of the watercourses in both the Warsak Lift and Las Bela subprojects would be renovated, which reflects the sizable areas under command in these subprojects that are not now receiving irrigation service. (See Table 12.5 for a summary on watercourses to be renovated, along with the percentages and lengths to be lined.) The estimated water savings from watercourse renovation varies from about 15 percent of the surface and public tubewell water delivered to the watercourse head in FGW areas (where 15 percent cement-brick lining would be installed) to over 25 percent in 6-R (with 40 percent lining) and about 30 percent in Las Bela (with 50 percent lining). It is estimated that about 236,000 ac/ft, or a little more than 18 percent, of the publicly generated water would be saved annually in the seven subprojects after these improvements are completed.

The length of the sakari khals (communal watercourses) in the Sehra-Naulakhi and Warsak Lift subprojects are about twice that for the area commanded, as compared with Punjab's subprojects. An effort would be made, working with formal water user associations in these two subprojects, to redesign the watercourses and chaks to achieve higher efficiency in watercourse deliveries. Assurances have been obtained from GOP and GOProvinces that the following criteria would be used by field teams in setting priorities for watercourse renovation (i.e., the same as for the OFWM Project, Cr. 1163-PAK):

- establishment of formally organized and registered WUAs (required prior to installing permanent structures)
- willingness of water users to improve their own watercourse branches and field ditches and to install appropriate OFWM practices (i.e., land leveling, improved field layout, etc.) on their individual farms.

Table 12.5 Watercourse renovation planned under the CWM Project

		Cement Brick Lining		Water Savings
Subproject	No.	%	ft (000)	(000 a/f)
Pakpattan	160	15	395	27.3
6-R	225	40	1,482	53.9
Shahkot	75	20	247	14.8[a]
Niazbeg	70	15	461	12.5[a]
Sehra-Naulakhi	360	15	1,541	97.9[a]
Warsak Lift	100	15	304	18.1
Las Bela				
Total	1,050		4,721	236.6

[a]Includes savings of public tubewell water.

Project Management Training and Technical Assistance

As noted earlier, we will discuss the theory and application of improved project management in the following major section of this paper (Managing Projects by Results Instead of by Inputs). This new management approach will require unique training and technical assistance components, however, and these will be outlined immediately below.

Staff Training and Orientation. Three short-term training activities important to initiating project implementation originally agreed to by the GOP are as follows:

- o A three-week training course sponsored by the Ministry is programmed to be held in Islamabad to provide project management training and orientation for administrative staff for each of the seven subproject offices. The course should include the following subject matter components: general organizational and management techniques, program budgeting and scheduling, coordination, personnel management, and general orientation to a command-area approach to irrigation management. These sessions were originally scheduled to be held in late fall 1984, but have been delayed pending appointments of subproject managers and their senior technical staffs.
- o Project orientation workshops of one-week duration would be held in each of the seven subproject management offices for officials of participating agencies. They would be scheduled upon completion of the three-week training course for project administrative staff. These workshops would be directed by each subproject manager.
- o Early in the life of the project, all extension personnel working in or to be assigned to the seven subproject areas would be given a six-week training course in irrigation water management. These classes would likely be held at the OFWM Training Institute in Lahore. The instruction would cover extension methods, extension organizational development and participation, irrigation water management and scheduling, agronomic practices and cropping patterns, and farm management. All such training is scheduled for completion within the first year of the project.

During the initial World Bank supervision mission for the CWM Project carried out during October 1984 (Fairchild et al., October 29, 1984), discussions with GOP and provincial officials surfaced considerable confusion about the philosophy of the CWM Project and the organizational framework required. It was agreed, therefore, that a CWM project orientation seminar for senior GOP and provincial administrators and donor agency personnel would be held in Islamabad in mid-January 1985.

In addition to the three formalized training activities outlined above, in-service training activities would be scheduled throughout the project period. For example, one such training activity, needed early in the life of the project, is a workshop for the computer programmers to be assigned to the subprojects, so that they can incorporate needed water scheduling and socioeconomic data evaluation routines and software for their respective subproject offices. Also, specialized training would be scheduled for WUA officials. It is envisioned that members of the expatriate technical assistance team would make a major contribution to these training activities.

Technical Assistance. Technical assistance inputs would consist of two interdisciplinary teams, one to be headquartered in Lahore to service subproject management offices in Punjab and NWFP, and the other to be headquartered in Karachi to service subprojects in Sind and Baluchistan. The team leader, headquartered in Lahore, would be an agricultural economist and would oversee both field teams. These expatriate teams, funded by USAID, would involve about 252 man/months of service (World Bank Staff Appraisal Report, April 30, 1984) as follows:

Profession	No.	Man-Months
Team Manager	1	36
Water Management Spec.	2	48
Farm Management Spec.	1	24
Extension Spec.	2	48
Agronomist	2	48
Computer Spec.	1	24
TDYs	-	24
Total	9	252

Per the World Bank Staff Appraisal Report (April 30, 1984), USAID would also provide additional technical assistance in support of the CWM Project that is not covered by the project budget. This component would include assistance in the planning and design of the currently undeveloped irrigation facilities in the Las Bela subproject, assistance and design of other civil works when needed, and strengthening of planning and related management capabilities. This assistance would be provided through ongoing, long-term technical assistance contracts under the USAID Irrigation Systems Management and On-Farm Water Management projects.

The CWM Project would also provide funding for about 877 man/months of local-hire supervisory consultants, plus supporting staff as follows:

Profession	No.	Man/Months
Project Manager	1 (part-time)	13
Project Engineer	2	108
Field Engineer (Civil)	2	108
Junior Engineer	4	216
Inspector (Sub-engineer)	4	216
Accountant	2	108
Ass't. Accountant	2	108
Total	17	877

The local-hire supervisory consultants would be responsible on a day-to-day basis to the subproject managers; however, they would report administratively to the project coordinator in the CWM cell in the GOP Ministry of Water and Power. The local-hire team's major responsibility would be to give direct assistance to the subproject managers in carrying out their responsibilities. Services to be performed by these local consultants include (a) reviewing plans and designs for civil works to assure compliance with agreed criteria or specifications; (b) spot-checking civil works during construction and checking all completed works to determine quantity of work completed and if quality meets PID specifications; and (c) assisting subproject managers and ministry in preparation of IDA reimbursement applications, consolidated Quarterly and Annual Progress Reports and the Project Completion Report, and consolidated Annual Project Work Plans.

Project Implementation Schedule

Project implementation would be scheduled over five years, FY85-89. During the first year, major project activities would be (a) establishing and training the project management staff, (b) organizing WUAs and initiating the implementation of the OFWM component, (c) procuring equipment, (d) appointing consultants, and (e) preparing plans and designs for the annual remodeling and drainage components. Construction of civil works for the canal rehabilitation and remodeling component and the surface drainage subcomponent would be initiated in the second year and would continue throughout the rest of the project period. Civil works related to the subsurface drainage component would be started in the third year. This schedule provides WAPDA and Punjab's PID the opportunity to fully coordinate the installation of subsurface works in the 6-R subproject with that being planned for the remainder of the Fordwah-Sadiqia drainage area.

World Bank funding approval was obtained in June 1984, with project implementation to proceed on July 1, 1984, or as soon as possible thereafter. As will be shown in the concluding sections of this paper, organizational delays have already put the project seriously behind schedule. But first, we will turn to a discussion of the remaining project component: project management.

MANAGING IRRIGATION PROJECTS BY RESULTS INSTEAD OF BY INPUTS

So far, we have discussed the role of project management without defining the concept explicitly. Before proceeding to explore the application of an improved management theory--e.g., "Management by Results" (MBR)--a more specific explanation of our terms of reference is in order.

The roots of this new management concept emanate from the original identification of the four major factor inputs in a production process--land, labor, capital, and management--as set forth in classical economic literature. Considerable effort is expended in international development efforts such as irrigation projects to insure that appropriate levels of land, labor, and capital (including both monetary and technological resources) are available at a project site. Heretofore, little effort has been devoted to development of a viable management input and the institutional framework in which it can operate. Yet, it is the management input that ultimately determines the efficiency with which the other factor inputs--land, labor, and capital--are converted into desirable outputs, such as increased physical products, higher financial returns to private sector participants, and improved well-being for society as a whole.

Drucker (1974) has noted that "management is not so much a technique as a culture" and has repeatedly emphasized that the culture of management arises from a particular kind of methodology employed. We differ from Drucker primarily in that he placed major emphasis on "management by objectives," while we favor major emphasis on a "management by results" (MBR) approach.

The MBR approach was designed for use in public agency projects because implementing agency objectives and government's overall public policy goals may differ significantly, whereas private sector management issues, as addressed by Drucker, seldom take such distinctions into account. For example, a national planning commission may designate high priority to increasing a nation's output of agricultural products and may promote funding of a major irrigation project to help achieve this goal. Once such a project is built, however, government may turn it over to an irrigation ministry or department to administer. That agency, in turn, may specify its objective in terms of delivering a design level of canal water to the head of a watercourse. In such cases, the agency's objective may well be achieved, while the national policy goal of increasing agricultural output may not materialize. MBR appears to be a useful methodology for addressing this problem.

How important is the management factor in the production process? A macro-example from the U.S. development experience will serve for illustration. Dennison (1974) has estimated that during the 1929-69 period, two-thirds of the growth in U.S. Gross National Product resulted from improvements in the output-to-input ratio (an efficiency concept), while only one-third of the expansion was attributed to increased amounts of land, labor, and/or capital inputs. This miraculous increase in efficiency has been attributed primarily to rapid improvements in available technology and in the quality of human capital. In reality, a critical element has likely been the corresponding improvement in management systems capable of effectively developing and applying these technological improvements. (Seckler and Nobe, 1983, p. 293):

> In sum, economic progress is essentially an increased rate of output per unit of time. This objective can be achieved only by two means:

(a) by more intensive application of inputs per unit of time and/or (b) by increased efficiency of transformation of inputs into outputs. The latter is a rather mysterious effect: more output with the same input! Whatever management is, that is what it does.

Case studies evaluating the success of private sector firms are replete with examples of firms with more or less identical levels of land, labor, and capital input endowments; yet, some firms prosper while others fail. In most cases, the deciding difference lies in the quality of management. In the same vein, in public agency efforts to improve sector output (for instance, via a publicly sponsored irrigation project), success or failure is, in most cases, dependent primarily upon the quality of the management input. Unfortunately, in most developing countries, the failures of public irrigation projects to achieve production goals far outnumber the successes. This is true primarily because the public agency involved in implementation (e.g., an irrigation ministry) is organized in such a way that its objectives are met when a particular input (e.g., irrigation water) is delivered to the head of a watercourse rather than when a more desirable public goal, such as increased agricultural output, is achieved by farmers who use this water. It is for this reason that our focus on project management is on an MBR approach rather than on the more limiting concept, popularized by Peter Drucker (1974) and his contemporaries, of "Management by Objectives."

The Theory of Management by Results (MBR)

The MBR concept, as far as we can determine, emerged only recently in the development literature (Seckler and Nobe, 1983). Its application to the Command Water Management Project in Pakistan will be its first real test. As with Drucker's "Management by Objectives" approach, MBR traces its roots to the fundamentals of cybernetics, particularly as applied to men rather than machines (Ashby, 1976). Fundamental to cybernetics is the concept of a "transformer," as illustrated in Figure 12.1, whereby inputs are "converted" into outputs. In applying the MBR concept to a publicly sponsored irrigation project, a deliberate effort is made to modify the behavior of public agencies responsible for "programming" the transformers

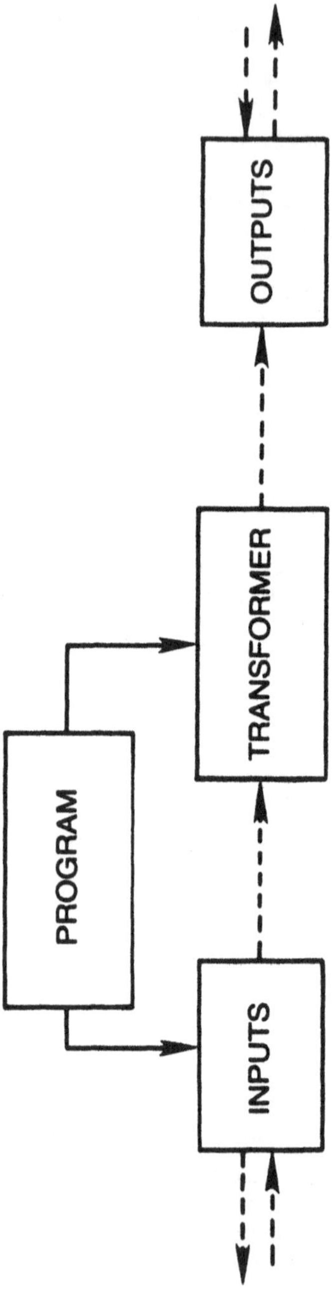

Figure 12.1 The transformer concept of cybernetics

(farmers) so that the objectives of both are more responsive to outputs than to inputs. When considering the potential for improving input/output relationships, such as embodied in large-scale irrigation systems, there are basically two types of transformer mechanisms to choose from: deterministic and cybernetic.

In a deterministic mode, a bureaucracy such as an irrigation department operates as if all the critical inputs are under its endogenous control; therefore, it tends to focus only on delivery of its water input, as dictated by an invariant routine based on engineering design specifications. In like manner, an agency responsible for some other key input, e.g., an extension delivery system, provides this input in isolation of the water input of the irrigation ministry and independent of those inputs supplied by the private sector, such as fertilizer and credit. In such an uncoordinated delivery system, it is pure accident if significant increases in output result from expanded supplies of inputs.

In contrast, the key characteristic in a cybernetic transformer-controlled system is the fact that such a system is subject to endogenous control of the inputs that it delivers. Therefore, its control program can adapt by responding to signals from the transformer that the expected results are not being achieved. If such response capability is expanded to include error signals for all input deliveries which produce less than expected project outputs, appropriate correction signals can be transmitted to the input suppliers, so as to improve results, regardless of which of the input signals is initially in error or which input is out of balance with the other inputs.

A common clock is the classic example of the use of an endogenous transformer mechanism. For example, when a timing clock is used to regulate a simple sprinkler irrigation system, the same level of delivery will result, regardless of whether there is a drought or if there are periods of more than normal amounts of rainfall. If the operational mode of the public agency that merely delivers water to the head of a watercourse limits its efforts to fulfilling only this objective, it operates in the context of an endogenous transformer system--ticking endlessly away without knowledge of its impact on the local agricultural production environment.

Conversely, a thermostat connected to a home furnace is an excellent example of a system controlled by a cybernetic transformer. With the thermostat set to operate within a given range of temperature control, the signal to increase or decrease the heat output of the furnace is responsive to signals of changing room conditions, e.g., sudden heat loss through an open door, an increase in the number of live bodies or level of their activities within the room, etc. By simple analogy, the clock-regulated simple irrigation system used in the endogenous transformer example can be converted to a cybernetic-controlled system by adding moisture sensors in the seedbed or devices used to measure moisture levels in the plants and connected to the transformer mechanism. A bureaucracy which has control responsibilities for one or more factor inputs into an irrigated production system that is aware of what kinds and delivery levels are needed in response to changes in the local production environment, and the ability to respond thereto, can be said to be operating in the cybernetic transformer mode.

The above examples of the clock versus the thermostat concepts, as applied to the organizational mode of public agencies responsible for sending program signals to input suppliers and transformer units, suggest clearly that shifting from the former to the latter will require a certain degree of administrative reorganization. Clearly, just improving diagnostic analysis capability will not suffice; implementing institutional change elements is needed as well.

In the case of public agencies heavily involved in publicly directed irrigation projects in developing countries, such as irrigation ministries, the usual operational mode is to focus on a delivery of inputs rather than on the impact its delivery program is having on production outputs. In the case of attempting to implement a Command Water Management Project in Pakistan, for example, such an endogenous approach was clearly embodied initially in the organizational and operational elements of the Federal Ministry for Water and Power and in the provincial departments of irrigation. To a large degree, the same was true for the federal and state counterparts of the ministries and departments of agriculture who are responsible for research and extension delivery systems. Such agencies were clearly concerned with meeting organizational objectives, but the objectives were stated in terms of input delivery, not production response by farmers, who--in an irrigated agricultural system--are the ultimate transformers of input into output. We will refer to such agencies as administrative organizations (AOs), as contrasted to management organizations (MOs), which are organized so as to utilize an MBR approach. The principle organizational components of AOs are identified in Figure 12.2.

In contrast, a managerial organization is an improved form of an administrative organization, in which feedback on output results, positive or negative, is directly utilized to modify successive rounds of input program implementation. A fully operational managerial organization (MO) is structured so as to perform six essential functions in specified operational sequence, as shown in Figure 12.3. These are (1) policy formulation and implementation; (2) project management (transforming policy goals into project objectives and designing programs to achieve specified objectives); (3) program implementation; (4) cybernetic transformation (converting inputs into desired outputs which, in the case of irrigation systems, is done by farmers, not the input suppliers); (5) performance monitoring (with particular emphasis on outputs being achieved); and 6) performance evaluation (a function shared by policymakers and program managers), followed by marginal adjustments in the policy and management functions if expected results are not being fully achieved.

The ideal MO mode would encompass these six elements for all farm input suppliers, whether public or private. As a minimum, some central management control should be achieved for the inputs supplied by the public agencies, particularly the irrigation canal water supply and the extension delivery system, as well as fertilizer, pesticides, etc., if distributed by public agencies. Of these present or potential public agency input suppliers, recent experience has shown that extension is the least equipped to respond to an MO approach (Nobe, 1984). Yet, of all potential public agency involvements, the extension field agents remain the most critical to successful implementation of MBR because of their direct contact with farmers. This situation should come as no surprise, given that MBR rests on a cybernetic foundation, that people are the ultimate "black boxes" of cybernetic theory, and, in the case of irrigation systems, that farmers perform the function of transforming inputs into desired outputs.

Translating MBR Theory Into Pakistan's CWM Project Design

At the time the World Bank/USAID CWM Appraisal Mission[9] visited Pakistan in October 1983 to evaluate the Command Water Management Project proposal, high priority

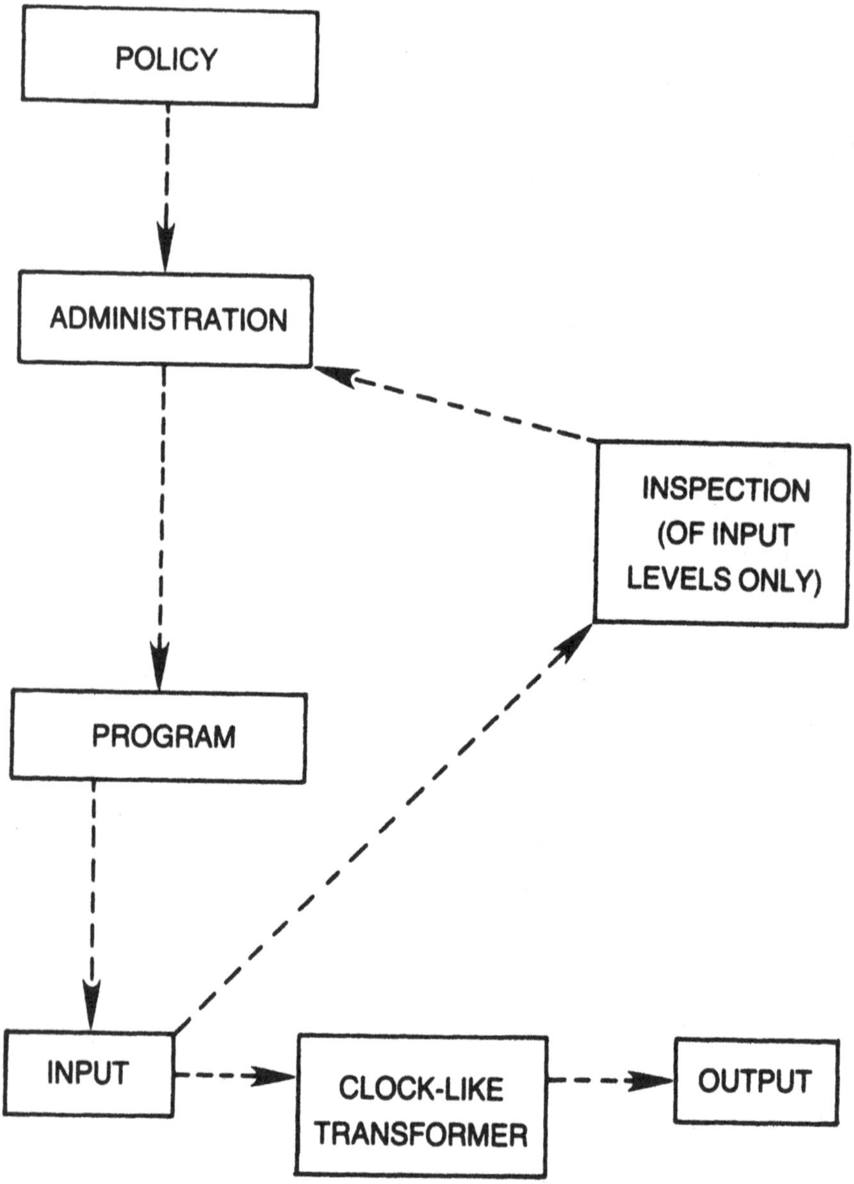

Figure 12.2 Structure of an administrative organization

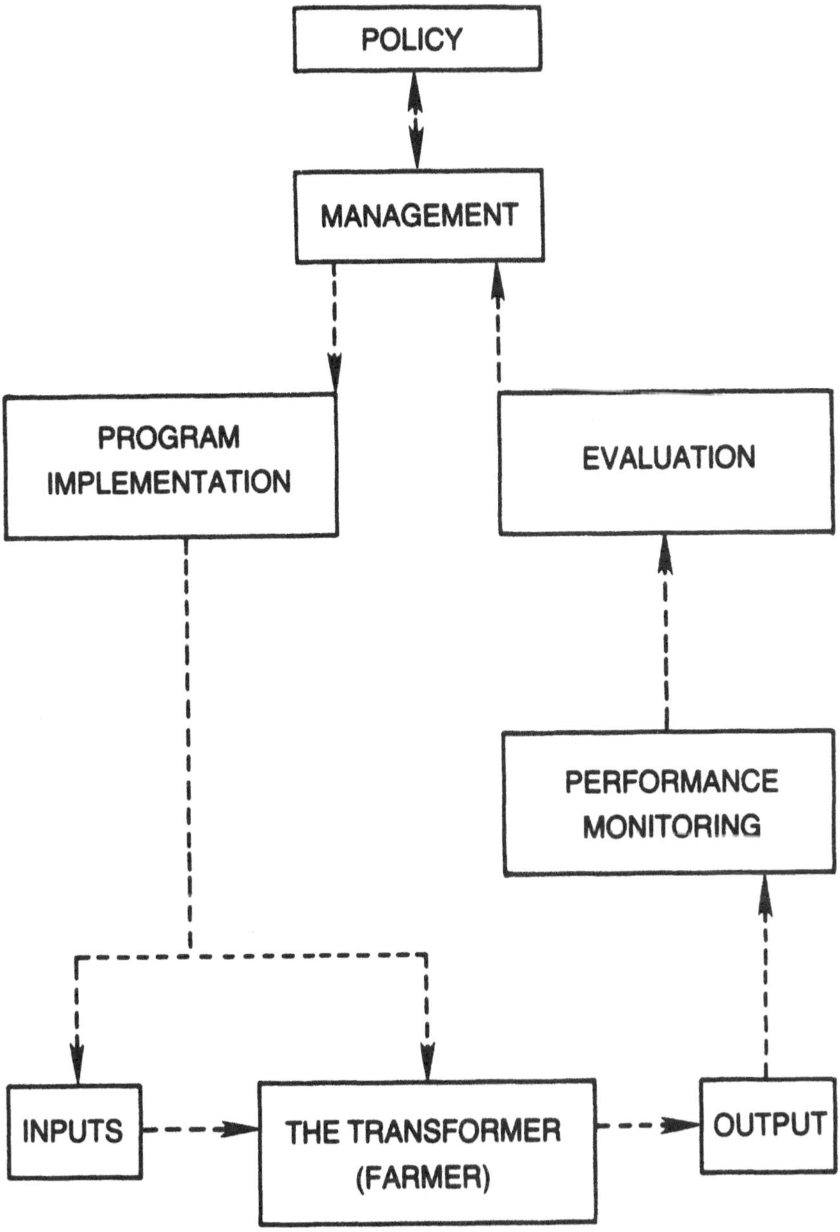

Figure 12.3 Structure of the managerial organization

had already been assigned to this project by the Pakistan central government. But considerable difference of opinion about the approach to use existed between central government policymakers, on the one hand, and the heads of key participating provincial government agencies, on the other. Central government spokesmen were leaning in favor of adopting the India approach, as embodied in command area development authorities (CADA). In the CADA approach, coordination is achieved by creating single-project management agencies that take over the normal functions of departments of irrigation and agriculture (particularly extension) within specified project areas. Conversely, at the provincial level, such agency heads in Pakistan favored retention of control over their traditional functions, e.g., irrigation canal water and extension deliveries. The institutional framework for the Pakistan CWM Project proposed by the World Bank/USAID Appraisal Mission was structured so as to mesh workable components from these two diverging points of view (Fairchild et al., 1983a). Specifically, line agencies such as Irrigation and Extension were to retain their traditional functions but would be required to respond to management decisions from CWM subproject managers. These managers would obtain decision-making leverage over the line agencies working in their project areas via reporting to a provincial policy committee composed of the Chief Additional Secretary for Planning and Development (Chairman) and three other key agency heads--the Secretaries of the Departments of Agriculture, Irrigation, and Finance.

The primary project management design objective was to induce farmers to increase agricultural output. As input suppliers, both public and private, make input delivery changes, response by farmers, as measured by output performance, would be evaluated and translated into delivery signals to input suppliers. As noted in a consultant report to the Pakistan USAID Mission in Islamabad (Nobe, 1983):

> The fundamental concept underlying the Command Water Management (CWM) Project is that a desired objective for an irrigation command area (e.g., all area to be irrigated from a single canal distributary, with or without supplemental tubewell water) is to increase agricultural output through management of water <u>and nonwater inputs</u> available to farmers in the command area.

> The process envisioned is to alter farmers' behavior so that the timing, sequence, and level of inputs result in an increased value of output which exceeds the costs of those inputs to the farmer. In essence, then, the focal point for the CWM Project is <u>management</u>.
>
> . . . And, that is the objective of the CWM Project--to take an essentially limited set of inputs, particularly water, and to prioritize and "mix" water with the nonwater inputs in such a manner that more agricultural output will result than if these agricultural inputs were to continue to be supplied in an uncoordinated manner.

The original design of the institutional framework for the Pakistan CWM Project corresponded to a remarkable degree to that of an MBR organization, as shown previously in Figure 12.3. Figure 12.4 shows the CWM Project's proposed institutional linkages, using the proposed organizational charge for the subproject in Sind Province as an illustration (Fairchild et al., 1983b). The six major MBR functions, as identified in the project's organizational management structure and as agreed to in principle by the provincial and central government policymakers, were described in a Pakistan USAID Mission/Islamabad consultant's report (Nobe, 1983) as follows:

1. <u>Policy</u>

   Each province has agreed to establish a Provincial Policy Committee, to be chaired by the Chief Additional Secretary, Planning and Development Department (or Board). The Secretaries of Agriculture, Irrigation, and Finance are members of the Committee in all provinces, and in Baluchistan, the Secretary of the Livestock Department is an additional member. The policy function is to set broad parameters for the CWM subprojects in each province. The parameters are set at the policy level, while the variables are left to the managers. In military terms, by analogy, policy is equivalent to "strategy," while management is responsible for "tactics."

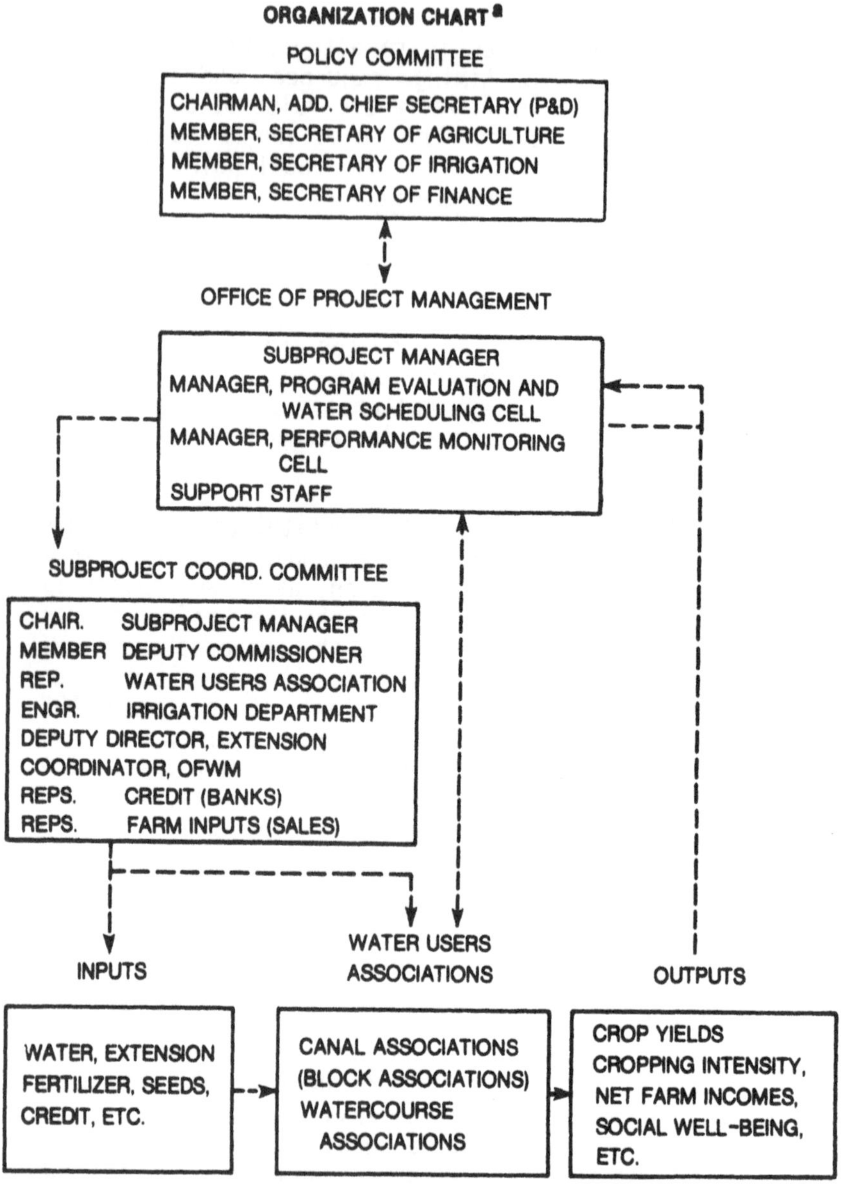

[a] DOES NOT INCLUDE INSTITUTIONAL ARRANGEMENTS FOR IMPLEMENTING AGENCIES (i.e., DEPARTMENTS OF IRRIGATION AND AGRICULTURE), WHICH WOULD FOLLOW CURRENT ORGANIZATIONAL STRUCTURE OF SUCH DEPARTMENTS AND AGENCIES.

Figure 12.4 Organization chart, Command Water Management Project at the province level (e.g., Sind)

2. Management

An Office of Management will be established for each subproject area, with each subproject manager responsible to the respective Provincial Policy Committee Chairman. Operating under an MBR approach, one primary function of the subproject manager is to transform general policy goals specified by the Policy Committee into specific objectives, defined as desired project area outputs at specific times, places, quantities, and qualities. The second major management function is to design, prioritize, and schedule program elements which are expected to produce the output package specified in the objectives. Since monitoring and evaluation will provide continuous information about how well the project elements are doing, the manager will be in a position to adjust program inputs, or 'fine tune' the project efforts, over time, by scheduling water and nonwater inputs in varying proportions. Since the Las Bela subproject is a new irrigation area, limited basic and farm-level research activities should be added to that Office of Subproject Management.

3. Evaluation and Scheduling

To a degree, the evaluation function is shared by the subproject manager and the policy committee. The latter may only engage in periodic evaluations, while the manager does so continuously as a basis for carrying out his input scheduling function.

4. Program Implementation

Each province will have a Program Coordinating Committee for each subproject, chaired by the subproject manager. Membership would include representatives of participating line agencies (e.g., Irrigation and Agriculture), suppliers of nonwater inputs and credit, and representatives of local farmers' water user organizations. Delivery of the total mix of water and nonfarm agricultural inputs would necessarily emanate from membership of the Coordinating Committee.

5. Input-Output Transformation

The farmers in the subproject areas are the ultimate transformers of agricultural inputs into desired project outputs. As a condition for receiving project assistance, farmers' water user associations will be established at the watercourse level in all subproject areas, initially to provide the labor and some cash cost-sharing of watercourse improvements. Over time, however, such farmer groups will be the contact point for Extension workers and they can work jointly to receive credit and nonwater input deliveries by demonstrating sufficiently large chunks of demand to attract suppliers. Ultimately they can offer their produce to the market in like manner.

6. Monitoring

Basic to the MBR approach is the monitoring of project performance in terms of outputs emanating from the total project efforts, in order to provide more timely, reliable and sufficient supplies of water and nonwater agricultural inputs. Therefore, so that management will know the results of its efforts (e.g., MBR), not only increased water yield, but also increasing cropping intensities, crop yields, net farm incomes, and improved well-being will be monitored. The monitoring activity will be under the direction of the subproject managers and will be carried out in part by their own staffs, area extension personnel, and, in the case of Punjab and Sind provinces, with assistance from personnel in the evaluation cells in Planning and Development Departments or Boards (e.g., The Punjab Economic Research Institute in Lahore).

Although the MBR approach to designing institutional arrangements for Pakistan's CWM Project was agreed to, in principle, by Pakistan government policymakers in October 1983, further program modifications were made as a result of Pakistan and World Bank financial negotiations, which extended through May 1984. Initially, modifications were made in the organizational chart (but not in the basic MBR principles) so as to clarify intent for Pakistan government policymakers and World Bank and USAID administrators. The modified organizational chart which appeared in the final World Bank Staff Appraisal Report (1984) is shown in Figure 12.5.

During final project financing negotiations held at World Bank headquarters in Washington, D.C., in May 1984, the Pakistan government requested and was granted an important modification in the organizational structure for managing participating subprojects in Punjab Province.

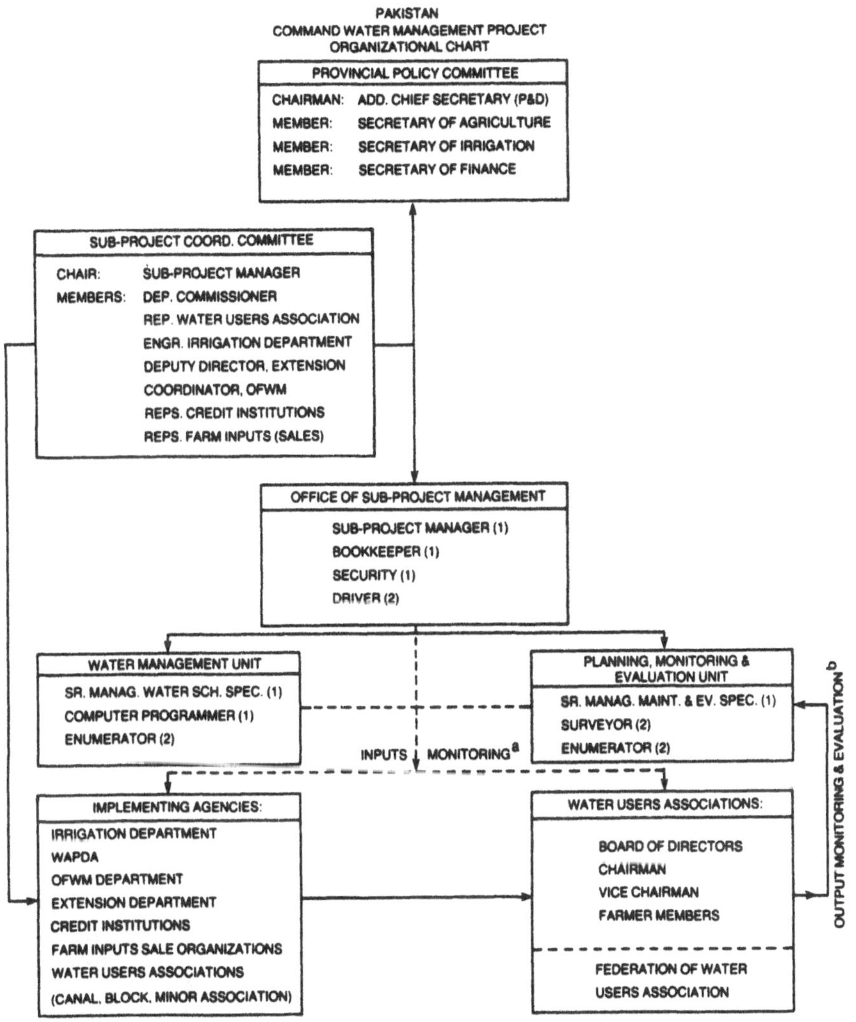

Project Management Information System

[a] Will monitor project activities (in physical & financial terms) such as staff recruitment; procurement; implementation of civil works; input distribution & usage, e.g. irrigation water, fertilizer, seeds, pesticides; & provision of agricultural services, e.g., extension & credit

[b] Will monitor & evaluate, among others, increased input distribution, services, crop yields, intensities, net farm income, employment, etc.

Figure 12.5 Pakistan Command Water Management Project organizational chart

Unlike the other three provinces of Sind, Northwest Frontier, and Baluchistan, each of which each has a single participating geographic project area, the Punjab Province has four participating subprojects, geographically separated by considerable distances. In the case of Punjab Province, therefore, the Pakistan central government requested that a single-project management office be established in Lahore and that all field programs be under the in-line supervision of the Secretary of Agriculture. The Chief Additional Secretary of Planning and Development would, however, continue to chair the Provincial Policy Committee.

These requested modifications were agreed to by World Bank negotiators on condition that, in addition, small scheduling and monitoring cells be established at each of the four participating project sites. While the needed modifications to USAID's original project paper to incorporate the above-listed changes were minor, its revised project paper, originally promised for completion in June 1984, had not yet been completed by the start of the agency's next fiscal year on October 1, 1984. Therefore, in retrospect, it is clear to us that modifying the Pakistan institutional framework to embody an MBR approach in the CWM Project will be far easier than speeding up the USAID bureaucracy in its project modification and approval process for its funding involvement in this project.

ANALYSIS OF INITIAL PROJECT PERFORMANCE

Since the Command Water Management Project was only recently approved (June 13, 1984), it is too early to make any definitive analysis of project performance. However, there have been several reportable activities and, in addition, there are numerous critical and sensitive matters already identified that will require constant monitoring.[10] These are as follows:

Project Preparation and Appraisal

Project design preparation was done by WAPDA, with the assistance of USAID-supplied experts. Because of WAPDA's considerable experience in this subject area, selection of it for project preparation remains a logical decision. However, two improvements could be suggested for the next project preparation: (1) supply additional expatriate experts--especially in the areas of institutions, management, and training and (2) WAPDA should be encouraged to have closer working relations and coordination with the provinces. Nonetheless, both World Bank and USAID officials appeared to be generally pleased with WAPDA's effort, as reflected in their joint appraisal of the project conducted in October 1983 (WBSAR, April 30, 1984).

Project Effectiveness

Originally, the credit effectiveness date for this project was September 11, 1984. Because of delays in processing the project agreement for USAID's co-financing and approval of PC-1s (pro formas--authorizing GOP documents), this date was then extended to December 7, 1984. Delay in credit effectiveness because of the delay in USAID's approval for project co-financing did not prevent start-up of the project on July 1, 1984, as scheduled. But it does prevent IDA disbursement for project-approved activities until the total credit component becomes effective. USAID draft project agreements were scheduled to be available for GOP and GOProvincial review no later than September 1984, but the timetable was further delayed. USAID officials have not stated that signing of these documents should be possible in December 1984[11]. Therefore, project effectiveness by the extended date of December 7 still appeared possible but not probable.

Local Financing

GOP and GOProvinces have prepared PC-1s consistent with the estimated costs included in the Bank's staff appraisal report. Therefore, commitment and follow-through by GOP and GOProvinces was highly critical. This potential constraint was removed when the PC-1s were approved by the Executive Committee of the National Economic Council, GOP, on October 24, 1984. Considerable effort was expended by World Bank to develop a workable project-management organizational framework that was acceptable to GOP and GOProvincial personnel. Time will tell as to their commitment and support of this institutional modification effort. In the meantime, it is important that expatriate assistance, such as institutional experts, be made available to the project for leadership development and training efforts. As noted earlier, implementing of the continuous monitoring and evaluation program is absolutely essential to the success of this project, which is based on an MBR concept.

Role of Water User Associations (WUAs)

Formal WUAs in Pakistan have only been in existence for about three years. But, so far, their role has been limited to installation and maintenance of renovated watercourses in the On-Farm Water Management Project (Cr. 1163-PAK). It is too early, however, to determine their effectiveness based solely on these activities. In the CWM Project, additional responsibilities are being placed on these local farmer organizations. There remains much nurturing of WUAs if they are to act as a vehicle for greater farmer involvement in the planning and implementation of this project. The subproject management offices and provincial OFWM directorates must be willing to spend an inordinate amount of time with the WUAs during the early phases of this project.

Technical Assistance and Training

Generally, USAID is co-financing this program component. They are authorizing initial temporary duty staffing of technical assistance from an earlier negotiated project (Irrigation System Management Project). This should be helpful. In addition, 255 man/ months of expatriate technical assistance are to be provided under the CWM Project. The GOP agreed to the 252 man/month level in the CWM Project with the proviso that, after nine months of operation, the program will be reviewed for the purpose of making any mid-course correction, as appropriate. The World Bank, on the other hand, is fully committed to this level of technical assistance and views this scheduled project review as an opportunity for expansion of the technical assistance component. In the meantime, USAID and the Bank must ensure that the very best available expatriates are assigned to this project.

Management Organization

The most critical element that can spell success or failure of this project is the effectiveness of the proposed project management organization. This organizational framework, based on an MBR concept, is an untried pilot approach in Pakistan. This institutional framework will only be as effective as the commitment of provincial officials in training and encouraging the subproject management personnel and the farmers' water user associations to employ it properly.

LESSONS LEARNED AND RECOMMENDATIONS FOR FUTURE PROJECT DESIGN AND IMPLEMENTATION

Given 20/20 hindsight as a result of our recent experience with the Pakistan CWM Project, we have learned some important lessons, which should also be of value to others who will be involved in designing and implementing irrigation projects in the future. Our observations and recommendations can be grouped under six major categories: (1) improving project feasibility studies, (2) use of joint World Bank/USAID Project Appraisal Missions, (3) more lead time for the project approval process, (4) meshing of MBR project staff with existing host country irrigation and agricultural agencies, (5) arranging for training of host country project managers, and (6) providing for an intensive on-farm water management training program for participating extension workers. Each of these will be discussed briefly in turn.

Improving Project Feasibility Studies

Our recent experience with the Pakistan CWM Project has strongly reinforced our view that irrigation projects will not function properly unless they are based on both sound engineering design and modern management principles. In this case, project preparation was carried out by WAPDA as primary consultants to the Ministry of Water and Power (MWP). The project feasibility report was released in August 1983 (MWP, 1983). Funding for this effort was provided under a UNDP project, with the World Bank as the executing agency and with expatriate experts provided by USAID to assist WAPDA. The CWM Project proposal had emerged as a major recommendation in an earlier WAPDA report (WAPDA, 1979) and had been included by the government of Pakistan in its Sixth Five-Year Plan--FY84-88 (MPD, 1983).

The basic WAPDA planning and design documents for the CWM Project are replete with shortcomings, as critics have been quick to point out. Among these were (1) inadequate or inaccurate data sets, particularly as they related to irrigation project delivery problems and leading to some cases of poor engineering design; (2) a reluctance to seek out and utilize data and judgments from within provincial-level irrigation and agricultural departments, agricultural colleges and research institutions, and farmers in the designated project areas; and (3) a proposed project management framework that was viewed by some as primarily self-serving to WAPDA interests.

Critics have suggested that these kinds of problems could have been avoided if the engineering design and project preparation assignment had been granted to an expatriate team composed of highly qualified technicians and institutional development specialists. We strongly disagree with such a proposal because it misses the basic point of how effective project design and implementation can be achieved in developing countries. The key point is that, ultimately, regardless of who designs and funds such projects, they will become the responsibility of the host government upon completion. We feel strongly that such projects will be more readily accepted as a host government responsibility if its irrigation-oriented agencies (e.g., WAPDA, PIDs and Agriculture Departments) are responsibly involved in the project design and implementation process. In like manner, we feel strongly that successful public irrigation projects will require farmer participation in the total process, ranging from project design to implementation and management to cost-sharing.

What steps would we have taken to improve project design? First, we would have taken great pains to provide WAPDA with the best interdisciplinary expatriate assistance (not management control) available. Rather than resorting to the competitive bidding process which USAID used to recruit its technical assistance input, we would have obtained a predominant capability waiver and then proceeded to develop a list of qualified expatriates with prior field experience in working with irrigation issues in Pakistan. Over the past twenty years, many such personnel were employed in Pakistan by Harza Engineering International, Inc., Tipton and Kalmbach, Inc., Colorado

State University, and other agencies and firms that could be identified. Personnel to be recruited for the expatriate team to assist WAPDA in the feasibility study would then have been selected from this list, regardless of their present professional affiliation. In our opinion, if USAID had used this approach, it would have had the expatriate team in the country much earlier than actually occurred and the team members would have been the best-qualified personnel available.

A second step that we would have taken at the outset would have been to have the Minister of Water and Power or his designee meet personally with key agency heads at the provincial level to provide the terms of reference for the proposed feasibility study. Specifically, it should have been made clear that responsibility for project planning and design would rest with the central government and that implementation and management of the project components would be the responsibility of the provincial governments. Above all, it should have been made clear that, while WAPDA had been delegated responsibility for preparation of the feasibility report, it would not have a major follow-up role in project implementation and management. Further, the central government's relationship with one or more international funding agencies involved--including respective areas of responsibility, time frame, and related matters--should have been set forth and agreed upon in advance of the project design phase. A clear demarcation of responsibility for all participating agencies should be made at the outset of any project, regardless of which developing country is involved.

Use of Joint World Bank/USAID Project Appraisal Missions

From the outset, it was mutually understood that funding for the Pakistan CWM Project would be a joint World Bank and USAID effort. In a broad sense, it was agreed that World Bank would provide funds for construction and equipment for the canal modification and drainage components, while USAID would primarily fund the on-farm water management component, technical assistance, and training, along with some equipment. Each agency proceeded independently to obtain funding approval for their portions of the total program, and, had both agencies remained on the same time frame, their joint planning and implementation efforts could likely have remained only loosely coordinated. As it turned out, however, while World Bank remained on schedule, USAID lost its "window of opportunity" to lock in funds during FY84-85 for its areas of responsibility. Rather than delay the total effort until October 1, 1984, when presumably USAID funding would again become available, World Bank then invited USAID to participate in the World Bank's project appraisal mission scheduled to be in the country during October 1983. USAID accepted this invitation and a joint World Bank/USAID team successfully carried out the field appraisal phase on schedule. This effort, in turn, provided the necessary information for preparation of the World Bank's staff appraisal report (WBSAR, 1984). This document was used as the basis for the final World Bank/Government of Pakistan project negotiations in May 1984, which led to World Bank Board approval of the project (on schedule) in June 1984, for implementation on July 1, 1984.

Within USAID, the project approval and implementation phases were scheduled on the following timetable: Some of the USAID participants in the appraisal mission would finalize a CWM Project Paper, drawing on relevant sections in the World Bank Staff Appraisal Report, and submit it to USAID's contract office in Washington, D.C., by June 10, 1984. USAID/Washington personnel, in turn, would obtain project approval and meet the conditions of project effectiveness by September 10, 1984, and would prepare a request for proposal (RFP) to be circulated among prospective universities and/or private consulting firms before that date. It was expected that a university and/or private consulting firm participant would be selected no later than October 1, 1984, when funds in USAID's new fiscal year budget would become available to begin a major training program for host country participants. Finally, it was expected that the selected expatriate team would be in the country no later than January 1, 1985.

Unfortunately, because of a series of administrative constraints that need not be detailed here, USAID was not able to hold to its timetable. As of late October 1984, the Project Paper had not yet been forwarded to USAID/Washington for approval and review by the GOP and the provinces, and the date for project effectiveness had been advanced to December 10, 1984. As a result, it was by then highly unlikely that USAID would have its expatriate team in the country before July 1985--a full year after the World Bank Board had approved its funds for project implementation! Needless to say, these unforeseen delays have been extremely frustrating to all parties concerned, but particularly so to Pakistan government officials who were anxious to proceed with project implementation at the earliest possible date after approval of its funding documents (PC-1s), which were approved on October 24, 1984. Although the joint World Bank/USAID appraisal mission came into being by accident rather than by design, it nonetheless did appear to have worked reasonably well. (The major problem with the joint planning and implementation effort developed only after the appraisal phase had been completed.) Therefore, we recommend that, in instances where joint funding is anticipated for future projects, joint World Bank/USAID appraisal missions should be used.

Planning for Adequate Lead Time for the Project Approval Process

Full implementation of this $80-M project is likely to be delayed a full year or more as a result of a breakdown in USAID's timetable for project approval. Unless this timing problem is resolved, the likelihood of future joint-funded World Bank/USAID irrigation projects will be considerably reduced. One option to consider is having USAID start its approval process a year or so in advance for such joint-funded projects.

Meshing Project Staff With Existing Host-Country Irrigation and Agricultural Agencies

Without detracting in any way from the technical and administrative problems that have developed with the CWM Project, we nonetheless feel that the most important and delicate aspect of project design dealt with formulating and gaining acceptance from the government of Pakistan for some institutional modifications needed to effectively manage this project. With full knowledge that it would be impossible to refine the entire government civil service system to accommodate our proposed MBR approach, we proceeded instead to design a project management framework that would retain the primary responsibilities of the existing irrigation and agricultural line agencies. At the same time, however, "clout" was provided for the subproject managers to call for changes in the input delivery systems in response to management decisions, as reinforced with results of monitoring system outputs. At the outset, we were prepared to accept a wide variation in the institutional framework that would be found acceptable among the central and provincial government agencies involved. Therefore, we were gratified and, indeed, surprised at the degree to which the institutional framework which was accepted by the GOP negotiators, as shown in Figure 12.5, resembled the theoretical framework shown in Figure 12.3.

In retrospect, we feel that our proposed project managerial framework was accepted primarily because the planning and oversight functions of central government, on the one hand, and the implementation and management functions of provincial governments, on the other hand, were retained. In the final analysis, the only major change that we introduced into the existing governmental organizational framework was a mechanism by which these organizations could focus on achieving desirable levels of project outputs instead of merely being content with monitoring their respective inputs. In essence, by providing a mechanism for responsible agencies in government to know the results of their actions (by monitoring outputs instead of only delivering inputs), these traditional administrative-type organizations are now evolving into management-oriented organizations that can employ the MBR approach.

As noted earlier, central government decision makers were already convinced that organizational changes were needed because their existing irrigation systems simply were not achieving expected agricultural output targets. Therefore, they readily responded to our MBR proposal, once we had convinced them that adoption of a self-contained organization approach of the India CADA type was not the only option available for needed organizational change. At the provincial level, irrigation and agricultural agency administrators responded because the threat of having a self-contained agency that would duplicate their agency efforts had been removed.[12] For them, giving up some degree of self-control of their delivery systems in order to respond to delivery signals emanating from an external source (e.g., from the subproject managers) was preferred by far to having their agency functions totally removed from the project areas.

Whether the MBR mechanism can be successfully implemented in this project effort remains to be seen. To a large degree, success or failure will be dictated by the degree to which project managers can be trained to manage instead of merely administer and by the degree to which extension workers can be trained to provide relevant management advice to farmers. Assuming these training objectives can be achieved, then the institutional framework introduced in this pilot project effort can be replicated throughout the irrigated areas of Pakistan and in other developing countries as well. And, this can be achieved without having to face the difficult, if not impossible, task of radically restructuring an existing civil service system or superimposing self-contained project personnel cadres (e.g., CADA) upon the existing system.

There is no denying the fact that this project is devoting a greater share of the total project budget to institution building and management training than has been the case in other World Bank projects with which we are familiar. But, since we truly believe that deficiencies in the management facet of the production process are the primary cause of the poor performance level of most irrigation development efforts during the post-World War II era, we would argue that the potential benefits to be gained from improving project management will far outweigh the extra costs entailed.

We believe that, in the longer term, it will be far more cost effective to manage projects in such a manner that they will be productive indefinitely rather than to continue to put huge amounts of capital funds into construction efforts for poorly managed projects that will fail within a few years and then have to be redone. Government decision makers in developing countries and in international donor agencies alike are encouraged to monitor the Pakistan CWM Project as it unfolds to ascertain whether our initial optimism for project success merits their further attention. We certainly think this project will succeed, but it will take longer than we had originally anticipated.

Training of MBR-Oriented Project Managers

Within the organizational framework for irrigation projects in Pakistan, the term "project manager" (or indeed, even "project administrator") was not being used. Traditionally, such positions were always filled by engineers, so the term that usually appeared at the top of their project organization charts was simply "chief engineer." We might well argue between ourselves about whether an engineer or an economist might be best suited for a project manager position, and others have argued that it should be an agronomist or even a person from the elite administrative cadre. Instead, however, we have identified such positions as nondiscipline specific (although, most likely, most of the first subproject managers will still be drawn from the engineering ranks of irrigation departments). Our position description for the project manager slots simply specified that incumbents must have organizational and management skills sufficient to operate MBR-oriented systems.

Such personnel are in extremely short supply in Pakistan (and indeed in all developing countries), and those that are present are usually gainfully employed at highly competitive wage rates in the private sector. The only solution to this dilemma is to engage in a retraining program to make MBR-oriented managers out of selected personnel previously trained in specific disciplines, such as engineers, agriculturalists, economists, and the like.

During the project design phase, we recommended to USAID decision makers that they provide funds for such training far enough in advance of the expected project implementation date so that even the first set of project managers would be adept in modern management skills. For a number of reasons, these training efforts will be delayed at least until early 1985. In our view, the training of project managers should have been started as soon as it became apparent that donor funding would likely be approved for the CWM project, which in turn would have required the provinces to identify such prospective subproject managers at an early date. Project approval appeared to be highly probable by late 1983--in time to have allowed for a selected cadre of project management personnel, perhaps 10 to 15 persons, to be placed in an intensive project management training program, even if the selected participants would have been required to go abroad for such training. Yet, by late 1984, only one subproject manager had been identified (for Las Bela in Baluchistan Province), and the subproject management training component was still in the design stage. In sharp contrast, the newly created India Irrigation Water Management Training Project, a training program in the U.S. for the first group of 20 persons who will staff in-service training institutes to be established under that project, began in June 1984--about a full year in advance of the expected implementation date for the India project.[13] Clearly, we strongly favor an early training approach as used by the India/USAID Mission over the delayed response that has emanated out of the Pakistan/USAID Mission.

Providing On-Farm Management Training for Extension Personnel

Developing an extension cadre for service in the selected project areas, skilled in delivering on-farm management information and assisting farmers in establishing the water user associations that are required as a caveat in the World Bank funding authorization, will also require a concerted retraining effort. The task is further complicated by the large number of personnel who will require such training. Fortunately, there is presently a facility in Lahore that engages in training extension personnel for use in implementing the ongoing nationwide On-Farm Water Management Project. Its administrators were willing to assist in training extension personnel for the CWM Project. Initially, the core of the extension staff needed for the CWM Project could have been adequately trained at this facility, with only minor curriculum modifications. That program could then have served as a prototype for setting up additional training facilities elsewhere in Pakistan, as now planned for under a separate USAID-funded On-Farm Water Management Training Project. As in the case of the need for creating a training program for project managers at an early date, the same recommendation was made to USAID for training extension workers. USAID, however, has delayed full implementation of this recommendation. Even if it were now to be initiated immediately under its related, fully funded training project, graduates in sufficient numbers to make a significant difference in the subproject areas is at least a year away.

Any hope that the CWM Project would remain on schedule and achieve early success has already been constrained by the undue delay in proceeding with meeting its personnel training needs. The failure to make a timely delivery on this key ingredient for project success (one which, paradoxically, requires an almost insignificant amount of total project funds) brings to mind the infamous allusion, "for want of a nail...the battle was lost." At this point in time, however, we can only plead with agency policymakers responsible for future irrigation projects in Pakistan and other developing countries not to make the same mistake of delaying needed manpower training efforts.

CONCLUDING COMMENTS

Seckler and Nobe (1983, p. 291), in discussing the problems likely to be associated with converting administrative organizations into management organizations in developing countries, had forecast most of the difficulties that we have now experienced with the Pakistan CWM Project design and implementation efforts. They stated, in part:

> The problem of instilling MBR techniques in traditionally focused public service organizations has no simple or quick answer. . . . (Nonetheless) a major revolution in such rigid systems is in order. What appear to be necessary are three factors. First, the man at the top must be sufficiently powerful and imbued with the MBR philosophy. Second, there must be organizational changes to model systems of information and control in cybernetic and scientific terms. Third, personnel throughout the whole organization must be exposed to and understand modern management principles before lasting changes can be achieved.
>
> In sum, while Drucker is right in describing management as a 'culture' not a 'technique,' this statement should not be interpreted in the sense of 'cultural relativism'--that, so to speak, any management system 'appropriate' to the local culture will do. On the contrary, while management 'styles' may differ from place to place in accord with local standards of interpersonal relationships, the culture of management itself is absolute and universal; it is the objective culture created by and necessary to MBR. Thus, while every organization or society always has the choice of whether it wants to manage effectively or not, once that choice is made, it does not have a choice over which kind of management that culture must create. The conflict between the cultural demand of the MO (Management Organization) and traditional cultures as expressed in . . . [existing administratively oriented organizations] . . . is perhaps <u>the</u> basic cultural conflict in the developing world today.

In addition to the problems and difficulties alluded to above, we were faced with one unexpected and serious additional problem. Relative to achieving desired organizational changes within the participating Pakistani agencies, effecting the necessary corresponding changes in USAID's Pakistan Mission modus operandi was a monumental task that has not yet been satisfactorily resolved. Presently, this Mission remains primarily a rigid AO, while the relevant Pakistani agencies are beginning to evolve into MOs. As Seckler and Nobe also wrote (1983, p. 285): "The 'litmus test' to determine the degree to which any particular organization is 'managed,' versus simply 'administered,' is the degree to which the organization knows the results of its behavior." In this context, we found that the USAID Mission in Pakistan simply was not yet ready to modify its procedures so that it can "know the results of its behavior" (e.g., undue delay in its CWM project approval process). This situation is a serious obstacle to implementing the project in a timely manner, given that USAID funding approval is a condition of effectiveness for the total World Bank/USAID-funded project effort (World Bank, 1984, p. 41).

Assuming that the USAID Mission in Pakistan is not all that unique within their system, it is our conclusion that others who may be interested in applying the MBR approach to irrigation projects in developing countries will necessarily have to include USAID (and/or other relevant donor agency decision makers) in the MBR indoctrination process. Without that extra effort, the chances of achieving success in such endeavors are likely to be considerably reduced--or at best, unduly delayed.

NOTES

1. World food consumption increased from about 850 mmt in 1960 to about 1400 mmt in 1979, or about 3 percent annually (Food data source--USDA).

2. Yet, it is estimated that the number of people consuming below the FAO/WHO recommended caloric intake has increased from 900 M in 1964 to 1,100 M in 1972-74.

3. For a comprehensive treatment of the CWM Project, refer to the World Bank Staff Appraisal Report (1984).

4. See, for example, The P.I.D.E. Economic Model of Pakistan's Economy (FY60-79), Pakistan Institute of Development Economics (PIDE), Islamabad, 1982; summary of field

surveys in Draft Sixth Five-Year Plan: Role of Fertilizer, National Fertilizer Development Center, Islamabad, January 1983; K. C. Nobe, An Overview of Pakistan's Current Agricultural Development Policy Options (USAID Contract Report), Islamabad, Pakistan, December 1982; and Forrest E. Walters, Pakistan Fertilizer Policy: Review and Analysis, USAID Contract Report by Chemonics International, Washington, D.C. and Chemical Consultants, Ltd., Lahore), Islamabad, Pakistan, September 15, 1984.

5. The lowest order command, covering (on the average) about 400 ac and 35 farm units.

6. The watercourse is the communal irrigation delivery facility within each chak.

7. Indus Basin Salinity Survey, WAPDA, 1981, indicates that 22 percent of the Indus Basin (about 9.0 M ac) has a groundwater table within 6 ft of the surface and 42 percent (about 17.0 M ac) within 10 ft.

8. A systematic rotational schedule of irrigation deliveries to farmers served by a single watercourse.

9. Warren Fairchild of the World Bank served as team leader for the joint World Bank/USAID appraisal mission. Other World Bank team members were G. Lituma, C. P. Cheng, J. Mohamadi, U. Qamar, and A. Colcini. USAID provided the services of R. Backus and P. Muligan of their direct-hire staff; L. Busch, W. Andrews, and K. C. Nobe served as consultants.

10. The authors returned to Pakistan in October 1984 to carry out an initial project supervision for the World Bank and to assist the GOP in implementing the CWM Project (Fairchild et al., 1984). Some of the bottlenecks alluded to above were partially alleviated as a result of that mission effort, but other constraints related to USAID involvement, such as the conditions of project effectiveness, were not resolved at that time (see section on "Lessons Learned").

11. As of late October 1984, the USAID project paper for the CWM Project had not yet cleared the Office of the Mission Director, USAID, Islamabad, and the date for its submission to USAID/Washington for contract approval had been advanced to late November. Final USAID project approval for its components are, therefore, likely delayed until February or early March 1985, at the earliest.

12. During the initial World Bank supervision mission carried out during October 1984, however, it became apparent that considerable confusion still exists at the provincial level as to how the subproject organizational

structure is to be implemented. Therefore, GOP administrators agreed that a CWM project orientation seminar for GOP and provincial line agency heads and donor agency administrators should be held in Islamabad in mid-January 1985.

13. These Indian participants, funded by USAID, were enrolled in a special seven-month, nondegree training program provided through the International School for Agricultural and Resource Development at Colorado State University, Fort Collins, Colorado.

REFERENCES

Ashby, W. Ross. An Introduction to Cybernetics. London, England: Methuen & Company, Ltd., 1976.

Dennison, Edward F. Accounting for United States Economic Amity 1929-69. Washington, D.C.: The Brookings Institution, 1974.

Drucker, Peter F. Management: Tasks, Responsibilities, Practices. New York: Harper and Row, 1974.

Fairchild, Warren (Chief of Party), K. C. Nobe, and others. Discussion Paper: Sehra and Naulakhi Subprojects, Command Water Management Project, Sind Province, Pakistan. Karachi, Pakistan: World Bank/USAID Appraisal Mission, October 23, 1983a.

Fairchild, Warren (Team Leader), K. C. Nobe, G. Lituma, C. P. Cheng, J. Mohamadi, A. Dolcini, W. Andrews, L. Bush. "Pakistan: Proposed Command Water Management Project," World Bank/USAID Appraisal Mission, Aide Memorie/Issues Paper, Islamabad, Pakistan, October 31, 1983b.

Fairchild, Warren (Leader), K. C. Nobe, and Usman Qamar. "Pakistan: Initial Supervision of the Command Water Management Project," World Bank Supervision Mission, Aide Memorie/Issues Paper, Fort Collins, Colorado, October 29, 1984.

Ministry of Planning and Development (MPD). The Sixth Five-Year Plan (FY84-88). Islamabad, Pakistan: Government of Pakistan, 1983.

Ministry of Water and Power (MWP). Feasibility Report: Command Water Management Project. Islamabad, Pakistan: Water and Power Development Authority (Consultants), Government of Pakistan, August 1983.

National Fertilizer Development Center. Draft Sixth Five-Year Plan: Role of Fertilizer. Islamabad, Pakistan: NFDC, 1983.

Nobe, K. C. An Overview of Pakistan's Current Agricultural Development Policy Options (CID Contract Report for USAID). Islamabad, Pakistan: Office of Agriculture and Rural Development, USAID, December 1982.

Nobe, K. C. Organization and Management of the Command Water Management Project (Contract Report for the Agency for International Development). Islamabad, Pakistan: USAID Mission, November 1, 1983.

Nobe, K. C. "Organizational Constraints to Greater Extension Involvement in AID-Funded Agricultural Programs in Lesser Developed Countries," Knowledge Transfer in Developing Countries: Status, Constraints, Outlook, J. B. Clair and L. W. Watts, eds., proceedings of a Conference on International Extension at Steamboat Springs, Colorado, July 1983), INTERPAKS, Office of International Agriculture, University of Illinois at Champaign-Urbana, Illinois, and Office of International Programs, Colorado State University, Fort Collins, Colorado, 1984.

Pakistan Institute of Development Economics. The PIDE Economic Model of Pakistan's Economy (FY60-79). Islamabad, Pakistan: PIDE, 1982.

Seckler, David W., and K. C. Nobe. "The Management Factor in Developing Economies," Chapter 12, Issues in Third World Development, K. C. Nobe and R. K. Sampath, eds. Boulder, Colorado: Westview Press, 1983.

Walters, Forrest E. Pakistan Fertilizer Policy: Review and Analysis (USAID Contract Report by Chemonics International, Washington, D.C. and Chemical Consultants, Ltd., Lahore). Islamabad, Pakistan: USAID, September 15, 1984.

Water and Power Development Authority (WAPDA). Revised Action Program for Irrigated Agriculture. Lahore, Pakistan: WAPDA, May 1979.

Water and Power Development Authority (WAPDA). Indus Basin Salinity Survey. Lahore, Pakistan: WAPDA, 1981.

World Bank. Pakistan: Command Water Management Project Report No. 4971-PAK. Washington, D.C.: South Asia Department, Irrigation I Division, The World Bank, April 30, 1984.

13 The Management of Paddy Irrigation Systems: A Laissez-Faire, Supply-Side Theory

David W. Seckler

INTRODUCTION

Peter F. Drucker, one of the pioneers of management science, urges managers to ask themselves continually: "What is our business? And what should it be?" These questions are basic to the specification of objectives, assignment of tasks, and monitoring of the results of effective management systems.

If these questions were asked of the management of paddy irrigation systems, a candid answer would have to be, "We do not know." As Wickham and Valera (1978, p. 61) correctly observe:

> While it is generally agreed that better water management is needed, it is not clear what is required to achieve it. What do we really mean by improved water management, and how can it be attained?

The problem is indeed worse than this, because it is not generally recognized that "We do not know," even in theory, what the objectives, tasks, and results of alternative management systems in paddy irrigation should be. Instead, there is a strong propensity to apply theories of irrigation management developed for other irrigated crops--maize, wheat, sugar cane--to paddy, as though the two were the same. But if there is one certainty in this field, it is that paddy irrigation is fundamentally different from that of other crops.

Paddy irrigation is different in at least two basic physical parameters that substantially affect the design and operation of irrigation management systems. First, in paddy, water is primarily stored on and drained from the surface of fields, whereas in other crops, water storage and drainage is in the subsurface profile of the root zone. Second, at least in the high-yield varieties (HYVs), paddy yields are ultra-sensitive to water stress, when compared with other crops. For all practical purposes, the paddy production function is like a binary switch--either on, with no water stress and high yield, or off, with water stress and drastically reduced yield.

The theory to be advanced here is that because of these two physical parameters, paddy irrigation systems have a self-regulating property that leads to a reasonably optimal allocation of water supply between farmers. Thus, in complete contrast to other crops, it is doubtful if management improvements in the form of rationing and rotation of water supply to farmers would result in cost-effective improvements over the allocation achieved by naturally functioning, laissez-faire systems.

For purposes of this discussion, this theory is restricted to the small (less than 1,000 ha), rivertine paddy irrigation systems characteristic of Southeast Asia and to the HYVs of paddy. Large systems in flat, alluvial plains differ mainly with respect to drainage effects, as discussed below. The deep-rooted, more drought-resistant "local" varieties of paddy form something of an intermediary case between the HYVs of paddy and other crops. These cases are not discussed further in this paper.

Of course, these conclusions regarding the allocative effectiveness of laissez-faire paddy irrigation systems should not be interpreted as denying the need for well-designed, constructed, maintained, and operated headworks and canal systems. However, it does appear that once water is delivered to something like a 10-15 ha block, the net returns to terminal systems, over field-to-field irrigation, may be negative. Hence the supply-side part of the theory. Instead of expending resources on improved management and terminal systems, it is likely to be more cost effective to build more and better main systems.

If this theory is correct, it will help explain why the few empirical studies of improved management in paddy irrigation systems generally show such poor results (see the review by Lazardo, Taylor, and Wickham, 1978). If the theory can be refuted, the refutation will help form the answers to Drucker's basic questions.

DIFFERENCES IN THE IRRIGATION PRODUCTION FUNCTION AND RELATED MANAGEMENT FUNCTIONS BETWEEN PADDY AND OTHER CROPS

Figure 13.1 illustrates a hypothetical production function for other crops. The point of maximum water-use efficiency (E), in terms of yield per unit of water, is defined by the tangency between the vector and the production function. The essential feature here is that E lies considerably below the point of maximum yield (M). Therefore, a major task of irrigation management in other crops is to ration irrigation water supplies to each and every user in the system so that they will produce at E rather than at M.

By inducing farmers to produce at E, both the objectives of efficient water use and a more equitable distribution of water are served. In the case of Figure 13.1, for example, if one set of farmers (A) changes production activities from M to E through rationing, the amount of water saved is WM-WE. The cost of the change to them, in terms of yield, is YM-YE. The amount of water saved by rationing can then be reallocated to another set of farmers (B)--who, it is assumed, had no water or production before. Figure 13.1 has been drawn, for simplicity, so that WM-WE = 1/2 WM. Thus, the gain to B will be WE amount of water, with a yield of YE. The net efficiency gain in the system is 2YE-YM. In addition, the number of beneficiaries in the system has doubled. In many cases, farmers may even use water beyond M to R--the point where yields are reduced due to waterlogging and/or salinity--or even beyond R. In this case, the benefits of rationing water (to A) and reallocation (to B) are even greater than shown above. Such are the enormous potential gains of improved irrigation management in other crops.

Of course, it is very difficult to ration water use from M or R to E in other crops. It is in the private economic interest of farmers at the head of the system to produce at M as long as the private marginal cost of water is zero. Rationing irrigation water requires a tightly managed and highly disciplined irrigation management system, backed by the legal powers of the state. This is why so few irrigation systems are well managed. But, at least in other crops, the objectives are clear, even if the practice is muddled. In paddy irrigation systems, oddly enough, it appears that the situation may be reversed. That is, the objectives of would-be managers may be muddled, but the results of actual practices in naturally

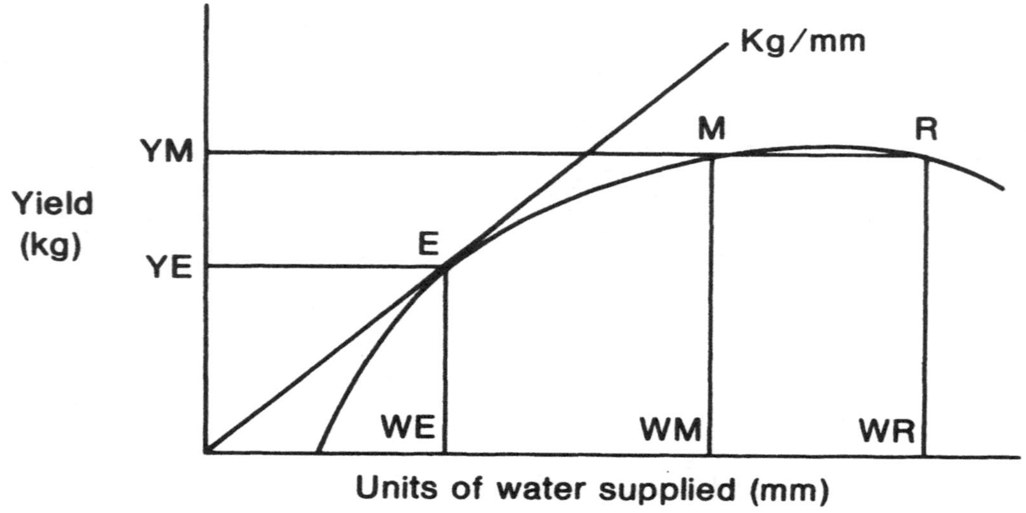

Figure 13.1 A hypothetical production function of other crops

functioning paddy systems may be as good as could reasonably be expected in an imperfect world.

Figure 13.2 (Thavaraj, 1978) shows a characteristic production function for paddy. The point of interest here is not in the specific values--which will, of course, vary from place to place--but in the shape of the function. For all practical purposes, this function divides into three linear segments. From 0 to 4 mm per day (pd), there is no yield response. Between 4-5 mm pd, yield jumps to 4.5 T/ha. Between 5-7 mm pd, yield increases linearly to 9 T/ha. After 7 mm pd, there is no yield response to increased water application.

The essential point is that the point of maximum water use efficiency (E) coincides, for all practical purposes, with the point of maximum yield (M). Thus, in paddy irrigation, unlike other crops, the management system does not have to ration water below M. At most, it only rations the surplus water from R to M. Thus, in principle, the rationing task in paddy is much easier than in other crops because it does not adversely affect the economic interest of the farmer.

The theoretical property that E = M in paddy irrigation has been empirically corroborated in a study by Wickham and Valera (1978), shown in Table 13.1. With no stress, at the maximum yield, kg/mm of water applied was maximized.

Since E = M, the most that a paddy irrigation management system would have to do is prevent over-irrigation--the application of surplus water which has no value to farmers. But here a rather obvious question arises: If the management objective in paddy irrigation is only to ration surplus water from R to M, why would farmers have R in the first place? The answer, of course, is that this water is only surplus in the static case, in terms of a point in time. In the dynamic case, over time, it is of value as reserve for future water supply. But if this is so, then the water is not truly surplus. Reducing water supply will increase the risk of a future loss to the user. This subject is addressed in the next section.

OVER-IRRIGATION IN OTHER CROPS AND PADDY

In other crops, the definition of over-irrigation is quite straightforward. It means application of water beyond field capacity (net of leaching requirements) with consequent deep percolation losses below the root zone of

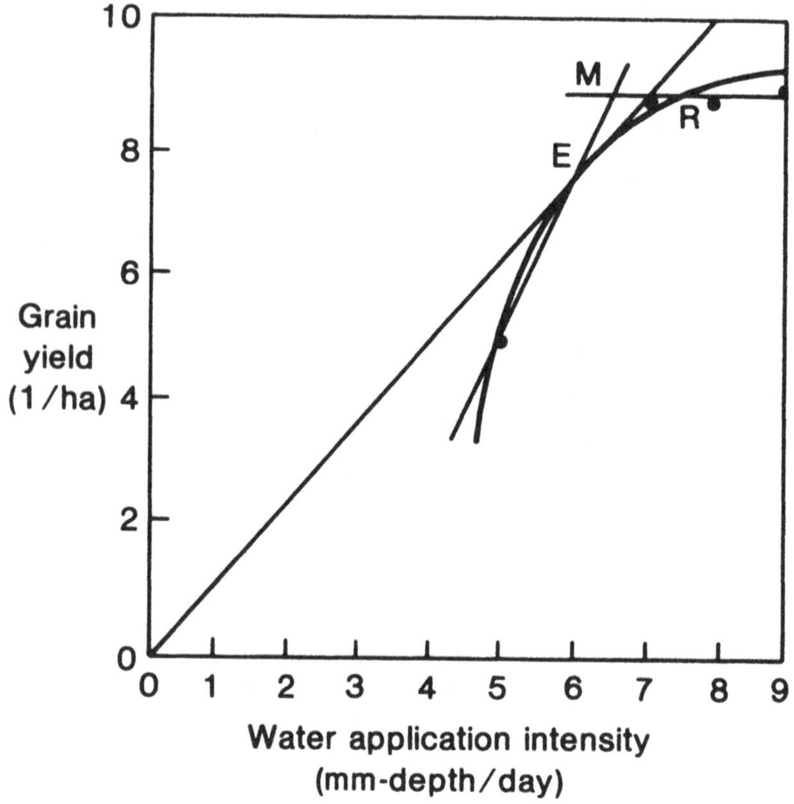

Figure 13.2 Yield response of IR8 to treatments of varied water application intensity, IRRI, 1969 dry season

Table 13.1 Water use and grain yield of IR20 under four water treatments, IRRI, 1972 dry season

Water treatment[a]	Days drained (no.)	Average yield (T/ha)	Water use[b] (mm)	Yield productivity of water (kg/mm)
No stress	0	6.2	773	8.1
Early stress	38	4.4	788	5.6
Late stress	39	2.0	806	2.5
Late stress to harvest	54	0.5	338	1.5

[a]No stress = flooded throughout crop growth; early stress = no irrigation from 43 to 81 days after seeding; late stress = no irrigation from 63 to 102 days after seeding; late stress to harvest = no irrigation from 63 days after seeding to harvest.

[b]From transplanting to about a week before harvest. Includes all rainfall.

the crop. Unless this deep percolation water is subsequently recaptured by pumping or downstream recharge, it is lost to beneficial use. In paddy irrigation, however, over-irrigation is extremely difficult to define, because once there is some standing water in a paddy field, neither percolation nor evaporation losses are significantly affected by additional depth of water in the field. If a farmer applies more water than he needs to, it will simply remain there for use in the future. Thus, in paddy--unlike other crops--there is no physical loss of water due to high rates of water application.

The only loss, then, is a possible economic loss due to excessive saving of water for future use. This economic loss can only be established by showing that the future value of water to a given farmer is significantly less than the value of the present use of water to other farmers. These trade-offs between future and present values form the rationale of rotational water delivery programs. However, the case for rotational programs is not easy to establish in paddy irrigation, as the following discussion shows.

A commonly perceived problem in paddy irrigation is high water depth in paddy fields at the heads of the system, while tail enders are subject to water stress. First, it should be noted that, while this perception is taken as the basis of the following analysis, in at least some cases the perception itself is erroneous. Wickham and Valera (1978, p. 63), for example, found no difference in the incidence of stress days between the heads and the tails of the systems they studied. As they note, this is probably because the tails are at lower elevations, and thus gain from surface and subsurface drainage what they lose from direct irrigation inputs--a subject discussed in the next section.

Second, in any well-designed irrigation system there must necessarily be "tail-ender" phenomena somewhere in the system. Since water supply and demand conditions continuously vary, some area must be available to absorb the variation--to act as a kind of "surge tank." This area will prosper with plenty and suffer with little. This area need not be at the tail of the system, rather it may be, as in other crops, on the land each farmer is able to irrigate--as in the warabandi system of India (Malhotra, et al., 1983). However, the potential for allocating the tail problem to each farmer, as in warabandi, is probably not feasible in paddy systems because of border losses from small plots with surface water surrounded by dry land. But in any well-designed irrigation system, a tail must be there; the only question is the equitable distribution of the tail.

Third, in considering the extent of possible over-irrigation, paddy differs from other crops in that there are both physical and economic constraints on the maximum depth of water in the field. With other crops in well-drained soil, there are no practical limits to over-irrigation. Also in paddy, the amount of over-irrigation can be directly observed by the farmer as field water depth, whereas in other crops, over-irrigation is invisible, i.e., below the root zone of the plant. The physical constraint on over-irrigation in paddy is the fact that the HYVs are subject to excessive submergence losses. The limit varies from about 5 cm depth shortly after transplanting to about 15 cm for a mature crop. A farmer would not rationally choose to go much beyond these limits, even with free and limitless water availability.

The economic constraint on over-irrigation is the labor and land cost of building and maintaining dikes for water storage. Since dikes are in the form of rectangles or triangles and, given soil characteristics, must have fairly constant height-to-base proportions, the cost of dikes increases with the square of height. Thus, for example, a 15 cm dike will cost about ten times as much as a 5 cm dike--or 2.25 times as much as a 10 cm dike. These costs are accentuated by the fact that, for structural reasons, the spillway from the dike--and thus the level of water storage--is only about 60 percent of the height of the dike itself. Thus, to save labor costs of constructing and maintaining dikes and to save land, farmers would be induced to store no more water than they perceived they would need under conditions of risk of future water supply.

While farmers may under- or overestimate their need for field storage, it is not at all clear that a management group could do better (especially given the quality of soils, terrain, and climatological data available to them in most cases). For these reasons, one can hypothesize that the distribution of heights of dikes in a paddy irrigation system reasonably corresponds to an optimum allocation of field water storage facilities--at least from the private point of view of the farmers themselves. If so, then rationing-rotation programs to redistribute storage capacity will entail private loss, which can only be justified by greater social gain.

Table 13.2 shows the average dimensions of dikes in three irrigation systems in the Philippines. The proportions of height to width and spillway height to dike height area are also shown. Unfortunately, the distribution of spillway heights in the irrigation area were not presented in this study. Spillway heights (apparently not related to management systems) ranged from 5.3 to 10.3 cm, with an average of 7.3 cm. Irrigation water requirements (after transplanting) were around 4.75 mm pd in both the wet and dry seasons. As noted before, it does not appear expedient to let fields dry out, so a minimum depth should perhaps be 2 cm. Thus, on the average, farmers could store about 5.3/0.475 = 11 days' supply on their fields. In drought periods, where rationing would be most necessary, ETA would be higher, at about the 6 mm pd maximum requirement shown in this study. This would amount to a 9-day drought contingency supply. This does not leave much room for maneuvering in rationing and rotation programs.

Assume, for example, an extreme case, where the command area is divided between head enders and tail enders and that, without a rotation program, the head would have the maximum of 15 cm of water in the field, while the tail would have 5 cm. With rotation, 5 cm would be taken from the former and given to the latter, so that everybody has 10 cm total, or 8 cm net water storage. With a 6 mm pd requirement, everybody would have about 13 days' supply. If the drought lasted no more than 13 days or so, the returns to the rotation program would be very high; half of the total crop (that of the tail enders) would have been saved from water stress. However, if the drought lasted much more than 13 days, the results of the rotation program could be catastrophic-- all of the crop, instead of only one-half, would be lost.

Thus, the returns to rotation programs in paddy irrigation are highly sensitive to the ability to predict the duration of drought periods. It should be emphasized that, in paddy, rationing water supply through rotation does not create additional supplies of water which would otherwise be lost to beneficial use, as in other crops; it only changes the time of beneficial use of a given supply of water and, hence, is sensitive to the accuracy of predictions regarding present versus future needs. In the above example, it is doubtful if the return to rotation would justify the risk, under reasonable assumptions regarding predictive ability.

Table 13.2 Average paddy parameters in research sites served by similar water delivery methods. Selected turnout areas in three national irrigation systems, Philippines, 1972-73 (Miranda and Levine, 1978)

Delivery method	Sites (no.)	Paddy area as dikes (%)	Paddy spillway height* (cm)	Dike dimensions Width (cm)	Dike dimensions Height (cm)	Area per paddy (ha)	Topography
			Santa Cruz River Irrigation System				
Simultaneous supply	3	1.6	5.3	24.1	9.3	0.09	Relatively flat
Farm-ditch rotation	1	2.0	6.0	23.0	11.1	0.07	Slightly undulating
			Angat River Irrigation System				
Simultaneous supply	5	2.3	7.7	35.1	12.9	0.19	Flat
Lateral rotation	3	4.7	8.3	40.8	12.9	0.08	Rolling
Farm-ditch rotation	3	2.9	6.7	41.9	11.5	0.11	Relatively flat
			Peñaranda River Irrigation System				
Lateral rotation	4	5.0	10.3	36.7	18.3	0.07	Rolling
Farm-ditch rotation	3	2.1	7.0	32.6	11.5	0.15	Flat

*Excluding those without spillways

Of course, these observations on rationing and rotation at the level of individual farmers should not be interpreted as denying the desirability of rotation in the main system among minor canals. Rotation is necessary to economize on the size of conveyance systems. It may also be necessary to provide a better distribution of water to groups of farmers between the heads and the tails of the systems. However, unless there are malicious obstructions of water flows toward the tail (and thus, less total water supply in the system) or real drainage losses, rotation between groups will not generally improve the efficiency of water use--only the possible equity of water distribution. Even with respect to equity, there is the problem that with a given supply of water, more water to the tail of the primary canal will simply create more tail enders toward the head. In paddy, the effective command area can be either lengthened or widened by rotation but not, unfortunately, both. This is one of the major problems in the reports on "improved management." The gains to the target group are noted, but the possible losses to others are ignored.

A NOTE ON THE REUSE OF WATER IN PADDY IRRIGATION SYSTEMS

The performance of the small, rivertine paddy systems characteristic of Southeast Asia can only be appraised in terms of the system as a whole--in terms of the use and reuse of water through an entire river basin. Drainage and percolation losses from one part of the system are commonly reused for irrigation in adjacent fields or by downstream systems. Thus, improving the efficiency of any particular part of the system (for example, by minimizing drainage losses) may simply reduce the productivity of another, unobserved part that depended on drainage for irrigation. Thus, the concept of field irrigation efficiency, so important in other crops, is virtually meaningless in paddy irrigation. Of course, some percolation and drainage losses are true losses of water to deep aquifers or to the sea. But perhaps the best solution to these losses is more reuse downstream--more projects, rather than attempts to improve the management of existing projects.

CONCLUSION

The subtitle of this paper stems from a famous dialogue between the eighteenth-century Minister of France, Jacques Colbert, and some Parisian businessmen. Colbert asked, "What can I do for you?" The businessmen replied, "Laissez-faire passer!"--roughly, "You can leave us alone!"

The laissez-faire philosophy, at least in its more enlightened form, depends on what is called "the presumption of the market." This means that it is presumed that people pursuing their own interests (which is likely to include the interests of others) will reach a reasonably satisfactory state of affairs without intervention by the state or any other collective entity--other than those collectives which they voluntarily choose to form. This presumption is quite often wrong. There are many instances of externalities, collective goods, monopolistic powers, and failure to assure basic needs which justify collective intervention in naturally functioning, laissez-faire systems. However, the burden of proof is on the interventionist. Unless there are good grounds for believing that intervention will result in a net improvement over the previous state, the system should be left alone. Such appears to be the case in the management of paddy irrigation systems.

NOTE

A condensed version of this paper has been published in Overseas Development Institute Network Paper 11b, May 1985, with comments by several irrigation experts. The writer is grateful to Ernest M. Thiessen and R. K. Sampath for detecting an error in the explanation of Figure 13.1, which has now been corrected.

REFERENCES

Lazardo, R. C., D. C. Taylor, and T. H. Wickham. "Irrigation Policy and Management Issues: An Interpretive Seminar Summary," Irrigation Policy and Management in Southeast Asia, International Rice Research Institute (IRRI), Los Banos, Laguna, Philippines, 1978.

Malhotra, S. P., S. K. Rahaja, and David Seckler. "A Performance Monitoring Study of the Warabandi System in

N.W. India," Colorado State University, Fort Collins, Colorado, 1983.

Miranda, S. M., and G. Levine. "Effects of Physical Water Control Parameters on Lowland Irrigation Water Management," Irrigation Policy and Management in Southeast Asia, International Rice Research Institute, Los Banos, Laguna, Philippines, 1978.

Tabbal, D. F., and T. H. Wickham. "Effects of Location and Water Supply on Water Shortages in an Irrigated Area," Irrigation Policy and Management in Southeast Asia, International Rice Research Institute, Los Banos, Laguna, Philippines, 1978.

Thavaraj, S. H. "The Importance of Integrating Nonengineering Aspects in Irrigation System Design," Irrigation Policy and Management in Southeast Asia, International Rice Research Institute (IRRI), Los Banos, Laguna, Philippines, 1978.

Wickham, T. H., and A. Valera. "Practices and Accountability for Better Water Management," Irrigation Policy and Management in Southeast Asia, International Rice Research Institute (IRRI), Los Banos, Laguna, Philippines, 1978.

14
Improved Irrigation Management: Why Involve Farmers?

Max K. Lowdermilk

> Engineering is not the fundamental problem underlying irrigation development in the LDCs. Engineering principles are known and can be adapted, but the major problem, however, is to discover ways to utilize farmer clients more effectively in operations and maintenance and development programs which will create rural transformation. Rural transformation essentially requires changes in farmers' behaviour, motivations, and expectations, which is hardly possible until institutions exist to provide them with improved production possibilities and incentives . . . (Wiener, 1976).

INTRODUCTION

An engineer stated in serious humor that there are six phases in every irrigation project. The first phase is the <u>high enthusiasm and highly publicized expectations</u> of the designers. Second, is the <u>disillusionment</u> of the implementors, who discover that the designs are greatly inadequate. Third, is the <u>panic</u> of the operational staff, who discover that the system won't operate as designed. Fourth, is the <u>search for the guilty</u>, which results in a vicious cycle of blame between designers and implementors, operators and extension workers. Fifth, is the <u>blame of the innocent</u>, i.e., the farmer, who had nothing to do with developing or operating the system. (Reports often conclude that ignorant and obdurate farmers won't cooperate with project authorities. Instead, they sometimes destroy structures, steal water, and create other problems.) Sixth, if a system works at even 40 to 50 percent of designed efficiencies, the praise and honor usually go to the nonparticipants, who are not the engineers, but the politicians who used pressure to get the project. This humorous satire says volumes about the different actors in the irrigation drama, especially the farmers, who usually play a rather passive role in the entire process. However, it is the farmer who is often hurt most when systems perform poorly.

The most underrated and misunderstood dimension of irrigation development today is that of the individual and collective irrigation behavior of farmers. Much is known about irrigation technology, the design and construction of dams and canals, crop water requirements, and irrigation practices, while the social and organizational aspects of irrigation continue to be the Achilles' heel in system development, improvement, and operation. The increasingly higher costs and the low performance of irrigation projects in India and elsewhere are a serious national and international concern. Government officials and donors are slowly realizing the high economic, social, and political costs involved when farmer end-users or beneficiaries play only a passive, spectator role in irrigation projects and programs which affect them directly. Seldom is it realized that on millions of farms in India, the irrigation drama is often a tragic one for farmers, for whom irrigation is a matter of success and failure--or even life or death. Questions are now being asked at every level about how the lack of effective farmer involvement in irrigation projects is related to the high costs of systems, the perennial problems of operation and maintenance, low irrigation efficiencies, and resulting low crop production.

Dr. Y. K. Murthy (1980), former chairman of the Central Water Commission and now a consultant for the World Bank, summarized the problems of the Indian irrigation in a lecture entitled "The Irrigation Engineer and the Farmer." He states:

> The farmer, who is the kingpin in the process of agricultural development of the country, is faced with a number of woes which he is not able to overcome unless there is a revolutionary change in the attitudes of the engineers.

Dr. Murthy says that usually the farmer does not know when and in what volume water will be delivered at the chak outlet. He states that the farmer is strangely expected to improve and maintain chak conveyance channels without technical assistance or adequate incentives. The farmer is also expected to distribute water with equity and to abide by outdated water codes. The farmer is to do all of this while receiving little or no technical assistance in water use from the agricultural and irrigation departments which, due to lack of cooperation and coordination, pull in different directions and often blame each other for low crop yields. It is usually the farmers near the public outlet who utilize almost all the water they want, while tail enders and overall crop production suffer. Dr. Murthy further states that farmers usually face extreme difficulties in trying to organize themselves for the distribution of water.

Much progress has been made in many aspects of India's irrigation development. Most observers, however, agree that there has been miniscule progress in the development of viable farmer-controlled organizations for distribution of water, settling conflicts, making improvements, and doing regular maintenance of the productive end of the system--the chaks and farms serviced by public outlets. The purpose of this paper is to examine three basic questions which are increasingly raised by concerned officials and donor agencies:

- Why involve farmers in irrigation development and improvements anyway?
- Why don't farmers cooperate with irrigation authorities more effectively?
- What are some useful lessons about farmer involvement which may have relevance for India?

For the purposes of this paper, farmer involvement or participation means farmers playing an active role in decision-making regarding planning, implementing, operating, maintaining, and evaluating projects and programs designed to improve the productivity, equity, and effectiveness of irrigation projects. The following situations are not considered to be positive participation by farmers: manipulating farmers for political purposes or to please outside agencies, using tricks or coercion to make a short-term project look good, or involving only the elite or special classes of farmers. Canal authorities who administer water rather than manage water often confuse coercion with farmer participation.

WHY INVOLVE FARMERS ANYWAY?

This question is intentionally phrased to indicate the doubt and skepticism which still exists in many quarters. Historically, farmers around the world have planned, built, and operated their own private systems or parts of public systems. Without active participation of farmers, irrigation systems can never be efficient or cost effective. There are many things governments cannot do effectively for farmers. Farmers need a countervailing power and a voice to assure that their needs are met by those who should be more accountable to them.

Historically Farmers Have Been Involved

If irrigation is for the purpose of agricultural production, then it is axiomatic that farmers must be viewed as the key building blocks and central actors in irrigation systems. It is the farmer who, when facing great risks and uncertainties, must manage the water and all other inputs to produce food and fiber. This simple truth is often ignored by engineers who have not received adequate training in irrigation management and its role in the agricultural production process.

In India and other countries, farmers have probably played a much greater role in irrigation development in the past than they do today. For example, Pant and Verma (1983) quote sources showing that, from the second century in India, farmers built systems, organized for repairs and maintenance, and collected fees for operations and improvements. Today in India, there are communal systems and tanks serving significant areas of land, which are owned and operated by farmers. For example, in the state of Himachal Pradesh, about 70,000 ha of a total of about 100,000 ha are irrigated by private or communal systems known as khuls. In Tamil Nadu State in South India, there are about 40,000 tanks irrigating about 1 M ha. Of these, an estimated 50 percent are private or communal tanks (Palanisami and Easter, 1983). In all states of India, about 4 M ha are irrigated by small tanks. There are about 25 M ha in India which are irrigated by over 2 M private tubewells and about 8 M private open wells. It is not known how many of these tubewells and open wells are informal private cooperatives, usually based on caste and/or kinship group membership. In North India, the famous warabandi system is probably one of the most successful water distribution and allocation systems in the world for large public gravity systems where water supplies are scarce. It has been wrongly concluded that no organization is involved in the warabandi system. To the contrary, farmers and irrigation authorities work together to establish a system of continuous supplies of water provided on a regular turn system of units of time per unit of landholdings on fixed rotations. The warabandi system has its own norms, wherein the responsibilities of farmers and irrigation authorities are clearly defined. Sanctions exist for violations, and farmers on chaks "police themselves" in informal organizations for regular cleaning and maintenance (Malhotra, 1982; Pant and Verma, 1983) describe several cases and small experiments with farmer involvement in India, including "outlet or pipe committees" in Andhra Pradesh, water cooperatives in Gujurat, the phad system in Maharashtra, the sattadar system in Bihar, and other localized attempts at various types of farmer participation. With the exception of some studies (e.g., Jayaraman, 1981; Pant and Verma, 1983), few of these indigenous organizations have been adequately investigated for identifying lessons and principles which might prove useful in India.

In other countries, farmer involvement ranges from complete ownership of irrigation systems, through active involvement in management of public systems, to little involvement in operations on completely government-controlled systems. Maass and Anderson (1978), in a study of three irrigation systems in Spain and two in the United States (Colorado and Utah), show that all these systems had much local control with formal organizations of farmers to manage and operate the systems, collect fees, resolve conflicts, and provide order and equity in the distribution of water. In many of the U.S. systems the farmers of irrigation districts hire and fire the engineers and management staff. Coward (1980) describes several communal systems in Southeast Asia. He finds that the indigenous private water user organizations (mostly informal) have three basic characteristics: accountable leadership, division of the organizations into many functional subunits, and channel-based (rather than village-based) organizational forms. Water user associations in

Korea, Taiwan, and Japan play a major role in irrigation development. Though associations vary in structure, the farmers are usually responsible for construction, improvement, operation, and maintenance of irrigation and drainage facilities; prevention and relief of damage to irrigation and drainage facilities; financing the association and staff; and coordination with government policies on land, agriculture, industry, and rural development (Asian Productivity Organization, 1980; Lin Chun-Huei, 1976). In these systems, governments usually provide policies, loans, and technical assistance as incentives for creating and strengthening associations. Since 1945 in Mainland China, decentralized irrigation associations are stated to be in charge of construction, improvement, operations, management, maintenance, and drainage facilities. These organizations also prepare irrigation plans, regulate water distribution, settle disputes, collect fees and engineering costs, take and repay loans, and conduct studies for system improvements (Asian Productivity Organization, 1980).

Farmer Involvement Is Cost Effective

Experience indicates that active farmer involvement is cost effective in terms of mobilization of local resources, improvement and maintenance activities, reduction of irrigation department staff time, provision of local wisdom for better design and planning of systems, reduction in the destruction of facilities, fee and fine collection, resolution of disputes, and provision of an organized means for extension and farmer training (Food and Agriculture Organization, 1982).

First, farmers can and should pay for some of the costs of operation and maintenance and a share of the improvement costs. They can contribute resources (in cash or kind), labor for construction and regular maintenance, land for rights-of-way, or they may provide a combination of these. In a large \$42 M effort for rehabilitation of the chaks in Pakistan, farmer labor alone amounted to about \$7.6 M. In a recent World Bank project in Pakistan, farmers are providing an estimated 30 percent of project costs in labor and cost recoveries for improvement activities. Early (1982) reports that with heavy farmer involvement in improving the management and operations of a Philippine system, the cost/ha per year was only about Rs 30, which resulted in an extra ton of rice/ha. Also in the Philippines, in an irrigation management project, the irrigation association contributed the following over a 14-month period: maintenance and improvement activities, 571 work days; water distribution and fee collection activities, 980 days; management activities, 308 days; and cash outlays for canal repair, materials, equipment, and supplies, $470. This amounts to about $12/ha/yr (Bagadion and Korten, 1983).

Second, with farmer cooperation, there is a large saving in the time required for implementing projects. In Pakistan, where farmers were not organized properly prior to improvement activities, 60 percent of the total government staff time in implementing an irrigation rehabilitation program was used in trying to convince factions or farmers to agree on rights-of-way for conveyance channels, removal of trees, and the provision of labor for rehabilitation activities.

Third, farmers' local wisdom and experience are resources often neglected in irrigation rehabilitation and management improvement programs. At the planning and design stage, farmers can contribute knowledge about topography, soil types, depth of soils, location of outlets, and information about possible social problems resulting from certain actions. Often, engineers collect data and develop designs without any useful information from farmers. This often results in faulty designs and systems which are not maintained by farmers. The common response of farmers is, "The government built the system without us; let them also maintain it without us." To bypass farmer leadership, local wisdom, and potential farmer contributions can be costly in both the short and long runs.

Fourth, it has been found that, where farmers are organized and have a sense of ownership in the system, they will not only maintain the system, but also assure that structures and facilities are not damaged. Also, where systems provide predictable water supplies, experience shows that farmers are more willing to pay increased water rates over time. In several countries, the increase in water rates or a portion thereof is returned to water user associations for regular repairs and improvements (Agricultural Development Council, 1981). This is similar to taxing local people for schools and using the taxes only for local school purposes. Though adequate field studies have not been made to test the hypothesis that effective farmer organizations lead to higher crop production and levels of living for rural families, it is logical to believe that this is the case. Improved irrigation efficiency resulting from improved management and operations by farmers should lead to higher crop production, as well as equity and conservation of soil and water resources. There is much indirect confirmation of this from the Philippines, Taiwan, Korea, and Japan, where rice yields/ha range from over 2 to 6 tons. A Japanese observer of irrigation in Asia (Asian Production Organization, 1980, p. 17) states:

> When once farmers have a financial stake in a project, they develop a deep sense of involvement for it. Farmers in neighbouring countries do not seem to have as great a financial stake in the irrigation undertaking, hence, are less prone to developing the interest and involvement in the management of water.

What Governments Cannot Do Effectively for Farmers

It is obvious that there are many things in irrigation development which farmers simply cannot do or cannot provide for themselves. They are not planners, policy-makers, designers, or engineers. They require much more technical assistance and a continuous stream of services to create improved production possibilities. Seldom do engineers and planners stop to think about the other side of the coin of what the government cannot effectively do for farmers in irrigaton improvement. For example, governments can seldom bear the complete costs of irrigation projects without contributions from farmers. Neither can they effectively distribute water below the outlets, do regular maintenance below the outlet, settle disputes and resolve conflicts, supervise collective decision-making, punish offenders, enforce cropping decisions, select the leadership, form effective associations from the top down, monitor farm systems regularly, nor do a host of other things. Without active farmer participation, irrigation efficiencies will never be increased. As Corey (1981) has stated from his experience in India:

> Governments should not do the farmers' business . . . development experience worldwide indicates that field channel construction, operation and maintenance, on-farm improvements, and general

management of the irrigation water conveyence system below the outlet can be accomplished but with direct farmer involvement.

There are practical reasons for this. One is the 168-hour week, irrigation field staff are at best present in the chaks only about 20 to 25 hours of a 40-hour work-week. It is not realistic to think that irrigation field staff can be at their posts continuously (consider the costs). Assume that by the year 2000, there are 50 M ha in India with gated outlets at the 40 ha level, with one field man servicing about 10 outlets. Such an arrangement requires 125,000 field men for the 2.5 M gates to be operated, plus thousands of extra supervisors. If there were gated outlets at the 8 ha chak level, the number of gates and field men would increase by a factor of five. What country could afford so much in recurring administrative costs? A serious question seldom raised about India's irrigation sector is: Is the government presently attempting to do too much for farmers in irrigation development and improvements? The concern among donor governments about the role farmer users should and can play in system operation, management, and maintenance is so great that a recent report to an agency recommended that, unless viable water user associations are established, no financing should be provided for on-farm improvements (U.S. General Accounting Office, 1983). Chambers (1983) suggested the following guideline for determining the degree of adequate government intervention:

> In general, government should unambiguously avoid doing that which communities can do for themselves in their own interest, but should intervene when exceptional problems are beyond a community's power to overcome. Is there such an appraisal which can be made in India? What new policies are needed to implement such a proposal?

Farmers Need a Countervailing Power and a Voice

Farmers not only have a right to take part in most of the decisions that affect their lives directly, such as at the outlet level, they also need an organized means to make their needs known. Especially on government-controlled gravity systems, farmers seldom have a voice in the projects and programs designed for them. Unlike farmers in most gravity systems in South Asia, farmers in some countries--through federated farmer organizations--have developed a strong voice and countervailing power to make irrigation and agricultural authorities accountable (Maass and Anderson, 1978). In most South Asian countries, the authorities who administer water do not feel that they are accountable to farmers. There are good reasons for this. At the time when the first irrigation bureaucracies were established during the nineteenth century, little attention was given to client participation in development. Also at that time, all public irrigation systems were administrated and not managed in the modern sense. Farmers in India, as a result of the Panchayat Raj and over 35 years of grassroots democracy and rural educational programs, are now becoming more aware of their rights. They are becoming more vocal regarding irrigation needs, agricultural price policies, and other decisions which affect them directly. The time is almost here when engineers and other professionals will have to become more accountable to the clients they serve. Increasingly, rural leadership is improving where some farmers and local government leaders have education's equivalent to professionals in agriculture and irrigation. Through political or other means, they will develop a strong countervailing power to irrigation and agricultural authorities.

Tukase (1982) predicts that the 1980s will be a period of irrigation development whose emphasis will not be on the construction of large dams and canals. He sees, instead, a growing and widespread interest evolving in the institutional aspects of the whole system; human management issues, such as farmer organizations; training of professionals and farmers; and ways to improve project design, implementation, and management. The major question being asked today is: How do we go about implementing these new concepts, which are more cultural and site specific than other aspects of irrigation development? We can begin by examining some of the major reasons why farmers are currently not playing a key role in irrigation projects.

WHY ARE FARMERS AND IRRIGATION AUTHORITIES NOT COOPERATING MORE EFFECTIVELY?

Experiences in many countries indicate that there are several common priority constraints in attempting to gain more farmer participation. These include systems too poorly designed and operated to provide predictable water supplies, problems of large bureaucracies, donor and host country pressures for quick results, and a dire lack of training for farmers and engineers.

Most Public Gravity Systems Don't Work Well for Farmers

Many studies in South Asia, including India, document the weakness of public irrigation systems. To date, few provide farmers with a known or predictable supply of water (Bottrall, 1981; Wade and Chambers, 1980; Johnson, et al., 1977; Reidinger, 1974; Murthy, 1980). Much of the noncooperation, and even destruction of structures, is the farmers' response to a system which is not dependable. When systems do not work, farmers will naturally attempt to modify them to obtain water by any means. The causes of the problem are seldom farmers themselves or their actions, which are usually only symptoms of the real problems of poor design, faulty operation and maintenance, and weak institutional arrangements. How often are problem symptoms confused with problem causes, due to lack of adequate diagnosis of systems and farmer needs? Hashim Ali (1982), in a survey of irrigation systems serving 232 villages, documented over 557 defects in system design and operation. Corey (1981) and others in India indicate that without the expectation of a fairly predictable water supply from the main system, it is next to impossible to do on-farm development or establish rotational water-supply systems with active farmer involvement. Recently in Gujurat (Mahikadana Project) and Rajasthan (Gamberia Project) states, training workshops were conducted to teach how to diagnose irrigation systems (Water Management Synthesis Project, 1981, 1983). In Gujurat, at the Mahi Kadana Project, farmers were willing to pay over six times the cost of canal water for private tubewell water because the public system was not dependable and could not provide water control. At the Gamberia Project in Rajasthan, due to faulty design and poor operation practices, farmers had to fight each other and the irrigation officials to get water. On a lined distributary of 5 km in length, there were 30 or more illegal outlets, plus sections where large boulders were used to check-up water for faulty outlets. This was due to poor design of the distributary, resulting in critical flow where adequate discharges from outlets were not possible without checks. A state of "water anarchy" best describes such systems.

Large Bureaucracies and Response to Local Needs

All large, centralized bureaucracies face serious problems in trying to respond to the diverse and complex needs of local management, especially these of local organizations (Korten, 1981). Any large organization has long-established and entrenched procedures, norms, and codes for carrying out its work. Such approaches and rigid procedures do not provide adequate flexibility for meeting the physical and management needs of specific local irrigation systems. It must be realized that irrigation management and farmer organization are the most indigenous aspects of irrigation systems. No single model or method can be applied universally, even in the same state or region. There is, therefore, a real danger in recommending the warabandi system, small chak sizes, or a particular organizational model for all of India, as some often do. There is simply no quick fix or universal solution for most irrigation problems. The most efficient systems known are those where decentralized process approaches are used to diagnose and solve problems at the project and farmer levels of the system.

Large bureaucracies in irrigation usually want rigorous adherence to fast-paced and often inflexible implementation schedules. In terms of project completion, pressures to hurry up usually result in long delays, often costing millions of dollars.

It is not known to what extent centralized planning and inflexible procedures result in long delays in project completion. A recent review of 67 irrigation projects in India shows that 66 were not completed at the time planned or within the budget. The range and median years of delay were respectively 2 to 22 years, with a median of 7 years (Economic Intelligence Service, 1983). Also, a Public Accounts Committee report (1983) to the Lok Sabha showed that there were 8 major irrigation projects 15 to 20 years behind schedule. This study also identified 42 major projects with cost overruns of 500 percent. As a result, the annual revenue costs for irrigation based on 1981-1982 budget estimates was about Rs. 425/ crore (Public Accounts Committee, 1983). What nation can face for long such high costs resulting from poor planning and action?

Another problem is the pressure donor agencies often bring to bear on host country irrigation bureaucracies. Under such situations, farmer involvement and management issues are usually left marginal or ignored due to the inordinate pressure to move too much money too quickly. It is well known that it is much easier for a host country to acquire loans for physical irrigation infrastructure than for software items like water management improvements. There are many reasons for this, which may include: the ease of administration of large loans for certain categories preferred by some donor agencies; the desire of politicians and engineers to build large irrigation monuments instead of systems which can be managed; and the well-known fact that design and construction are more prestigious, glamorous, and profitable than mundane O&M activities (Early et al., 1982). Jayaraman (1981) has documented, through an interesting study, the preferences of engineers for design and construction. A major problem in obtaining loans for management, training, action research activities, and assistance for evolving farmer organizations is that these components seldom move much money fast. The real reward systems of many donor organizations are for project development which obligates too much money fast rather than quality work in project implementation.

Another set of problems of large bureaucracies relates to those procedures, norms, and attitudes which become standard and frozen over time. In long-established irrigation organizations, one finds traditional design procedures, fixed attitudes about farmers, and paternalistic approaches to farmers who work directly against officials developing credibility with farmer clients. For example, many designs of irrigation systems do not include how the system will be managed or what the roles of the farmers in management will eventually be. Irrigation system design, especially in the areas of water delivery and distribution, is often very traditional and inflexible (Murthy, 1980). Many irrigation design methods, according to Early and others (1982), come from outdated textbooks using techniques from the western U.S., where there are vast, flat landscapes with large fields and farms. These conditions do not fit the unique vertisol soil conditions, such as those in Maharashtra, or the shallow soils found in many states. In much of India, there are rolling landscapes and very small farms with highly fragmented fields operated under complex land-tenure systems. Design and the procedures for design have built-in assumptions which are often strongly held. Once implemented in the project, the limits are virtually set in terms of what optional management systems and modes of farmer involvement can be best used. After a system has been constructed and found not to work well, a frantic search for a management mode then begins. Rather than the typical blueprint-type designs where everything is specified, there is need for much more flexibility in design to provide for a learning process for evolving the most appropriate mode of management. This often can be established prior to the design process by studies on similar systems nearby. This can also be established in new systems by including field experiments early in the life of the project, which provide feedback for improved design. The inflexible, blueprint approach in designing the management mode seldom works well in actual practice, though most design is still done this way.

Those professionals who work in long-established irrigation and agricultural bureaucracies develop traditional attitudes and values related to their particular professions. In some countries where I have worked, irrigation engineers not only work together, they socialize in clubs together and even live in colonies or separate zones together. There are a number of common beliefs and sentiments which often work against effective farmer involvement. First of these is the prevalent view that social and organizational factors are not really very important for irrigation improvement. An example is given in an extensive study that was done on the Indus Basin in the 1960s by Sir Alexander Gibbs and Associates. In the findings of this study comprising 24 volumes, a total of only 18 pages is given to social and organizational problems of irrigation systems in Pakistanz.

Second, there is a rather widespread view that farmers must be commanded or put under great compulsion in order to force cooperation. This approach was used recently in Pakistan's Punjab, where martial law authorities developed a water user's ordinance. This rigid legal approach lays out strict rules and regulations, with sanctions against farmers for noncompliance. Such a top-down legal approach, which provides no incentives for farmers, will not work in a free society like India.

Third, there is a view that, if only the right technology can be found, we can bypass the problems of farmers and their organizations. This is merely a half-truth and, as such, is dangerous. There are some technologies (such as the warabandi in northern India), small chak sizes, and underground conveyance systems which can help reduce farmer conflicts. There is, however, no technology which completely rules out the need for farmer organization of some type and at some level.

Fourth, there are many myths about what farmers will or will not do, usually held by those with little or no farm background and who have little data or experience in working directly with farmers. Some of the common myths which are often hotly debated include the following:

- o Farmers will use water efficiently only if it is delivered to the outlet on a predictable basis
- o Farmers simply don't know the value of water
- o Farmers on the warabandi system don't take turns because water is in shortage and all have the same need for it
- o Farmers will not irrigate at night
- o Farmers are ignorant; therefore, they don't know how to irrigate efficiently
- o Main system improvement separated from on-farm improvement, or vice versa, is what is really needed.

It is a fact that in many countries, those who plan, design, implement, and operate irrigation projects have never worked on a farm. This is especially true where the larger percentage of irrigation engineers come from urban rather than rural or farm backgrounds. Therefore, with no direct farm experience and no training about farmer behavior and relationships, it is difficult for them to acquire positive attitudes about farmers. It is certainly not their fault but a matter of circumstances.

There are many other negative attitudes which often influence the approaches made to farmers by irrigation field staff. Allied to such attitudes is the fixation on certain technologies or procedures which can be applied universally. The warabandi system, other rotation water supply systems, small chak sizes, and the command-area development approach have each, at one time or another, been viewed as the single solution to India's complex and diverse irrigation problems. As a rule, the advocates of a universal solution usually have had little field experience in India at the farm level. It is estimated that about 65 to 75 percent of all irrigation engineers there from non-farm backgrounds have never themselves irrigated a field.

A final problem facing most large irrigation establishments is the lack of communication between users and irrigation authorities. Even a one-way flow of information to users from irrigation authorities is greatly inadequate. In recent studies in Pakistan and the Gujurat (Mahi Kadana Project) and Rajasthan (Gamberi Project) states in India, it was found that about 70 percent of the farmers reported that they did not usually receive information about the closure of systems for maintenance and repairs, or even opening dates. In a similar study in Pakistan, Lowdermilk and others found that over 83 percent of 389 farmers did not know the published date for canal closures. It was also found that the irrigation officials could not keep to the published schedule the department had established. Farmers seldom meet officials other than their local patwaris and chakidars, who are not always at their stations. In the Gujurat study, it was found that farmer-reported views about how the system could be improved were almost exactly like those reported by the irrigation officials working at the Mahi Kadana Project. The key problem is that villagers seldom have a chance to meet with irrigation officials at the executive engineer level. However, I have known several executive engineers with agricultural backgrounds who make it a point to visit Panchayat and other groups to hear farmers' needs on a regular basis. One engineer in Gujurat stated that the most cost-effective method for improving relationships between farmers and irrigation authorities would be a special week declared each year by the government, during which executive engineers and their staffs had to go out to villages and simply listen to farmers express their needs.

Training of Farmers and Engineers

Until recently, the training of farmers and engineers has not been emphasized adequately. In a recent analysis of 63 projects assisted by AID and the World Bank since 1970, it was found that training was included in only 5 projects. Of 40 projects with adequate cost data, less than 1 percent of the total budget was allocated for training. The cost/ha for training professionals, extension staff, and farmers for these projects was about $25/ha (WMSP, 1981).

Experience with farmer involvement in Sri Lanka and the Philippines indicates that farmers and those working directly with them require much training. The fact is often overlooked that successful irrigated agriculture requires more knowledge inputs than any other known form of agriculture. Farmers need help in learning how to manage water in the chak and on their fields. They need to know improved irrigation practices and better forms of cultivation. Useful knowledge, when invested in people, makes land, labor, machines, water, and other factors more productive. Welch in the U.S. (1970), as well as Ram (1974) and Rosenzweig and Evenson (1977) in India show that training and education make agricultural production information more useful and less costly. The great need in India in the 1980s and beyond is to find better ways to train farmers and those who work with them directly and indirectly.

Dr. Murthy (1982) has also stated that there is need for more opportunities for professional development training for irrigation engineers. The government of India is now convinced that a critical mass attack must be made on training both farmers and irrigation officials. The World Bank and USAID are now assisting India with one of the largest training efforts in water management to date for farmers, engineers, and agricultural professionals. The new International Irrigation Management Institute established recently in Sri Lanka will also place much emphasis on training professionals and farmers.

MAJOR LESSONS LEARNED ABOUT FARMER INVOLVEMENT

There are a number of major lessons learned in Asia which may have much relevance to India. These lessons have been field tested in several countries of the region.

No Model as Blueprint Exists for Gaining Effective Farmer Involvement

No single model exists that can be transferred to India or within India to resolve the farmer participation problem. A major lesson is that the development of a management mode involving farmers is a process which must be evolved in each site-specific cultural setting. This requires a built-in learning mode for each project. One process approach is known as the action research-development process for irrigation improvement. For the purposes of this discussion, focus is primarily given to Phase I, diagnostic analysis. Farmers are actively involved in this phase in working with an interdisciplinary team to identify priority system constraints. Phase I is followed by an experimental phase known as search for solutions. At this phase, the focus is on testing and monitoring new technologies and procedures. Much more action research is needed in India which places a strong focus on farmer involvement. Useful action research examines the proper mix of both hardware and software for evolving improved management of operations. A lesson learned is that a team approach which examines both physical and institutional environments is more useful than a separate discipline which examines farmer involvement only.

Farmers Do Respond to Adequate Incentives and Interventions

There are ample data to show that small and large farmers alike respond to adequate incentives for improved irrigation management (Bagadion and Korten, 1983; Uphoff, 1982; Lowdermilk, 1972, 1978). Farmers in India have responded positively by the millions to new crop varieties, use of improved levels of inputs, and the development of private tubewells. The major disincentives to farmers participating more positively in irrigation systems seem to be the malfunctioning systems themselves and the lack of adequate technical assistance and essential services.

It is interesting that farmers usually want more rather than less intervention from irrigation and agricultural officials to provide more order in the system below the outlet. In a reconnaissance of irrigation systems in India in 1981, farmers indicated to team members that they wanted both more flexibility and strong outside intervention (WMSP, 1981). When farmers cannot resolve difficult problems among themselves, they often desire and need to have an outside authority to intervene.

Government Policies That Provide Incentives and Authority for Farmer Organization

There is much agreement that farmers' organizations are most effective where there are incentives for the farmers to organize. These incentives include a predictable water supply, inputs and services for improved production possibilities, and flexible attitudes of irrigation authorities (Easter and Welsch, 1983; Bagadion and Korten, 1983). Government policies should provide for the clear authority of the water user group and specify the roles and responsibilities of the group and the irrigation authorities. In each place, it must be determined how much authority should be provided. This will likely differ greatly in different settings. This authority may include resource mobilization, operation and management, regular maintenance, fee collection, assistance in planning, supervision of implementation activities, help in integrating services, conflict management, checks on bureaucracy, regular monitoring of the system, and a means to communicate their views to higher authorities. It is often overlooked that farmers need training for many of the tasks they are expected to undertake in evolving strong local associations. Bagadion and Korten (1983) stated the hypothesis that the less authority an irrigation group has, the weaker the organization. Just as it is important to spell out farmers' roles with clarity, it is also necessary to identify the roles of the action agency or agencies involved. A major problem in India is the lack of clarity in the roles of irrigation authorities, agricultural workers, and farmers.

Initiate Farmer Organization Where a Predictable Water Supply and Local Commitment Can Be Generated

As discussed earlier, it is important to be able to provide a predictable water supply or the assurance of such a water supply. Equally important is the careful selection of the initial sites where local commitment exists or can be generated. In the Pakistan and Egypt Water Management Improvement Projects, the chaks selected had to meet certain physical and social criteria before improvement activities began. These criteria required chaks where:

- the majority of the farms were small
- a predictable water supply could be provided and where irrigation efficiencies, yields, and cropping intensities had a real potential for improvement
- the farmers agreed to:
  - provide all labor for earth work and lining
  - provide some of the cash for small watercourse structures
  - settle disputes over land, water, and other significant problem areas
  - supervise local improvements
  - provide masons
  - clear rights-of-way and resolve all conflicts
  - provide part of the cost of land leveling
  - create a formal or informal organization, such as a water user association, for the project activities
  - set up a method for regular operations and maintenance of the systems and improvements.

The successful program in the Philippines of farmer organization for irrigation management showed that it was important for the irrigators to make regular cash contributions to the construction costs involved. The amount required in cash was 10 percent of the construction costs and a regular fee of about $61/ha for ongoing operation and maintenance expenses (Bagadion and Korten, 1983). In a study of 36 small farmer development projects around the world, it was found that the two factors which were most important for success were involvement of farmers in decision-making and commitment of labor and cash to the project (Morss et al., 1976).

It is important that irrigators develop a sense of ownership in their project. Recently, the author observed an improvement project in India where all the planning, structures, earth work, and materials were contributed solely by the Irrigation Department, with no farmer contributions. This system, as other improved systems, was designed and implemented without adequate input from the farmers concerned. Doing the farmers' business for them is certainly no way to build up local capacity for useful farmer involvement.

A Punjabi farmer years ago confided to the author how rural people often trick government workers into taking the first crucial steps in development projects. He stood up in a walking position and demonstrated how the first step a person takes creates an imbalance; therefore, a second step has to be taken, and so on, with the result that all is done by the government authorities. The farmer stated, "If you want to evolve successful projects with us, get us to take the first step and provide some support." This local wisdom is important. Where irrigators will not take the first steps, it is often better to bypass their chaks for others. Government organizations may find this difficult; however, it is based on the well-known fact that all groups are prepared or ready at the same time for development activities.

Build Upon Local Forms of Organization and Begin With Small Units

Often, farmers have developed informal organizations which can serve as the basis for building stronger organizations. Local leadership and existing organizational forms should not be ignored (Bagadion and Korten, 1983). Some general strategies which seem to work include the following:

- Work within the cultural context in selecting leaders, and help them improve their skills in organizing farmers, making decisions, and resolving conflicts
- Identify and assess local informal organizations and, where appropriate, build upon these
- Do not bypass group leaders and leaders of factions
- Allow leaders to organize committees their own way; to manage or supervise labor, materials, and equipment; to settle disputes; and to operate and maintain the system
- Use much caution in attempting to introduce new organizational forms. Where there is doubt about a particular organizational model, first try to use the local form of organization
- Use local organizations to reach and involve individual farmers
- Where possible, keep organizations small so members deal in face-to-face relationships with each other
- Find ways to reward local leadership. (Lowdermilk and Lattimore, 1981).

It has also been found that small chak groups with face-to-face and daily interactions should form the basic building blocks for organization. Until much experience is gained about how to best federate irrigators' groups, it probably should not be attempted. In most cases this hydrological unit, instead of the village or administrative unit, should be used for irrigator organizations (Bottrall, 1981; Hutapea et al., 1979). In Indonesia, Hutapea and his colleagues found that, where the hydrological units coincide with village units, irrigation systems are easier to manage because conflicts are less likely. Where irrigation groups have federated up to the main canal or system level, much time and learning was required. Federated groups, to be effective, must be based on strong local units.

Involvement of Water Users From the Beginning in Planning, Layout, and Implementation

Most of the literature on irrigation associations is based on experiences of rehabilitation or improvements of existing systems. What do we do about involving farmers in new projects? Earlier, we stated that work should begin where there is a high expectation of a predictable water supply. This still holds, but to delay the work of organizing farmers until the predictable water supply flows easily spells failure. Far too often, irrigation staff do all the planning, provide all the resources, and complete all the work without farmer involvement. Then they expect farmers to cooperate in something which those farmers perceive as belonging to the government. Earlier in this paper, it was emphasized that contributions can be made by farmers in planning, layouts, and implementation (see Bagadion and Korten, 1983; Coward, 1983). Experience in Sri Lanka and the Philippines suggests that from 9 to 12 months of preparation is needed to prepare farmers for new projects. Where farmers have not had prior experience with irrigation, it takes time for them to learn new forms of cooperation.

Build Up the Capacity of the Implementing Agencies

Time is also required to achieve a reorientation of the methods and approaches of personnel who work with farmers. Without changes in attitudes, procedures, and skills of the implementing agencies, little can be accomplished in mobilization of farmers. To achieve this, a strong policy with incentives for staff to work closely with irrigators is needed.

First, engineers and social scientists need to learn to work together in diagnosing systems to identify the physical and social constraints and how these interface. Little attention has been given to proper diagnostic analysis of systems to increase our knowledge of how systems actually operate. New skills are also required for monitoring irrigation systems using a team approach. We still know very little about how to monitor irrigation systems effectively and utilize the information for improved system management. Even where monitoring is done, the indicators of farmer performance are often not included (Clyma et al., 1977).

Along with learning new skills for diagnosing and monitoring systems, engineers also require some new skills to work effectively with farmer groups. Nowhere in present engineering training are there opportunities to acquire this knowledge. Some of the changes needed for engineering staff include (Uphoff, 1982):

- o flexibility in working with farmers and farmer groups
- o willingness to learn from farmers and incorporate local wisdom in planning and design
- o willingness to use the trial-and-error method of learning, such as experiments
- o willingness to let farmers make their own decisions about matters that impact them directly
- o acceptance of the fact that the local organization actually belongs to the members, not to the government
- o overcoming the disease of paternalism, and viewing irrigators as members of the team
- o showing appreciation and respect for rural people.

Bagadion amd Korten (1983) ask several useful questions to gauge the capacity of the agency to effectively implement a strategy for developing irrigation associations:

- o Does the irrigation agency have a rapid diagnostic analysis method to identify how systems are presently operating, including farmer irrigation behaviors?

- Does it use this knowledge in planning, designing, and implementation of systems?
- Does it have a method to accurately estimate the area which will actually be irrigated?
- Does the agency have social scientist personnel or special personnel who can develop irrigation associations working as a team member of professional technical staff?
- Is there a special training program which develops field workers with the abilities to organize irrigation?
- Is there a clear framework...where the social, organizational, and technical aspects are viewed as interdependent?
- Do field-level technical and organizational staff and farmers understand their roles clearly?
- Are there mechanisms to assure that this is accomplished?
- Do the present procedures of the agency and the attitudes and actions of staff act to encourage irrigators to organize?
- Are there training programs to help technical staff to acquire the attitudes, skills, and knowledge required?

Most irrigation agencies do not yet have this capacity. A major question is how to develop this capacity in old line institutions with up to 100 years of tradition. In South Asia, especially in India and Pakistan, are some of the world's largest public gravity systems. Changes are taking place, however, because it is being found that there are large economic, social, and political costs in doing business the usual way. India, unlike some South Asian countries, is a democracy to the grass roots level, where farmer users of irrigation systems will make their needs heard, with louder and louder voices.

The experiences in the Philippines and Sri Lanka suggest that, as farmers organize for irrigation, more positive ties are developed with the irrigation authorities. In the Philippines, the personnel who do the social organizing work are employed by the irrigation agency. In Sri Lanka, a separate agency is used for organizing farmers. It is not yet known which is the most effective approach. It appears that under one authority, the technical and organizational activities are more complementary, and it is easier to involve farmers in planning, design, implementation, management, and operations.

Dr. Y. K. Murthy (1980) provides a type of profile of the future irrigation engineer in India. The new irrigation engineer will be concerned with the total system. No longer will the engineering responsibility be limited to the headworks and canals. Future engineers will understand more about crops, soils, and farmers' problems and will utilize this knowledge for improved designs which fit the physical and institutional environments. The new engineer will have a management orientation and will include in the design the type of management system, including the role of the users in operation of the system. Future engineers will be held accountable for developing and operating systems which provide predictable water supplies with equity for all users. The new engineer will create more flexible designs and allow for future changes to improve the system. The future engineer will also know more about basin planning, systems analysis, drainage, canal network development, operation and maintenance, and soil-plant-water relationships. By the year 2000 or so, all engineers will be involved in water management for improved system operations. The design and construction role will gradually diminish over time. The engineer in the twenty-first century will work in a streamlined organization which develops effective liaison with farmers. Government organizations dealing with farmers will become more flexible and responsive to farmers' needs. It is likely that, by the year 2000, agricultural uses of water will be brought under one unitary control which is basically an irrigation water management organization. New engineers will have more opportunities for study and professional development. With all the advances in irrigation, the professional who does not take time to keep sharp technically will be left behind in a matter of years. Witness the rapid spread of the computer, and consider how this will be used in 1994. By the year 2000 or so, there will be improved policies and water codes which provide better incentives to farmers and professionals. Finally, the irrigation engineer will not look upon farmers as adversaries, but as team members who share the common goal of increased production and higher levels of living for almost a billion people in India.

Since we began this paper with Wiener (1976), we close with his statement, that in the future:

> . . . policymakers and project designers will have to ask (new) questions about location of

projects where small farmers will benefit; devising small-scale farmer operated systems; providing means for small farmers to participate in planning, and evolution of projects intended to benefit them; and making sure that along with technical feasibility studies, careful attention is given to economic, social, and environmental assessments.

NOTE

This paper was presented at a Central Water Commission Seminar on October 19, 1983. The author is a project officer and irrigation water management specialist of the USAID Mission in New Delhi, India. The views expressed in this paper are those of the author.

REFERENCES

Agricultural Development Council (ADC). "Mobilizing Local Resources for Irrigation," ADC Report No. 22, New York, June 1981.

Ali, Syed Hashim. Technical Advisory Committee Study on Water Management Research and Training to the Consultative Group on International Agricultural Research, TAC, Los Banos, Philippines, 1982.

Asian Productivity Organization. "Farm-Level Water Management in Selected Asian Countries: Report of a Multi-Country Study Mission," APO Project Code OSM/III/79 June 12-28, 1979, Tokyo, Japan, 1980.

Bagadion, Benjamin U. "Farmers' Involvement and Training in Irrigation Water Management and Farm Management," paper presented at the Workshop in Implementing Public Irrigation Systems, East West Center, Honolulu, Hawaii, August 18-31, 1976.

Bagadion, Benjamin U., and Frances F. Korten. "Developing Irrigators' Organization: A Learning Process Approach to a Participating Approach," Putting People First, Michael Cernea, ed. Washington, D.C.: World Bank, forthcoming (1983).

Bottrall, Anthony. "Water, Land, and Conflict Management," reprint from ODI Review, No. 1, 1981.

Carruthers, Ian. "Neglect of O & M in Irrigation: The Need for New Sources and Forms of Support," presented at the Organization for Economic Cooperation and Development, Paris, France, August, 1982.

Chambers, Robert. "Rapid Appraisal for Improving Existing Canal Irrigation Systems," Discussion Paper Series No. 8, Ford Foundation, New Delhi, India, August 1983.

Chun-Huei, Lin. "Activities of the Yun-Lin Irrigation Association of Taiwan, Republic of China," paper presented at Workshop on Implementing Public Irrigation Systems, University of Hawaii, Honolulu, Hawaii, August 18-31, 1976.

Clyma, Wayne, Max Lowdermilk, and Gilbert Corey. "A Research Development Process for Improvement of On-Farm Water Management," Water Management Technical Report No. 47, Water Management Synthesis Project, Colorado State University, Fort Collins, Colorado, June 1977.

Coolidge, P. S., D. R. Daines, G. H. Hargreaves, and D. W. Miller. "Irrigation Projects Document Review: Executive Summary," WMS Report I, Water Management Synthesis Project, Agricultural and Irrigation Engineering, Utah State University, Logan, Utah, February 1981.

Corey, G. L. "Irrigation Water Management in India--Farmer Involvement," unpublished report, World Bank Mission, New Delhi, India, 1981.

Coward, E. Walter, Jr. Irrigation and Agricultural Development in Asia: Perspectives from the Social Sciences. Ithaca, New York: Cornell University Press, 1980.

Early, Alan C. Farm Irrigation Constraints and Farmer Responses: Comprehensive Field Survey in Pakistan, Vol. 2. Fort Collins, Colorado: Water Management Synthesis Project, Colorado State University, 1978.

Early, Alan C., et al. "Collaborative Irrigation System Management Research Results from The Upper Pompunga Integrated Irrigation System," paper presented at Seminar IRRI - Irrigation Water Management Development, October 1982.

Easter, K. William, and Delane E. Welsch. "Socioeconomic Issues in Irrigation Development and Distribution," Economic Report ER83-5, Department of Agricultural and Applied Economics, University of Minnesota, Minneapolis, Minnesota, April 1983.

Economic Intelligence Service. "Economic Intelligence Service Report," Bombay, India, September 1983.

Food and Agriculture Organization (FAO). "Farmers' Participation and Organization for Irrigation Water Management," The International Support Programme for

Farm Water Management, Land and Water Development Division, Rome, Italy, April 1982.

Hutapea, R., Prajanta Dirjasanyata, and N.G.S. Nordhoit. "The Organization of Farm-Level Irrigation in Indonesia," Irrigation Policy and Management in Southeast Asia, Agricultural Development Council, Bangkok, Thailand, 1979.

Jayaraman, T. K. "Attitudes of the Irrigation Bureaucracy in India to Scientific Water Management Tasks in Irrigated Agriculture: A Case Study from Gujarat State, India," mimeographed paper, Gujarat, India, December 1980.

Jayaraman, T. K. "Farmers' Organizations in Surface Irrigation Projects: Two Empirical Studies from Gujarat," Economic and Political Weekly, September 1981.

Johnson, Sam H., III, Max Lowdermilk, and Alan Early. "Water Problems in the Indus Food Machine," Water Resources Bulletin 13 (6), 1977, pp. 1253-1268.

Korten, David C. "Rural Development Programming: The Learning Process Approach," Rural Development Participation Review, Vol. 2, No. 2, Rural Development Committee, Cornell University, Ithaca, New York, Winter 1981.

Korten, Frances F. "Building National Capacity to Develop Water Users' Associations: Experience from the Philippines," World Bank Staff Working Paper No. 528, 1982.

Levine, Gilbert. "Perspectives on Integrating Findings From Research on Irrigation Systems in Southeast Asia," No. 26, Teaching and Research Forum: A Workshop Report, Agricultural Development Council, New York, 1982.

Levine, Gilbert, and Henry C. Hart. "Mobilizing Local Resources for Irrigation," Agricultural Development Council Report No. 2, Agricultural Development Council, June 1981.

Lowdermilk, Max K., and Dan Lattimore, eds. "Farmer Involvement," Planning Guide No. 2, Water Management Synthesis Project, Agricultural and Irrigation Engineering, Colorado State University, Fort Collins, Colorado, 1981.

Maass, Arthur, and Raymond L. Anderson. . . . And the Desert Shall Rejoice: Conflict, Growth, and Justice in Arid Environments. Cambridge, Massachusetts: The MIT Press, 1978.

Malhotra, S.P. "The Warabandi and Its Infrastructure," Pub. No. 157, Control Board of Irrigation and Power, New Delhi, India, 1982.

Morss, Elliott R., John R. Hatch, Donald R. Mickelwait, and Charles F. Sweet. Strategies for Small Farmer Development, Vol. 1. Boulder, Colorado: Westview Press, 1976.

Murthy, Y. K. "The Irrigation Engineer and the Farmer," The Fifth Bhaikaka Memorial Lecture, February 19, 1980.

Palanisami, K., and William K. Easter. "The Tanks of South India," Department of Agricultural and Applied Economics, University of Minnesota, Minneapolis, Minnesota, June 1983.

Pant, Niranjan. "Utilization of Canal Water Below Outlet in Kosi Irrigation Project: Administrative and Community Level Solutions," Economic and Political Weekly Review of Agriculture, Vol. 16, No. 39, September 1981.

Pant, Niranjan, and R. K. Verma. Farmers' Organization and Irrigation Management. New Delhi, India: Ashish Publishing House, 1983.

Public Accounts Committee (PAC). "1982-83 Report Planning Process and Monitoring Mechanism with Reference to Irrigation Prospects," Lok Sabha (Secretariat), New Delhi, India, April 1983.

Ram, Rati. "India's Agriculture During 1950-70: An Exercise in Growth Analysis," Agricultural Economics paper, University of Chicago, Chicago, Illinois, 1974.

Reidinger, Richard B. "Institutional Rationing of Canal Water in North India: Conflict Between Traditional Patterns and Modern Needs," Economic Development and Cultural Changes, 23, 1974, pp. 79-104.

Rosenzweig, Mark R., and Robert E. Evenson. "Fertility, Schooling, and the Economic Contribution of Children in Rural India: An Econometric Analysis," Econometrics, Vol. 45, July 1977, pp. 1065-1079.

Small, Leslie E. "Investment Decisions for the Development and Utilization of Irrigation Resources in Southeast Asia," Teaching and Research Forum: A Workshop Report, No. 26, Agricultural Development Council, 1982.

Tukase, Kunio. "Food Production and Irrigation Development in Asia," ADB Quarterly Review, Asian Development Bank, 1982.

U.S. General Accounting Office (USGAO). "Irrigation Assistance to Developing Countries Should Require

Stronger Commitments to Operation and Maintenance," report to the Administrator, GAO/NSIAD-83-31, Agency for International Development, Washington, D.C., August 29, 1983.

Uphoff, Norman. "An Overview of Concepts and Strategies for Farmer Participation in Water Management," mimeographed paper, Rural Development Committee, Cornell University, Ithaca, New York, 1982.

Uphoff, Norman. "Report on Institutional-Organizer Programme, Gal Oya," mimeographed paper, USAID Mission, Colombo, Sri Lanka, January 5, 1982.

Wade, Robert. "The System of Administrative and Political Corruption: Canal Irrigation in South India," The Journal of Development Studies, 1982.

Wade, Robert, and Robert Chambers. "Managing the Main System: Canal Irrigation's Blind Spot," in Economic and Political Weekly, Review of Agriculture, September 1980.

Water Management Synthesis Project (WMSP). "Irrigation Development Options and Investment Strategies for the 1980s India," Report No. 6, Water Management Synthesis Project, Utah State University, Logan, Utah, 1981.

Welch, Finis. "Education in India," Journal of Political Economics, Vol. 78, January-February 1970, pp. 35-59.

Wiener, Aaron. "The World Food Situation and Irrigation Programmes," ICIDD Bulletin, 34, 1976, pp. 21-25.

15
Legal and Institutional Aspects of Irrigation Water Management

George Radosevich

INTRODUCTION

This discussion of the legal and institutional aspects was the last of the seminar series, "Current Issues in and Approaches to Irrigation Water Management in Developing Countries." As indicated by a number of previous speakers (Fairchild, Nobe, Skold), while this topic is generally the last to be considered among the factors affecting the development, use, and management of water by agriculture, it is often the one factor that ultimately causes great concern in the implementation of such projects and programs. Perhaps there is an ulterior motive for putting this topic at the end of the series, as indicated by the quote, "Water lawyers are like beavers--when they get into the mainstream, they dam things up." That is not the intent of this paper. Basically what will be discussed is the institutional arena and alternative legal interventions that exist or might be employed in the area of irrigation water management.

THE INSTITUTIONAL ARENA

The institutional arena or institutional framework referred to throughout this paper is broadly described by three components as they affect water and related resources development and utilization. These three components are (1) policies and laws, (2) organizations, and (3) the system of implementation through planning and management. Each component will be discussed subsequently in greater detail. Within the discipline of law, there are various systems or families of laws which establish the legal and administrative jurisdiction over a subject matter. Once having identified the particular system of law within a country, the "law" consists of policies (including goals and objectives) and the various specific laws in the form of acts, decrees, codes, and the administrative rules, regulations, instructions, and manuals adopted to carry out these legal enactments.

The second major component in the institutional arena is the group of organizations directly or indirectly involved with water resources. These organizations are classified into governmental and nongovernmental bodies. In the case of governmental organizations, this includes the government agencies and courts (or judicial system) at the national, provincial (state), and local levels. Nongovernmental organizations of concern to this topic include the individual water user and various types of water user collaborative efforts. These water user organizations may be private or public, with the latter often difficult to distinguish from government-controlled organizations.

A third component consists of the planning and management activities involved in implementation of the laws by organizations. In this context, planning is synonymous with a "blueprint" for action developed by the organization. It should consist of three major steps that are continuous and dependent upon each other: (1) develop plans, (2) carry out specific actions, (3) evaluate achievement (Brusco and Wright; 1984). Management is the "process" that everyone in the organization from top to bottom is, or ought to be, involved in to implement the plan or blueprint. According to Lawrence Miller, a prominent business consultant involved in the development of management-improvement systems for many major corporations in the United States, management includes deciding how to do things and how much should be done, identifying the goals, providing feedback, and coaching the various levels of employees. In general, Miller maintains that management is a process that cannot begin or end at any one level within the organization (Miller, 1984).

A comprehensive examination of these three components --their interrelationship and the authority and power that can be generated to accomplish specified goals and objectives--provides the institutional insight to develop an efficient and effective irrigation water management program. It is, however, unfortunate that many problems faced in both developed and developing countries stem from the fact that the institutional framework is incomplete or inadequate. As Mr. Fairchild points out in his report based upon his many years of national and international experience, we must improve the institutional arrangements --it is essential to successful irrigation water management (Fairchild and Nobe, 1986).

THE SITUATION

Before going into a detailed discussion on the water laws and organizations as they relate to agricultural water use, we must be cognizant of certain situational factors that may have contributed to the institutional framework that exists in any given country. Also, too often, discussions are premised with "The problem in this country is ..." instead of "The situation in this country is...". The difference between the two is that the first already reflects a negative value judgment, frequently before understanding the events or conditions that have shaped or are shaping the institutional structure.

One of the most basic situational factors is the geoclimatic conditions. In most countries, the general direction of the water law is determined by these conditions. In arid areas, the concern is with the allocation of a scarce resource. In humid regions, the general concern is with drainage and flood control. Of equal concern is the seasonal and spatial distribution of water throughout the country, which may require special institutional attention directed toward construction of storage facilities, transbasin diversions, and conjunctive use of ground and surface waters.

Another factor is the state of development within a country. This would include an assessment of the water data; the extent of development and use of water for agriculture and other uses; historical information on water development, including past policies, plans, and programs; information on present policies, goals, and projected programs; as well as identification of problems of a technical, social, economic, and institutional nature. To understand the current situation, it is important to focus upon the "state" versus the "stage" of development. In a technical sense, "stage of development" implies progress from one period of time to another, which may not be the case. Adverse economic conditions may result from the introduction of a new technology, such as installation of deep tubewells, that may render useless shallow tubewells.

Unfavorable climatic conditions, such as drought, may cause a country to regress in spite of its efforts to maintain the status quo. Progress may require many changes, including formulation or alteration of policies, relaxation or tightening of government involvement and control, and introduction or restructuring of organizations to address short- and long-term needs or problems.

Historical bias is another factor that one must assess to fully understand the present situation. It is important to determine if the government is pro-agriculture, attempting to be self-sufficient, or interested in balancing international trade with the help of agricultural production. Further, it is important to assess if the government is pro-intervention or is committed to minimizing the degree of dependency of farmers (water users) on government by encouraging greater participation and sharing of responsibility by them.

In most areas of the world, there is an increasing demand for water, requiring the development of new supplies and/or a reallocation and more efficient use of existing supplies. Population increases and shifts create urban and rural domestic water supply demands, as do the emerging support systems and related infrastructure. The introduction of technological improvements increases the demand for water, as this enables certain developments to take place that require water as an ingredient or medium. In the area of irrigated agriculture, the acceleration of construction, operation, maintenance, and rehabilitation costs have effected the demand for water by requiring more efficient storage, delivery, distribution, and application systems.

In theory, water lost is water wasted, resulting in less than optimal production. In the last decade, there has developed a global awareness of the increasing and multiple-use demand for water and a focus upon more efficient and effective development and utilization of this resource. For a variety of reasons throughout a nation's history of development and use of water resources, there are identifiable cycles of progress toward meeting these demands. The cycles themselves are very often caused by the reordering of priorities at the national level, in which emphasis on water will at one point be a high priority, leading to greater than usual attention and funding of water programs, and at other times be of lower priority, leading to some slippage in the maintenance and rehabilitation of project facilities. This contributes to lower production rates, causing a reevaluation in the order of priorities, which may again place water management at or near the top. This cyclical phenomenon is also reflected in the laws and organizations where there has been a conscious recognition that the institutional arena exists as a constraint or may be used as a tool to implement desired changes.

SYSTEMS OF LAW

Generally speaking, from a comparative point of view, four major legal systems (or more accurately, legal families) have been identified and classified according to their theoretical natures. The first is the Romano-Germanic family, originating in Europe and, through colonization, spreading to other areas of the world. David and Brierly (1978) describe this family:

> Here the rule of law is conceived as a rule of conduct intimately linked to ideas of justice and morality. To determine and formulate these rules falls principally to legal scholars who, absorbed by this task of enunciating the 'doctrine' on an aspect of the law, are somewhat less interested in the actual administration and legal practitioners.

The second major family of law is the "common law" found in England and those countries modeling their legal system after the English law. This body of law is commonly referred to as "judge-made" law, based upon the standards of the community as interpreted by the court. In many countries, this body of common law has been supplemented by "legislative" law to set the standards and penalties, with the courts applying the law by interpreting it for cases brought before it.

David and Brierly (1978) identify as the third major legal system the family of socialist laws. This legal system has many of the attributes found in the Romano-Germanic family but has been especially tailored to the political philosophy reflected in socialist countries.

A fourth broad group of legal families includes those of a philosophical, traditional, customary, or religious nature, such as that found in the Islamic countries or at local levels where traditional or customary law is still prevalent. Over time, many countries have interjected portions of other legal systems in some fields, such as in water resources development, but retained the formal law over such matters as family relationships and property ownership.

A number of major legal systems specifically oriented to the development and utilization of water resources can be identified. Essentially, all water law systems dealing with the internal water resources of a country, as opposed to international waters, fall into one of two categories, based upon the primary political jurisdiction over the resource. A nation's water law system is either a national system or a federated system.

Countries in which the primary jurisdiction is placed at the highest level of government and in which the law adopted is applicable nationwide have a national system of water law. In such cases--as in Spain, Mexico, the Philippines, and Egypt--the federal government adopts a law and either promulgates regulations or delegates to regional offices of the national water agency the responsibility to formulate regulations independently or in conjunction with the state or province. On the positive side, this allows for nationwide uniformity of general principles, with flexibility to develop implementing rules and regulations regionally to reflect the different geohydrologic conditions. In most instances, this system avoids or minimizes interprovincial or interstate disputes over the allocation of water. On the negative side, it is often found that a national water law lacks specificity and comprehensiveness. Further, it is often found that national water agencies fail to develop rules and regulations reflecting different conditions found throughout the country. This may, in itself, lead to inadequate or impossible implementation of the law.

Under the federated system of water law, primary jurisdiction is in the state or province regarding the allocation, distribution, and administration of water, but with a certain degree of authority retained by the federal government for matters such as jurisdiction over interstate waters for navigation, flood control, hydropower site location, and commerce. Often the federal government must be involved in matters affecting interprovincial or interstate allocation of water. Countries having this system include Argentina, India, Pakistan, and the United States. On the positive side, such a system of law allows for the enactment of legislation reflecting the particular conditions found within the political subdivision. In so doing, emphasis can be placed where specifically needed, such as on groundwater development or utilization of water for municipal, industrial, and/or other needs. Because jurisdiction over water has been decentralized to the state level, it allows for promulgation of more specific rules and regulations under the laws. The disadvantages may be the lack of uniformity between the states or provinces in the legal systems adopted and conflicts arising over interprovincial or interstate water allocation. From an economic standpoint, there may be additional costs for fully staffed agencies within each state or province, as opposed to one central agency with regional or field offices.

Since most political boundaries or political subdivisions do not conform to hydrologic boundaries, a conscious effort must be made under either system to facilitate administration over water resources along hydrologic boundaries through the use of appropriate legal governmental entities. The river basin authorities and hydrographic confederations found in England and Spain are examples.

Major water law systems can be further classified as customary, traditional, or modern (Radosevich, 1976). The basis of this classification is the nature of the particular water laws or codes in the country. Customary water laws may or may not be written, generally allow for a great deal of local administration, and may be closely tied to a particular social pattern in the area. For example, Moslem water law is found in Saudi Arabia and other Islamic countries in which Sharia law is followed; the Hindu-Bali system exists in Bali, Indonesia, in which the subak, or village irrigation organization, operates. Traditional water laws can be found in Spain, Argentina, Pakistan, and India, where laws have been adopted addressing certain aspects of water use, generally providing for centralized administration of the law. Many countries have adopted modern water laws in the last few decades resulting from the recognition of the interdependent relationship between ground and surface waters, the integrated nature and need for water quantity and quality administration, and the multisector utilization of the resource. Modern water laws and codes, such as those adopted in the Philippines, some states in the United States, and England, contain policy statements to serve as general guidelines and comprehensive provisions covering conjunctive use of ground and surface water, integration of water quantity and quality control, and allocation and reallocation of water resources according to specified criteria and the demand on other and associated resources.

In 1975, the International Conference on Global Water Law Systems was held at Valencia, Spain. The purpose of this conference was to identify and describe the most significant "systems" of water law, based upon distinct principles and degree of influence through adoption in various areas or countries of the world (Radosevich, 1976). The conference produced a systematic description of many major systems and a methodology for assessing desirable characteristics of a water law system in any particular case. Among the water law systems reviewed were the Spanish, French, British, Romano-Italian, Soviet, Hindu-Bali, Moslem, Israeli, Latin American, and variations found in the United States. (For a brief summary of these systems refer to Radosevich, Global Water Law Systems: Summary Report, 1976, pp. 16-28.) Many of these systems have been introduced into other parts of the world. Figure 15.1 identifies the major legal systems and their paths of influence.

The analytical methodology developed for the conference to determine the most desirable water law system resulted from interdisciplinary, intersectoral, and intergovernmental contributions. The process that emerged requires the identification of inputs, formulation of policies based upon the national and/or regional objectives, and an examination of the doctrinal variations and organizational alternatives that exist in the major systems (see Figure 15.2). The paths identified by this methodology, of course, only served to develop a framework upon which the system of law may be built. Considerable effort must be made to tailor any particular model to the specific conditions and needs of the country or region.

WATER AND RELATED POLICIES

What is Policy?

In the context of this paper, policy is defined as a set of political and governmental pronouncements, generally in written form, that serve as directives or guidelines for achieving goals and objectives. Dunn has traced the etymological roots of the term "policy" from its Greek, Sanskrit, and Latin languages, as referring to the conduct of public affairs or the administration of government (1981, p. 7). Through his exhaustive work, he has developed a framework for policy analysis. He prescribed a

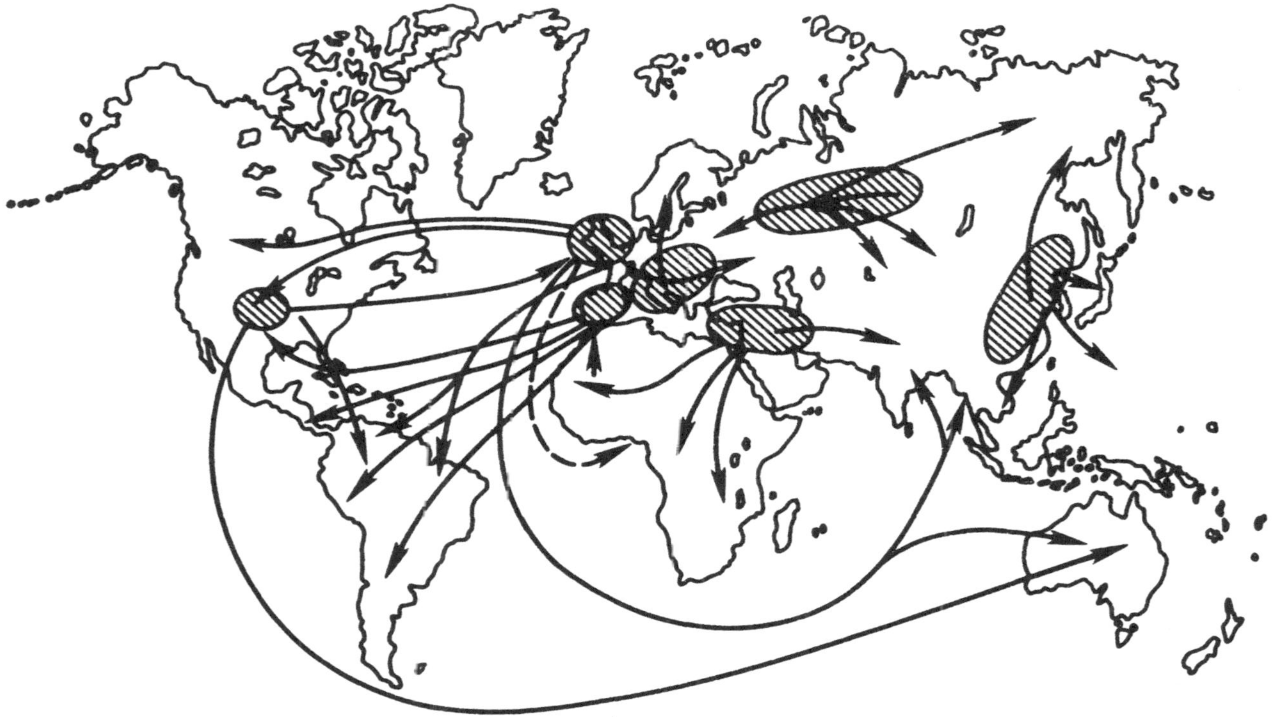

Figure 15.1 A descriptive map of major legal systems ( ) and their variations or paths of influence (⟶)

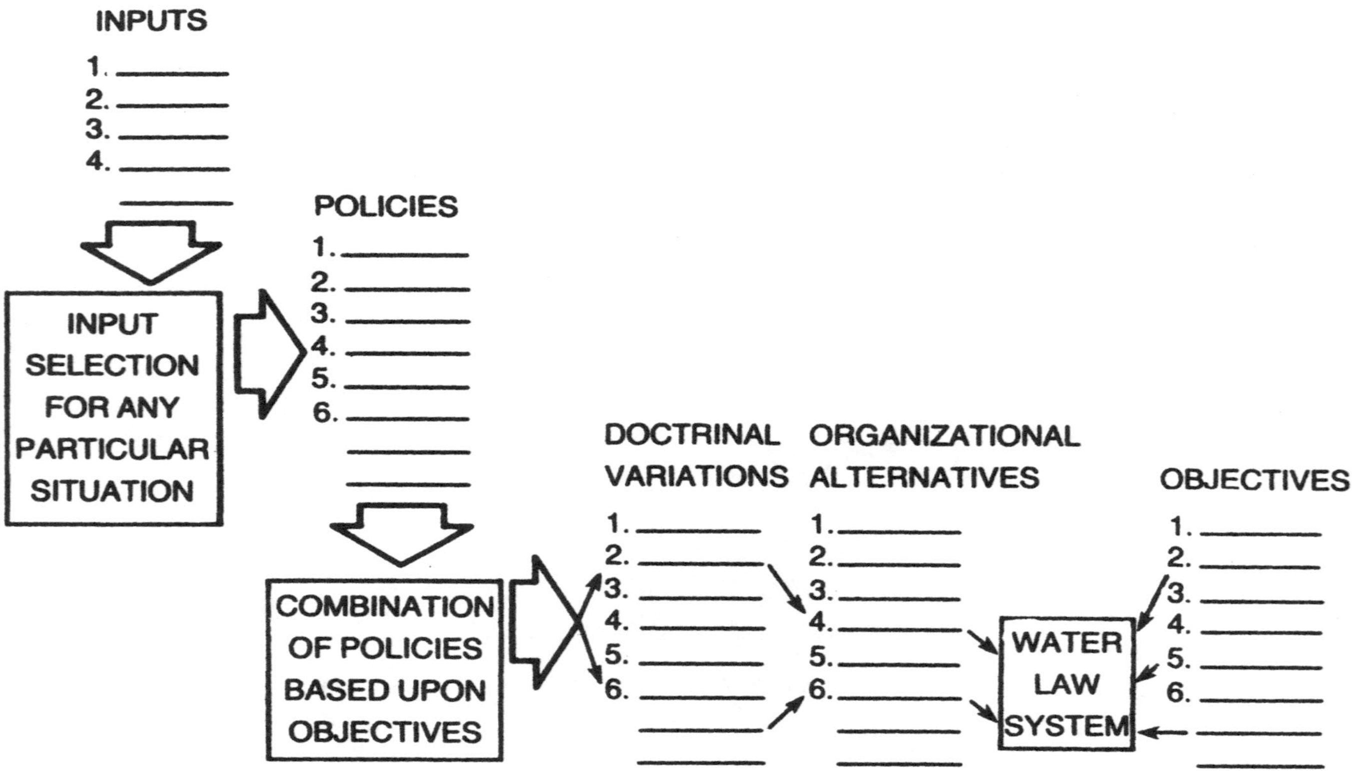

Figure 15.2 "Path of solutions" for determining the most desirable "water law system" for any particular case

range of analytical methods that include problem identification, forecasting, recommendations, monitoring, and evaluation. In the policy analysis approach that he describes, it is important to note that policy is not something that just is; often it is something that evolves from a recognition that a new direction must be taken or that particular problems must be solved with some fundamental guideline established to provide the thrust and direction for a particular process or program.

Where are Policies Found?

Policies may be generated in two specific areas. The first are "political" pronouncements formulated by legislative bodies or political leaders of a country. The second type of policy statements are "governmental," that is, those adopted by implementing organizations giving specific directions in carrying out certain tasks and duties relevant to the achievement of particular goals set by the political forces of the country. The first type of policy pronouncements may be found as separate government declarations, as preambles in particular legislation or decrees, or in such documents as five-year plans or other forecasting framework reports. The second type often are difficult to locate. They are generally not recorded in a systematic fashion. Often they are found in manuals or may exist only in letter form.

The Role of Policy in Water Management

Water policy and policy statements related to other resources or sector activities in which water is used as an input or medium are extremely important for the achievement of efficient and effective utilization of this resource. The primary purpose of water policy is that it adds both direction and flexibility to the planning and management process. One very interesting feature of policy is that it is subject to change according to national and regional needs and desires, even though the law and organizations remain the same. This is the major distinction between policy and law. Policy provides the focus and orientation; law provides the substantive parameters and procedural process for the implementing organizations.

WATER LAWS

Historically, discussions of water law have been confined to water quantity development and control in a particular area or nation. In the last decade, reference is to a much broader context, referring not only to water quantity laws, but also to laws regarding water quality, conservation, planning, geothermal resources development, weather modification, and a host of laws that may be enacted governing specific sectors or activities such as transportation, fisheries, coastal zone management, etc.

In general, water law refers to basic policies, substantive provisions, and organizational structures. Figure 15.3 illustrates a desirable approach to examining the formulation of water laws. It begins with expressed or implied goals as defined by people, translated into policies, and enacted into laws, decrees, or orders by a representative body or executive. These policies and laws are then refined by the implementing agencies, who promulgate internal policies, rules, regulations, and manuals that relate back to achievement of the goals. In the event of a dispute or question, courts are often called upon to test and interpret the water law relative to the goals and constitutional principles.

The main topics found in most water laws generally relate to the physical (quality) and chemical (quality) properties of the resource. Modern, comprehensive water codes generally begin with a declaration of objectives and principles, followed by statements on the ownership of water resources. In many cases, the nation's waters are owned by the state or public. However, in certain countries, there is a distinction between "public waters" that are owned or over which the government exercises jurisdiction and "private waters" such as springs and groundwater that may be privately owned or exempt from government control. The code may then contain sections or parts relevant to the quantity control and others to quality control.

On the quantity side, it is important to take into account the natural system in the formulation of a legal system (see Figure 15.4). Most often, there are specific sections that address surface waters and others directed toward groundwater. Modern codes contain sections dealing with the conjunctive use of surface and groundwater where there is a hydrologic connection between the two and where optimum use can be achieved by carefully managing the use

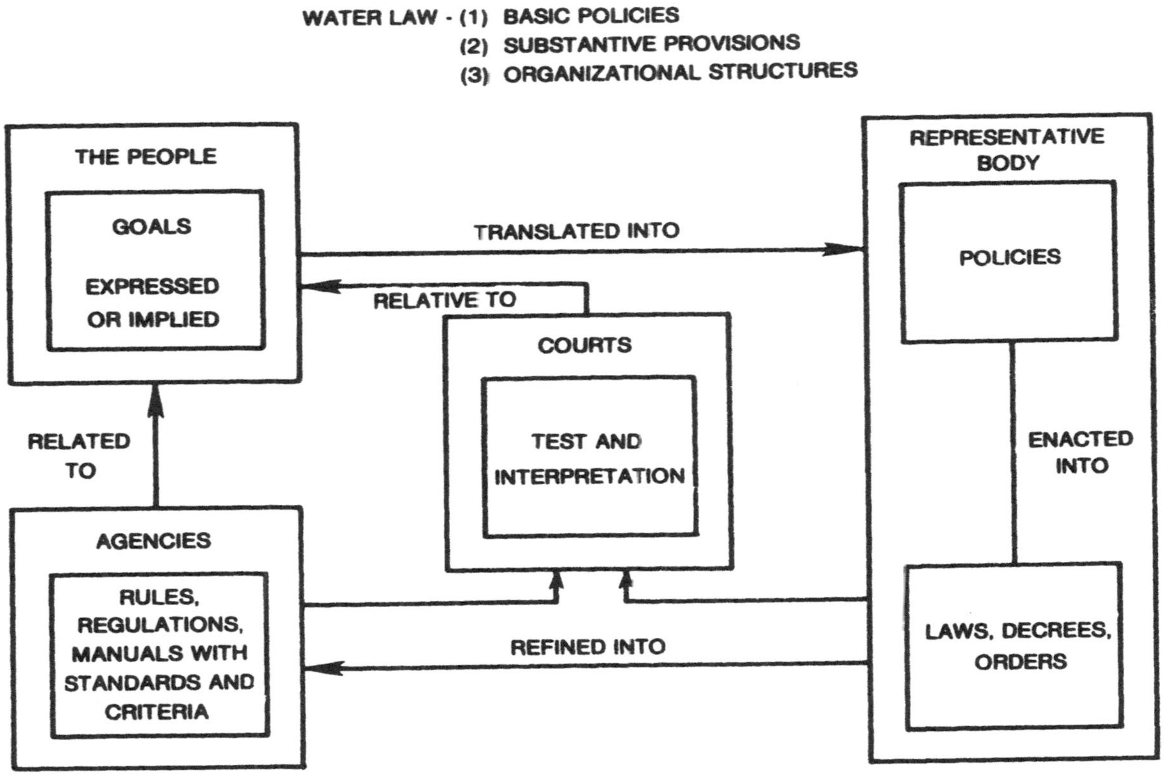

Figure 15.3 Formulation of water law

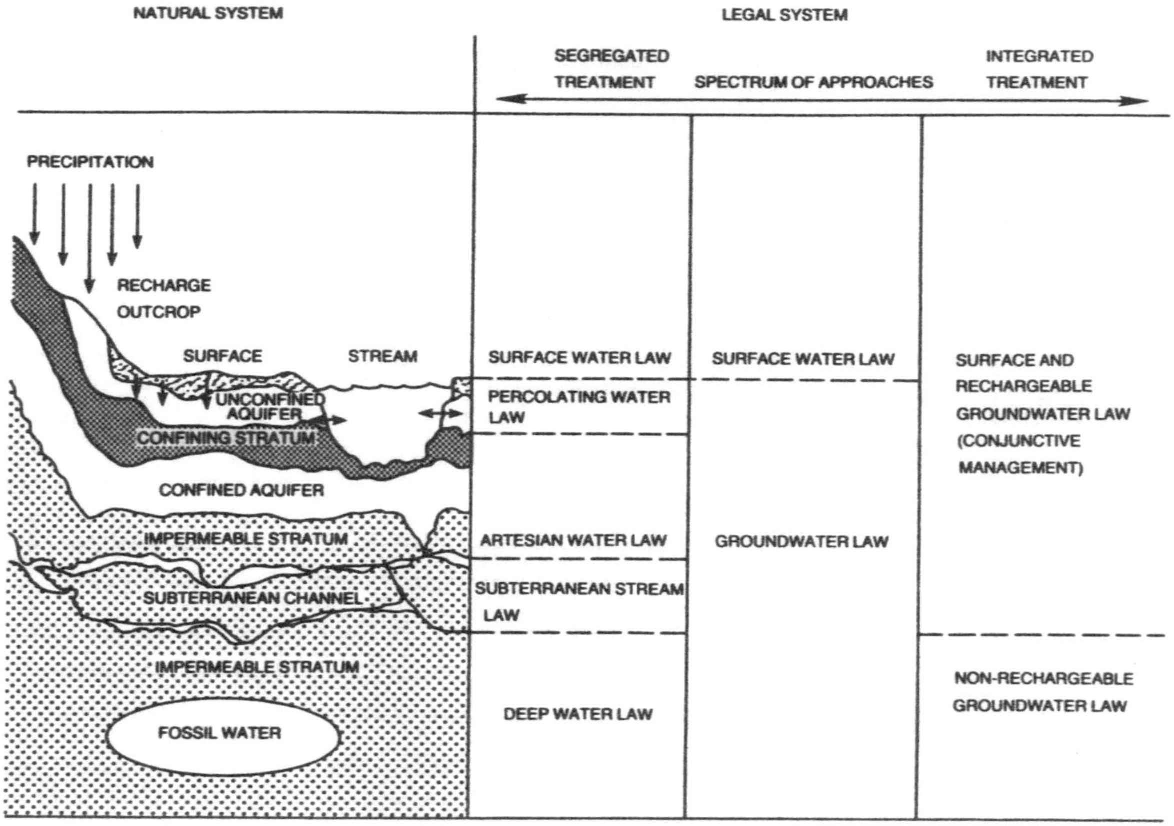

Figure 15.4 Natural system/legal system interface (John Muir Institute)

of these two sources according to time and geographic locations. In many instances, artesian and deep or nonrechargeable groundwater require specific legal interventions. Water quantity laws generally include provisions with respect to the allocations, utilization, conservation, and administration of water; coastal zone management and seawater intrusion protection provisions; interbasin and interprovincial transfer of the water; and enforcement of the laws.

Water quality laws are usually tied to environmental protection enactments or the solution to particular problems, such as contamination of domestic water supply and adverse effects upon other users or resources (e.g., agriculture and aquatic life). They are often under the jurisdiction of a separate agency. Water quality laws may give specific attention to surface and groundwater, provide for monitoring groundwater quality, establish ambient and discharge standards, require permits for certain types of discharges that may degrade water quality beyond acceptable levels, and provide for enforcing the water quality laws.

In modern water codes, there are efforts to integrate water quantity and quality control. An example of a recent water code that is comprehensive yet simple and pragmatic is the Philippine Water Code enacted in 1976. It consists of 9 chapters of 101 articles, totaling just 25 pages. Although it is certainly not the answer to every nation's or state's needs, it is useful to examine as a basic source document.

Analyses of water policies and laws often overlook or ignore the importance of agency rules, regulations, instructions, and manuals. Depending upon the particular legal system and the system of law adopted in a country, it is generally incumbent upon the implementing organization to develop rules and regulations that add specificity and functionality in the implementation of the law. These rules and regulations are most often the only guidelines used by local water administrators and officials. For this reason, they are often the most important (or should be the most important) documents for efficient and effective water management. These rules and regulations should be clear and concise, easily understood by the persons directed to carry them out, and refer to the policies and/or laws upon which they are based. Too often, field officers do not know, understand, or appreciate the importance of the job in achieving the goals and objectives set at the national or state level. Manuals should include not only technical specifications and design criteria, but also the legal pronouncements relevant to the administration of water and to the operation and maintenance of the particular system. What should and what does exist, however, differs considerably from country to country, depending on the role and attitude of government water agency(ies).

SYSTEMS OF ORGANIZATION

There are three distinct categories of organizations of concern to the matter of water development and management. These three systems are (1) government agencies, (2) judicial systems, and (3) nongovernmental entities. The role and significance of each varies from country to country, according to the degree of development in irrigated agriculture. Although their functions differ, as well as their methods of operation, the ultimate objective is (or ought to be) the most efficient and effective utilization of water resources for production-related activities.

Government organizations are generally formed at the national level and state or provincial level, with offices at the local level, depending upon the form of government and system of law. Under a national system of water law, the state and local agencies are usually under the direction and control of the national office. Under a federated system, the national agency may have regional offices or simply coordinate with state agencies, but the state agency is independent of the national office. At the national level, either ministries, departments, divisions, and sections or departments, bureaus, divisions, and sections are formed, with some variation in the hierarchy from country to country. Generally, one governmental ministry or department will have primary jurisdiction over water, with ancillary or additional jurisdiction delegated to other governmental organizations. Delegation of authority and/or jurisdiction over subject matter may be either comprehensive or specific. Regarding irrigation and agriculture, these two subjects may be in the same ministry, with jurisdiction delegated to two different departments, or a there may be a ministry for irrigation (or broadly, water) and another for agriculture, with departments or line agencies addressing the same topics. Because of the wide range of activities in which water is either the median or ingredient, interest and jurisdiction over water may be spread among a large number of government organizations. For example, in Sri Lanka, there are 9 ministries and 22 departments involved in some way with water resources. In the Philippines, prior to the adoption of the new water law in 1976, there were 50 different agencies with jurisdiction over water.

Two types of government agencies are of particular concern to irrigated agriculture. One is the development agency, such as the Water and Power Development Authority of Pakistan or the Bureau of Reclamation in the United States. The second is the operational agency, such as the irrigation departments found in many countries, or the state engineers' offices found in many western states of the United States. In some countries, these two activities are merged into one, such as the Royal Irrigation Department in Thailand and the National Irrigation Administration of the Philippines. They have responsibility for both the construction and the operation and maintenance of irrigation projects.

As a result of the delegation of subject matter jurisdiction to numerous government agencies, many practical problems emerge. The first concerns the difference between the delegation of authority versus the assignment of responsibility. Some agencies may be authorized to perform certain tasks but not be responsible to ensure that they are carried out. Others may be responsible for certain activities but not have full authority to ensure their successful implementation. The second major type of problem has to do with cooperation and coordination of agency activities. In many countries, there is no formal body at the highest level of policy-making to ensure that the actions and programs carried out by the agencies are coordinated or that, in fact, the agencies achieve any level of cooperation. Beginning in the early 1960s, a trend was started to establish water resources councils at the national level, consisting of the ministers or secretaries and heads of departments. A number of countries have formed or are in the process of forming this umbrella entity, which, in many instances, goes beyond policy formulation and coordination of activities to include the development of essential data banks for both supply and utilization of water.

The second major system of organization is the judiciary. The courts may be classified in a variety of ways. Generally, the organizational structure is some form of trial court--usually at the local level--with one or more appellate courts and, ultimately, the court of last resort. The courts are also subject to geographical venue, with the usual limitation that a judgment rendered in one jurisdiction may not be readily enforceable in another. Subject matter jurisdiction is also a classification that may or may not have relevance to irrigated agriculture. In general, the local trial court may hear matters pertaining to water disputes and interpretation of laws; however, in some countries special water courts have been established to avoid many of the complex procedural matters found to apply in general litigation of disputes and also to act in a timely fashion because of the critical nature of water itself. Perhaps one of the oldest and best-known special water courts is the Tribunal of Waters found in Valencia, Spain, which has existed for over a thousand years. It, however, does not consist of politically appointed or elected judges, but rather is an appendage to the water user organization system, in which the presidents of each of the eight irrigation communities in the Valencia Valley sit as judges to hear disputes or issues related to water. The Tribunal convenes at the Apostle's Gate of the Cathedral of Valencia every Thursday at noon. Attorneys cannot appear before it, and any party named in a complaint who fails to appear before the court may have delivery of his water suspended until he agrees to resolve the issue (Radosevich, July 1976). Other examples of water courts are those in the state of Colorado, having responsibility for not only resolving disputes and interpreting the law, but also for allocating water to applicants by decree (Radosevich, et al., 1976).

The third type of organization that is becoming increasingly important to efficient and effective management of water is the nongovernmental or (in some countries) quasi-governmental entity called the water user organization or association. The primary role of this entity is to provide the local infrastructure for improving irrigation efficiency and increasing agricultural production by operating and maintaining the local distribution system below government outlet. There are many examples of historically famous water user organizations, such as the subak in Bali, Indonesia; the Community of Irrigators in Valencia, Spain; the traditional irrigation associations of northern Thailand, and the mutual ditch companies in many western states in the United States. These organizations have evolved and been refined over time and, in recent years, have been considered to be an extremely useful and beneficial mechanism for encouraging and increasing farmer participation in water improvement programs (Radosevich, 1977).

There are various types of water user associations. Many have legal status; others are customary or indigenous organizations based upon tradition. The organizations may range from simple, single-purpose entities to complex, multipurpose organizations. For example, in the state of Colorado, there are single-purpose irrigation companies only responsible for delivery of water to shareholders; there are water user associations that consist of a number of irrigation companies working together to construct and improve a larger storage and delivery system; there are irrigation districts which have quasi-public status, enabling them to levy a tax upon landowners in the area in order to pay for the construction of storage and delivery works to an even larger but local area; and there are conservancy or conservation districts which are multipurpose and consist of jurisdiction over more than one watershed. Ultimately, there may be river basin authorities which are multipurpose, complex organizations having broad authority to develop and administer projects at the basin level.

Focusing on the local water user organization, some of the major purposes for its formation are to get farmers involved in decision-making; to get farmers involved in managing the local delivery system; to serve as a communication link between the water users and the government and as a form for the dissemination of information among the membership; to resolve disputes at the local level; to promote collective action among the farmers to improve the system; to assure there is a formal mechanism for administering government improvement programs; and to capitalize on the economies of scale gained from many farmers working together but under a centralized decision-making process that exists by having a committee or board as the contact point. These water user organizations may be formed from the bottom up, in which the initiative to work together comes from the farmers. This is essentially the method by which the irrigation companies in the western part of the United States and the system found in Bali, Indonesia, and Valencia, Spain, emerged. The second method is to form from the top down; in such cases, the government has concluded that water management could be improved if the water users entered into an organization of their own in order to undertake local operation and maintenance of their system and administration of water within their command area. Within the last ten years, a number of countries have introduced legislation granting legal status and setting procedures for creating water user associations, either at the initiative of the water users or at the direction of the government. Considerable success has been achieved in the Philippines through the National Irrigation Administration's (NIA's) ability to organize farmers. Pakistan has adopted water user association acts in all four of its provinces as the mechanism to implement the World Bank-funded On-Farm Water Management Program (WUA Pakistan, 1984).

INSTITUTIONAL PROBLEMS, CONCERNS AND RECOMMENDATIONS

Water Allocation

There are five general problem areas that have been identified from an institutional standpoint that effect improved water management of irrigated agriculture. They are related to (1) water allocation, (2) planning, development, and management of development programs, (3) cost recovery, (4) water user organizations, and (5) dispute resolution. These concerns and problems have to do with the adequacy and sufficiency of the law and with the capability or sufficiency of the organizational structures and their implementation approach.

The issue of water allocation concerns the identification of ownership of the resource, initially, and then with the process of allocating the water to the various users and types of uses. In most countries, all (or at least surface) water in rivers and lakes and groundwater below certain depths are under the jurisdictional control of the government. In many countries, exceptions are allowed for domestic uses of water, diffused surface water, and shallow groundwater. The amount of water consumed for domestic purposes or small households, however, is marginal, compared to that used for large irrigation projects. For this reason, ownership of water is very important to determining the nature of allocation of water by the government, particularly for medium- and large-scale irrigation projects. The problem that exists in many countries is the lack of commitment by the government to allocate water in a definitive fashion. As a result, there is little reliability for the water user to take the risk of his own investment. Another difficulty caused from the failure to allocate water to specific uses occurs in conflicts between types of uses, such as for agriculture, power, fisheries, and transportation. Included in conflicts that can occur are also quality degradation from return flows and differing requirements for multiple uses of water.

It is recommended that governments develop a system for allocating water to users and identify the type, quantity, source, and location of use in the form of some documentation. This raises the issue of "water rights" versus "licenses" or "ownership" versus an "interest" by the water user. In his paper for this series, Dr. Coward (1986) discusses the issue of "property rights" as they may affect water allocation, and particularly the perception that farmers may have as to the dependability of water supply for their crops vis-a-vis other water users. The point here is not to identify a "property right" concept in water, in the traditional sense (as the legal profession might), but rather to use the term in the broader context that one might find in developing countries. For example, in the western United States, a property right in water refers to a real, tangible, valuable, transferable, salable right in a certain amount of water, which is recognized and protected by the government.

In most countries, this concept does not, nor could, exist. In many developed as well as most developing countries, it is well understood that the government or the people own the water and that private ownership interests cannot exist, except perhaps in springs or certain underground water. Water may be allocated for use by concession or license but is still not considered a "real" property right giving the holder right to sell or transfer it to another. Licenses and concessions may be transferred upon approval of the government but are still not property rights that one might equate to the rights that exist in land.

The entire concept of "property" is one that is based upon the expectation a person has toward a particular thing. Social behavior is partially dictated by the extent of the expectation that exists in different countries. Perhaps this is one of the reasons why water users in many countries do not participate willingly in the development, operation, and maintenance of systems that deliver water to them. The source of their apathy may be that they have no interest or expectation in the use of water resources and, consequently, will limit their investment, time, and resources in order to conserve for purposes that may be more rewarding. For this reason, it is highly recommended that governments consider the adoption of tangible and definable means of identifying the rights or interests of water users. The process of adopting this mechanism allows for the establishment of a system of water allocation to individuals, groups, or projects, which can add to the sense of dependability and security for the water user to have available supplies, according to certain conditions as set out in the allocation. It will also assist the government in the planning process by knowing where, when, and how much water is allocated and how much may be available for future development.

Planning, Development, and Management (PDM)

As previously mentioned, in the context of this paper, planning is considered the blueprint, and management the process of implementation. Together, planning and management form a continuous process. Through the implementation process and evaluation of the results, it is found that some of the problem areas identified in the management of water resources stem from the lack or inadequacy of an accurate data base. Within the organizational structure should be located an entity with responsibility for collecting hydrologic data on supply and demand to include the amount, timing, and location. Within the last few years, discussions in several countries have revealed the fact that often one agency of government relies upon a set of figures on water availability and use for planning purposes, while another agency--whose activities definitely would affect the first--is relying upon data that is considerably different. Conventional means for data collection are necessary, particularly to obtain the "ground truth." However, use of more advanced techniques should be considered, particularly the use of satellite imagery and remote sensing techniques. For example, the use of data collection platforms (DCPs), which transmit data to a satellite that relays it back to a collection station established by the agency, has been of immense benefit to the U.S. Geological Survey and Water Resources Department of Colorado. The use of satellites can make a significant impact on the decision-making process of the government, particularly for verification and validation of the data base. It is maintained that without accurate data, accurate decisions cannot be made.

Another PDM problem consists of conflicting objectives within and among government organizations. These conflicting objectives may exist both at the geographical and sectoral levels. Geographically, a government may be interested in pursuing objectives that have little or no relevance at the regional or local level, and vice versa. Similarly, government efforts in the various sectors--such as in water and power development, agriculture, transportation, and fisheries protection--may create adverse impacts upon one another, unless the agencies cooperate and coordinate their activities.

There are three organizational issues or problems that hinder water resources development and management. The first is the bureaucracy of the structure and hierarchy of the organization. For decades, England has considered the pros and cons of centralization versus decentralization. In the '50s and '60s, water administration was centralized; in the late '60s and '70s, it was found that a more efficient means for managing water was to decentralize administration at the river basin level. The appropriate solution for any country, of course, depends upon its own needs, desires, and capabilities. Centralization has the benefit of a uniform policy formulation and dissemination. Decentralization allows for more water user participation because of the closer contact between the users and the decision makers. This issue is not one that is an either/or in most cases, but rather the proper mix or combination of centralization and decentralization, depending upon the issue and function to be performed.

The second organizational issue is concerned with the "functional" versus "objective" orientation of most government water agencies. This is considered a significant issue in attempting to develop programs of farmer participation and in improving the operation and maintenance of systems by the government agency. Most water or irrigation departments perform certain distinct functions such as construction, operation, and maintenance of storage, delivery, and distribution systems. From the top down, personnel are given certain tasks to perform, and, particularly at the local level, the duties are detailed and inflexible. The concern with this administrative approach is that there is, regardless of the level in the hierarchy, the responsibility only to perform the duties as designated, very often without consideration of the significance to the real objective in performance of these duties. This functional versus objective orientation is considered to be one of the reasons why improper operation and maintenance of some systems exists; it may contribute to water user disinterest in working with government officials and account for those officials failing to successfully operate and maintain the system.

Another organizational concern that has emerged within several countries within the last few years is the "subsidized" versus the "self-sufficient" operation of the water agency itself. In most countries, the irrigation department or water agency is centrally funded, thereby not relying upon the generation of any income for its own operation. In the Philippines, however, the NIA has been directed to become more self-sufficient, which may not only result in its receiving less funding from the central government, but may also require streamlining its own operations in order to eliminate marginal activities, reduce personnel requirements, and, in general, strive for greater efficiency in its operations.

It is suggested that government establish a PDM system based upon management by results (MBR) and adopt a process of communication through cooperation, coordination, and control (CCC) (Figure 15.5). Peter F. Drucker introduced the concept of management by objectives in 1972. Professors Seckler and Nobe expanded on this concept and relate it to improving the management capability of developing economies. In the process, however, they have concluded that a more effective method is to shift from management by objectives to management by results, in which one evaluates the difference between what is desired and what occurs in order to provide the feedback for improving the system. It is for this reason that MBR should be implemented through a process of communication in which the structural guidelines, operation and maintenance manuals, and performance of activities among and within a water organization be based upon cooperation, coordination, and exercise of the appropriate degree of control.

Cost Recovery

The issues of water pricing and cost recovery for construction and/or operation and maintenance costs involve not only economical and technical considerations, but also the adequacy of the institutional framework. Cost recovery in the United States has been tied to most of the Bureau of Reclamation projects constructed under

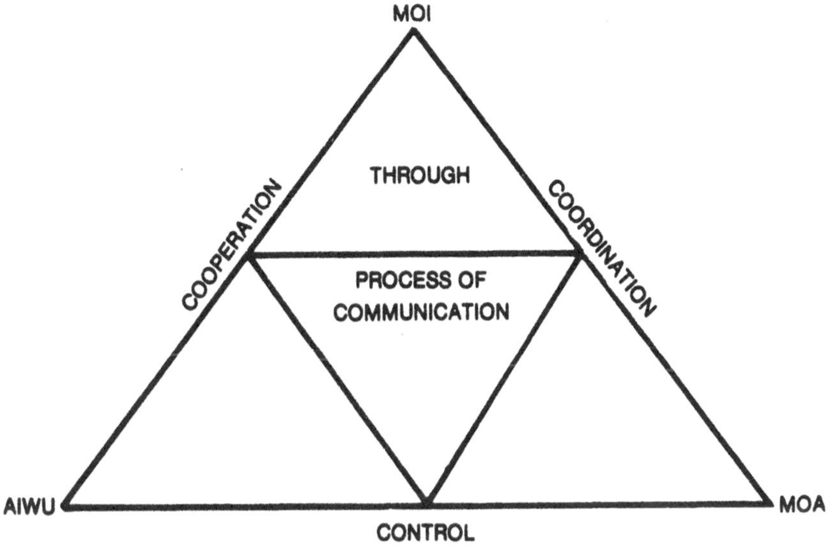

THROUGH
PROCESS OF COMMUNICATION
A JOINT EFFORT OF
MINISTRY OF IRRIGATION, ASSOCIATION OF IRRIGATION WATER USERS,
AND THE MINISTRY OF AGRICULTURE
IN IMPROVING WATER MANAGEMENT CAN
ACHIEVE INCREASED FOOD PRODUCTION

BY

COOPERATION — WORKING TOGETHER FOR ATTAINMENT OF THE NATIONAL GOAL OF INCREASING FOOD PRODUCTION

COORDINATION — OF RELATED ACTIVITIES, SUCH AS SYNCHRONIZED AND SYSTEMATIZED SETTING OF CULTIVATION AND IRRIGATION SCHEDULES, PERFORMANCE OF COOPERATIVE EFFORTS OF EXTENSION AND TRAINING CONTROL

CONTROL — NOT OF EACH OTHER, BUT OF THE RESOURCES:

WATER - ALLOCATION AND TIMELY DELIVERY BY MOI
- EFFICIENT DISTRIBUTION AND USE BY AIWU

OTHER INPUTS - TIMELY AVAILABILITY BY MOA

Figure 15.5 Improving water management for increased food production in Egypt

the 1902 Reclamation Act, an act establishing a land reclamation and settlement program for the western part of the United States. This cost-recovery concept is based upon the "ability-to-pay" principle, with repayment of a portion of construction costs over a 40-year period. Acceptance of the cost-recovery program is one of the prerequisites to government interventions. The water users must form a legal entity with a representative body to contract with the Bureau for project design, construction, and initial operation. This entity has generally been an "irrigation district," which has become one of the major institutions in the development of irrigated agriculture in the United States.

In the last 10 to 15 years, numerous countries have considered or attempted to implement cost-recovery programs for the construction, operation, and maintenance of irrigation systems. International donors, such as the World Bank, have stressed cost recovery in their grant and loan programs as a tool to improve water use efficiency. Laws in some countries, such as Thailand, now provide for irrigation fees to be placed into a revolving account for rehabilitation, operation, and maintenance purposes. Unfortunately, as is the case in Thailand, the law may make provisions for or require collection of water charges, but, in actual fact, few or no collections are made. Discussions with irrigation officials in several countries indicate that the rate for water charges was set far below the actual cost of collecting them; therefore, that portion of the law is not implemented.

A number of problems exist regarding the institutional aspects of cost recovery. The first is an inadequate legal framework and basis for the water charges. A clear distinction must be made between collecting for construction costs and collecting for operation and maintenance. Water users may not be able to relate to the total costs of constructing the project and, in most instances, do not feel an obligation to repay all or any of these costs, since they have had little or no input into the project. The situation is somewhat different with respect to collecting operation and maintenance charges; the farmer may readily see the benefit from an irrigation department's efforts to operate and maintain the system effectively.

The second problem is an inadequate organizational framework for informing water users of costs and for collecting the fees. A range of alternatives exists, from the revenue department collecting the fees as part of the land tax; to the irrigation department maintaining the records, collecting the charges in cash (or in kind); to the farmers collecting the monies through a water user association, with a proportion retained by the association for improvement within their own command area.

A good cost-recovery program should be contingent upon delivery of an adequate water supply. Unfortunately, this contingency cannot be met by the irrigation departments in all areas at all times. Often, this is because of the "functional" versus "objective" orientation of the agency. Other times, it is because of an inadequate supply of water. In still other situations, it is because of inadequately or poorly designed, constructed, maintained, and operated delivery systems. It is suggested that a government might (1) adopt a policy designed to achieve the objective of cost recovery, (2) adopt a law containing the legal mechanism for introducing a cost recovery program, and (3) develop an implementation approach which is realistic.

Water User Organizations

Perhaps one of the most talked-about institutional issues that has emerged in the last five years has been the role of and need for water user organizations to improve irrigated agriculture. These organizations may be nongovernmental or quasi-governmental, and are referred to by a variety of names in different countries. Generally, irrigators receiving water from a common point or from a project area come together to create a formal or informal organization for the distribution of water received from a government supply channel or directly from the source of supply (such as a weir or well) and to operate and maintain the distribution systems. In many instances, the impetus for this cooperative effort is the construction of a diversion and delivery system (Radosevich, 1976).

One reason water user associations have become a key topic of discussion in recent years is the fact that in many countries there are no or inadequate local organizations with primary responsibility for operation and maintenance of the on-farm water delivery system. There are many well-known and described forms of local water organizations, such as the subak in Bali, Indonesia, and the Community of Irrigators and Tribunal of Waters in Valencia, Spain. In more recent times, there are the irrigation associations in Taiwan and the various forms of the water user associations found in the Philippines,

Argentina, and the United States. The significant feature about these organizations is their apparent success in those situations in which the farmers can and do derive a benefit from their existence. In many countries, they have gained historical importance and have become "institutionalized" as part of the social fabric. In other cases, water user associations have been proposed and created where there was an initial need, but, as the benefits became less apparent or important, the organizations disintegrated. It is interesting to note that it has been difficult to replicate in northeastern Thailand the traditional irrigation organizations that have evolved and play a very important role in northern Thailand.

Although water user associations originally developed as "bottom-up" organizations, modern irrigation development of new lands and rehabilitation of old systems through government intervention in the planning and construction usually involves medium- or large-scale projects. Consequently, elements of certainty and uniformity in working with the farmers are required, leading to a "top-down" approach to forming the organizations. This can be done successfully if not forcefully imposed. Farmers should still be encouraged to understand the project and voluntarily form their local entity.

Another difficulty in forming water user associations is the historical bias against the farmers' ability to assume the role that the association would undertake. Officials often fail to give farmers credit for an instinct about and understanding of hydrology and hydraulics. The knowledge irrigators have acquired over years of experience could be a valuable input into any government or nongovernment program, but it is often wasted or even repressed by programs which tend to mold the irrigators to unfamiliar operational and social patterns which they do not fully comprehend. Formation of water user associations in order to get members of the irrigation community involved can have a very positive effect upon relations with irrigated agricultural programs and their implementation. In effect, the purpose of forming water user associations is to transfer to the water users some of the responsibility of participation previously carried out by the government. Benefits are derived by the government in freeing a number of employees to improve the operation and maintenance of the government canals, and benefits to the farmers are derived by providing opportunities for participating in decision-making and developing a sense of pride in their own system. Leaders of water user associations having historical records tell of the pride with which they carry out their duties. The historical legitimacy of their associations and the constitutional premise for doing the best job possible seems to encourage farmers into going "that extra mile" to do a better job. This pride in participation and accomplishment should be fully utilized in new irrigation programs.

Dispute Resolutions

The final institutional issue of concern is that of dispute resolution, which can occur at three levels. The first level involves interagency/interministerial disagreements, conflicts, or disputes that might arise. The resolution to this type of problem, of course, could be handled by the chief executive, minister, or other high official, as the case may be. At the interministerial or even interagency level, it is suggested that a national water resources council might be the appropriate vehicle. The council should consist of the secretaries of the various ministries having jurisdiction over water. The primary role of the council would be suggesting policy and providing guidelines for implementation, insuring coordination and cooperation among the various agencies involved, and containing a secretariat that has as one of its functions the maintenance of a water and related data base.

The second level, where disputes most often occur, is between the primary water agencies and the water users. Very often, irrigators complain that government officials do not take their interests into account, show favoritism to certain members of the community, are not uniform in their application of rules and regulations, indiscriminately cause canal closures or reduce the flow of water, etc. Agency officials commonly complain that farmers encroach upon reservoir and canal bank easement areas, expand their irrigated acreage beyond that amount of land calculated for the supply of the canals, steal water, cause obstructions in the delivery system, or, in some cases, cause the destruction of diversion and measuring devices. Resolution to these disputes occurs in one of three forms. The first is administratively, by government officials working with the water users and reaching a satisfactory settlement or the rendering of an administrative decision which is carried out by the government and followed by the water users. The second approach may be through the use of water user associations or federations of associations at the appropriate level, in which the water users themselves participate in determining the course of action that should be taken. The third is the traditional use of the courts to resolve disputes.

Obviously, not all disputes fit in any one of the three categories, nor can any one of the three categories resolve all disputes. Each has its own particular benefits in different types of circumstances. However, in most instances, traditional use of courts is not very satisfactory, the primary reason being courts often have long dockets and cannot operate in a timely fashion for the benefit of either the government or the water users. In many situations, the evidenciary and procedural requirements of a court are too time-consuming and costly for both parties.

The third level involves disputes between the water users themselves, and the use of water user organizations to resolve the conflicts, at least initially. This is particularly important to those problems occurring within the command area or even between the command areas of two or more associations. As an example, the Tribunal of Waters previously mentioned is a water user court that expedites resolution of disputes between irrigators with a minimum cost and with a high degree of fairness and understanding. Under circumstances where the water user association is not found to be a satisfactory mechanism, other possible mechanisms to use are a federation of water user associations, the government agency responsible, or the courts. It has been demonstrated in a number of instances that local and common organizations of irrigators can be effective mechanisms, not only in undertaking the performance of a specific task in the construction, operation, and maintenance of the systems, but also in dispute resolutions at the local level.

CONCLUSIONS

As initially mentioned, the institutional aspects of improving on-farm water management are often the last to be considered. One of the reasons may be that most of the players involved in water management tend to be technically oriented and so technical solutions seem more tangible, durable, and more easily implementable. Also, social and institutional solutions require considerable understanding of local conditions and culture, the taking into account of a wide variety of individual interests, and a much longer time horizon to implement. Institutional solutions involve both political and governmental considerations, requiring input from a wide variety of decision makers other than those in the substantive field such as Civil Service Commission, finance and revenue, training, balance of payments, etc.

It is important to create an atmosphere for progress and improvement within the lead agencies. There are several suggested ways to develop the proper environment. One is to articulate the purpose for carrying out the various duties and functions of the agencies. This may be done by identifying the goals or objectives, translating them into policies, and encouraging the organizations to become objective oriented, as opposed to function oriented. The second suggestion is to develop a system of planning, development, and management (PDM) based upon management by results (MBR) and the process of communication through cooperation, coordination, and control (CCC). The third suggestion is to develop and adopt a water allocation approach consistent with the PDM objectives. It may be useful to identify the allocation of water to projects or to water user groups in terms of water rights, licenses, permits, or concessions. But it is felt that the basis for the allocation of water should be clearly stated in the form of a document which the water users can understand. The document should make the water users feel that their interests have been taken into account and will not be indiscriminately cast aside for the sake of a new project. Water users should be provided an opportunity to assume a responsible role in improving water management at the local level. The concept of the water user association as an identifiable mechanism is highly recommended; further, it is suggested that governments might adopt a policy encouraging efficiency in water diversion, delivery, and use, as well as independency of the water user and self-sufficiency of the government agencies.

As time effects all elements of society, it is important to evaluate the legal and institutional structure periodically and to provide careful consideration of the effects of legal and organizational changes that are proposed. Experiences observed in many countries in which radical changes have been brought about indicate that implementation is difficult and enforcement is necessary. Three criteria should be considered in attempting to introduce legal interventions. The first is to minimize the disruption to the existing institutional framework, which requires examining the present structure and determining how it can be improved upon or, at least, how an intervention can be tailored to existing conditions. The second is to determine that any proposed institutional change is implementable and cost efficient. Many examples exist of workable solutions in one country or a part of the country meeting with failure when introduced elsewhere. It is essential to determine whether the lead agency has the capability and interest to carry out new assignments or whether, in fact, any new duties are simply going to add to an already overburdened agency. Also, if the cost of implementation exceeds the budgetary capability of the implementing agency, the likelihood of it being carried out is very minimal. The third criterion is that legal changes should strive for definiteness in the interest of the water users and the water supply, viability in the water supply, and flexibility to adapt to priorities with equity considerations. These three criteria must obviously take into account the involvement of many viewpoints, realizing that ultimately what emerges as policy, law, and organizational framework will be a political decision as to what should be done. If what ought to be done and what is done are fairly similar, then one can conclude that a high degree of agreement exists between the political, governmental, and private interests involved in water management. If the gap is wide, it will be necessary to evaluate the results against the objectives and process to determine where changes need to take place.

REFERENCES

Brusco, D., and D. Wright. "Planning and Management," Frontier, January 1984.

Coward, E. W., Jr. "State and Locality in Asian Irrigation Development: The Property Factor," Chapter 16, Irrigation Management in Developing Countries, K. C. Nobe and R. K. Sampath, eds. Boulder, Colorado: Westview Press, 1986.

David, R., and J. E. C. Brierly. Major Legal Systems in the World Today, 2nd ed. New York: MacMillan, 1978.

Dunn, William N. Public Policy Analysis, Englewood Cliffs, New Jersey: Prentice Hall, 1981.

Fairchild, Warren, and K. C. Nobe. "Improving Management of Irrigation Projects in Developing Countries:

Translating Theory into Practice," Chapter 12, Irrigation Management in Developing Countries, K. C. Nobe and R. K. Sampath, eds. Boulder, Colorado: Westview Press, 1986.

Miller, Lawrence M. American Spirit: Visions of a New Corporate Culture. Newark, New Jersey: Wm. Morrow Press, 1984.

Radosevich, G. E. Moslem Water Law and Its Influence on Spanish Water Law and the Irrigation System of Valencia. Fort Collins, Colorado: Colorado State University, July 1976.

Radosevich, G. E. Global Water Law Systems. Proceedings of the International Coference on Global Water Law Systems. Fort Collins, Colorado: Colorado State University, 1976.

Radosevich, G. E. Improving Agricultural Water Use: Organizational Alternatives. Washington, D. C.: USAID, March 1977.

Radosevich, G. E., K. C. Nobe, D. Allardice, and C. Kirkwood. Evolution and Administration of Colorado Water Law: 1876-1976. Longmont, Colorado: Water Resources Publications, 1976.

Seckler, David, and K. C. Nobe. "The Management Factor in Developing Economies," Chapter 12, Issues in Third World Development, K. C. Nobe and R. K. Sampath, eds. Boulder, Colorado: Westview Press, 1983, pp. 279-295.

Water Management Wing. Water Users Associations in Pakistan. Islamabad, Pakistan: Ministry of Food, Agriculture, and Cooperatives, Government of Pakistan, 1984.

16
State and Locality in Asian Irrigation Development: The Property Factor

E. Walter Coward, Jr.

INTRODUCTION

I begin with the observation that much of the history of irrigation development has been an interplay of state and locality initiatives and actions.[1] In some eras and in some places, local groups (and sometimes individuals) have seized opportunities to develop hydraulic facilities, sometimes in response to incentives (or at least the absence of disincentives) created through state policies. In other circumstances, the state has taken the lead and used its power and authority to mobilize resources and devote them to developing irrigation resources.

At the present time, there is a preponderance of state involvement in irrigation development: no modern nation-state in Asia is without an irrigation agency, for example. But this dominant state role must be seen in context with two additional features. First, even in state-initiated and controlled irrigation schemes, typically there remain significant local tasks to be performed--tasks frequently assigned to water user groups--real or imaginary. And second--apart from the state's involvement in irrigation, though sometimes supported by it--in many Asian countries, there continues to be a dynamic local irrigation sector supported largely by local resources. The basic proposition I wish to explore in this chapter is that irrigation development is the result of activities by both the state and the locality and that improving irrigation development outcomes is dependent on discovering (and using) better means for joining state and locality actions.

IRRIGATION INVESTMENT, PROPERTY, AND PROPERTY RELATIONS

There is value in beginning the discussion with a somewhat abstract conceptualization of irrigation development and then proceeding from that general notion to more specific discussions of practical deeds. As I have suggested previously (Coward, 1983), one can view irrigation development as a property-creating process. This property-creating process has two linked meanings. On the one hand, it means that irrigation development leads to (1) the creation of new objects of property (weirs, canals, water rights, etc.) and (2) the possibility of new property relations. The latter point refers to the fact that, as new objects of property are created, the relationships among people related to that object of property may also be adjusted or created de novo. To take a simple example, if an existing canal is extended to serve additional water users, the canal extension and the additional water rights created can be seen as new objects of property. Likewise, the relationships among the newly served water users and between them and the original users can all be seen as new property relations--i.e., social relations based on their joint (though not necessarily equal) rights to the canal and its water supply.

Beyond the matter of property creation, one can also consider property maintenance. Based on a review of the local irrigation sector in parts of Asia, I have argued that these experiences suggest that "ownership of and responsibility for irrigation works invariably coincide" (Coward, 1983). That is, based on the cases reviewed, one could see that the group making the original irrigation investment also had the responsibility for the upkeep of the facilities they had constructed. Their efforts to maintain these works could then be viewed as rational economic behavior aimed at protecting prior investments.

Conceiving of irrigation development as property creation, we can then move to a body of theoretical writings that deal with issues of property rights.

IRRIGATION AND PROPERTY

The Theory of Property Rights

For the last decade or more, there has been a growing amount of literature in economics dealing with the important topic of property rights. This work is concerned with theory for explaining fundamental economic activities by reference to the "interconnectedness of ownership rights, incentives, and economic behavior" (Furubotn and Pejovich, 1972). Or, in other terms, the key idea of the property rights literature is that "different property rights arrangements lead to different penalty-reward structures and, hence, decide the choices that are open to decision makers" (Furubotn and Pejovich, 1972).

On one point, the property literature is consistent and clear: property is about social relations; about relations among men, not relations between men and things. Furubotn and Pejovich (1972) describe property rights as:

> . . . the sanctioned behavioral relations among men that arise from the existence of things and pertain to their use. Property rights arrangements specify the norms of behavior with respect to things that each and every person must observe in his interaction with other persons, or bear the costs of nonobservance.

As discussed by Goody (1962), analysis of property can be clarified by employing the distinction between objects of property (valued things) and property relations. As Bohannan (1963, p. 102) indicates, property rights is the concept we typically use when social relations are of primary importance. That is, property relationships are a subset of social relationships which are based on the position of two or more individuals in relation to some property object. A basic assumption of the property literature is that the nature of this social relationship (i.e., the nature of property rights) will be an important determinant of how the resources that are the property object will be used.

This body of literature is also concerned with the basic function of property rights. Demsetz (1967) has offered an hypothesis regarding this matter:

> . . . property rights develop to internalize externalities when the gains of internalization become larger than the costs of internalization.

That is, when groups create things whose use can benefit other than the creators, the conditions are set for the group to attempt to form property rights for the purpose of excluding those others from the benefits of property objects they have constructed. When it is not possible to achieve this internalization, incentives to create or sustain the property objects will be absent.

The research agenda suggested for students of property rights includes the following large questions:

- o What is the structure of property rights in a society (or group) at some point in time?
- o How has this property structure come into being?
- o What consequences for social interaction flow from a particular structure of property rights?

As will be detailed below, these three property questions can have considerable utility in understanding irrigation behavior in particular contexts, including the context in which state and locality resources are being joined, successfully or not, for irrigation development. Understanding irrigation behavior in a particular place is greatly facilitated by examining it through the prism of property structure--giving attention to the distribution of property rights in land, water, and hydraulic facilities. Moreover, attention to the history of the investment process that created the existing structure of property can yield critical insights of importance to introducing irrigation changes. Finally, a critical matter to be understood is the relationship between property rights and the performance of fundamental irrigation tasks, such as water allocation, maintenance, and conflict management.

As will be discussed later, by improving our understanding of property, it may subsequently be possible to view property rights as a policy variable to be used in designing future irrigation development activities. Not that present irrigation development policies are devoid of property content; rather, current policies usually lack explicit property assumptions. Particularly in areas with irrigation experience, development efforts often result in the destruction of existing property arrangements, with consequent negative results.

Property and Irrigation Groups

In a paper prepared for a seminar in Thailand last year (Coward, 1983), I used the concepts of property and investment to review and order our understanding of the social organization of various traditional (or indigenous) irrigation systems in Asia. From that review, I concluded that collective social action in these systems is based on property relations. That is, that these irrigation groups formulated principles of action and acted out irrigation tasks in ways that reflected prior and continuing investments in their hydraulic property. It is this relationship of co-property holders that legitimizes and activates their solidarity.[2]

In this chapter, I wish to set the following question: How does the property factor relate to the respective roles of state and locality in Asian irrigation development? More specifically, what can we discern about the various structures of property rights that emerge from state, local, and joint irrigation investment activities, respectively? And, what consequences for social interaction and irrigation activities flow from these respective property structures? In short, whereas the previous property discussion (Coward, 1983) emphasized property creating and property relationships with small, somewhat autonomous irrigation groups, this discussion expands the topic to consider the impact of the state on these processes.

PROPERTY-EXPLICIT SYSTEMS--SOCIAL ACTION AND THE ARCHITECTURE OF OWNERSHIP

The structure of hydraulic property rights is not uniformly obvious across irrigation schemes. One category of systems with particularly explicit ownership patterns are those we can term "share" systems. In these systems, one's property rights in water are represented in more or less definable units, each of which represents some portion of the total water supply available (or sometimes a share of time which is, in fact, a proxy measure of a share of water).[3] This form of ownership appears remarkably robust, occurring in a wide range of contemporary socioeconomic contexts (from Nepal to Colorado) and having long historical continuity as well (as in Spain and various Arabic settings).

As I have discussed with regard to a Philippine case, property rights represented in shares can be an effective instrument for allocating an uncertain water supply. In this arrangement, one share is always a portion of the total water supply available and, thus, increases and decreases as does the stream flow being diverted. Social action for repairing and maintaining the irrigation works directly reflects this architecture of ownership. One is liable for labor, materials, and occasionally cash, in a ratio consistent with the number of shares owned.

In both the Philippine case and a Nepali case which has been described in detail (Martin and Yoder, 1983), the shares were derived from the original social action of investing in creation of the hydraulic facilities. Those who came after the establishment of the system gained rights to shares by paying some equivalent of the original investment and/or committing themselves to continuing investments to sustain the existing facilities.

The Philippine and Nepali examples both deal with situations of relatively small community irrigation facilities. However, some examples of share systems can also be found among larger, government-managed schemes. One case is the warabandi system found in parts of India and Pakistan (Malhotra, 1982; Renfro and Sparling, 1983). The theory of the warabandi arrangement is that each cultivator is assigned a turn, represented by a specific period of time--a time share--and the volume of water available during that slice of time is his to use.[4] This time share becomes a property right legitimized by the state through the creation of a formal and legal warabandi roster for the delivery channel in question. The warabandi share, as a property right, then serves to organize the social relationships of irrigation among the cultivators and between them and the irrigation agency. While the original warabandi holders may not have obtained their shares through investment in the hydraulic works, they may have invested in land development (which can be seen as a prerequisite to enhancing the water right with actual value).

The architecture of water ownership is perhaps seen most clearly in situations with a formal or informal "market" for selling or trading shares; Maass and Anderson (1978) have analyzed the formal water markets in systems in Spain and the western United States, and Renfro and Sparling (1983) have analyzed informal trading of warabandi turns in Pakistan. In these situations, property rights are sufficiently secure, and it is recognized that they can be transferred and exchanged as can other objects of property, such as land and cattle.

Non-Share Systems

While the property underpinnings of the social organization of irrigation are easily revealed in share systems, these property arrangements may be either unformulated or simply less conspicuous in other places. For example, it appears that in many state systems in Asia, while there is a generalized intention of the irrigation agency to provide water to designated users and a generalized right of users to obtain an irrigation supply, those users have no specific claim to a certain amount of water, portion of delivery time, or whatever. One result of this absence of specific property rights is a high level of ad hoc behavior involving the individual (sometimes collective) negotiation of impermanent and idiosyncratic use rights. Such transient rights foster a more negotiated form of social structure composed of temporary coalitions, numerous (though short-lived) dyads, and considerable atomizing of behavior. These features of social relations characterize relations both among cultivators and between them and the agency staff. In short, indefinite property rights are reflected in an ad hoc pattern of social organization.

The case of central Java is an interesting one to consider; it seems to refute the above line of thought (Oad, 1982; Duewel, 1982). In these public systems, cultivators have only a generalized water right; nonetheless, they have created quite structured patterns of irrigation organization. How is this to be explained? Perhaps the explanation lies partly in historical facts, partly in the contemporary scene. Many of the present public systems are comprised of small systems that previously were owned and operated by the community. While the state now owns and operates the diversion structures supplying these systems and some of the major works within the command area, local people remain the owners of, investors in, and operators of the ancillary facilities, which are often dense in this region of Java. Moreover, many contemporary villages continue to receive part of their irrigation supplies from small community works, which they own and operate, in addition to the water supplied by the state facilities.

Thus, one may hypothesize that the basis for social action in irrigation in these communities derives from the historical ownership of hydraulic property and water rights, as well as the contemporary ownership of both small community systems and the local appurtenances using the water supplied by the state works.

In sum, property-explicit systems, such as water share arrangements, reveal the property basis of much social action for irrigation. From them, we see that clear arrangements of hydraulic property rights can become the basis for calculating water apportionments, as well as assigning various operational costs. Moreover, we are able to detect the property undergirding associated with social relationships among cultivators and between them and the irrigation system authorities.

THE PROPERTY FACTOR IN STATE INVESTMENT AND LOCAL ACTIONS

In the opening of this paper, reference was made to the large and expanding role of the state in irrigation development and the corresponding demise of local involvement. Many state agencies have become frustrated by the phenomenon of lack of local support, or withdrawal of support, for state-built irrigation facilities. Agency explanations have usually suggested some form of laziness or passiveness on the part of cultivators. Here, I want to explore the property factor as a possible explanation.

I begin with a previously stated proposition--irrigation development is a property-creating process (Coward, 1983):

> . . . investments to create irrigation facilities always create, or rearrange, property relationships with regard to those new facilities.

The second proposition is that property creating (or the investment process) also creates a structure of property relationships (Coward, 1983). We have argued above that, in communal systems, this investment process establishes a social bond among those who have invested and is the social basis for subsequent irrigation behavior.

However, conventional state investments in irrigation exclude cultivators from the investment process. Therefore, what we need to examine are the following relationships resulting from state investments in irrigation:

- o the relations among cultivators with regard to the hydraulic property
- o the relationship between the cultivators and the state with regard to the hydraulic property.

Since with state investment, the cultivators typically are not made co-property holders, the irrigation investment acts of the state do not fuse the cultivators into a corporate group or reinforce collective behavior that may exist based on some other factor. In effect, in a state system, the water users resemble individual consumers more than a cooperative band. In most circumstances, the "consumers" will act independently of one another and focus attention on their relationship with the "owner" of the resource they require. In some circumstances, they may act collectively to try to ameliorate some common problem. There is little impulse to act collectively to use or protect a property object owned in common--in fact, there is no common hydraulic property deriving from state action. As is easily recognized, this atomized condition describes the situation in many public schemes.

Moreover, conventional state investment--making the state the owner of the hydraulic property--also structures relations between cultivators and the state. Perhaps the most fundamental consequence is that cultivators, as non-owners of the hydraulic property, are alienated from that property and may not act as though they are responsible for it (even though government wishes them to do so). Note that the usual pattern is not one in which state and local users are co-property holders (though there may be a certain degree of rhetoric suggesting this in some cases). Rather, cultivators are put in a subordinate relationship to the irrigation agency, and all uses of the hydraulic facilities are (in theory) mediated through the agency.

If state investment occurs in settings with existing community irrigation facilities (and an increasing number of such cases are occurring), the usual property consequence is the destruction of existing property relationships. That is, property relationships built around the prior investment process and the property objects that have been created are disrupted, confused, and muddled to the extent that they no longer serve to organize social action. This occurs because the state either ignores or discounts the ownership of existing facilities and water rights and lodges the rights to all new hydraulic property in itself. The result, as noted above, is to alienate the water users from the facilities and remove the basis for continuing collective irrigation activities.

Exceptions to this general consequence may be instructive to consider. One case is rather well-known success of irrigation associations in Taiwan (at least in the 1960s and early 1970s, when considerable writing was done on this topic). A second, less well-known case is the incorporated subaks of Bali.[5] Each of these are interesting exceptions and, therefore, puzzles to be explained.

The Case of Taiwan: Is Property Involved?

The uniqueness of the Taiwanese irrigation associations is that they were not mere appendages of the irrigation agency. They were the irrigation agency, and, through an elaborate system of elections and representatives, they hired and directed the technical staff which operated the irrigation facilities. They were, thus, more analogous to the modern corporation than to the atomized consumers typically found in large government systems. While the ultimate ownership of the irrigation facilities and the water was, no doubt, with the state, a serious transfer of these ownership rights had been given to the irrigation associations, in trust.[6] The result was that the property system that entrusted state-created hydraulic property to the irrigation associations established a property basis for those associations. That property foundation then provided a legitimate basis for organizing social action at the local and higher levels and for creating an administrative apparatus to manage the transferred property rights.[7]

It may also be the case that the irrigation associations were able to base their lower level organizational units on prior irrigation-owning groups and property principles. It is known that many of the present large systems in Taiwan are amalgamations of previously existing small systems, some of which reportedly were organized around property principles similar to those discussed above for contemporary community schemes (Wang, 1972). Thus, while speculative, there is reason to suggest that the early success of irrigation associations in Taiwan reflects a property factor.

The Incorporated Subak: Is Property the Factor?

For the past several decades, beginning during the colonial period but accelerated in recent years, the state has been investing in irrigation development and incorporating previously small, independent local systems (subaks) into irrigation networks built and managed by the state.

In general, the pattern has been to build more substantial diversion structures and to convey water from these diversions through large delivery canals to the existing subaks. In most cases, this supply is delivered in a manner that retains the physical integrity of the subak, so that the state canal system merely substitutes for the natural stream formerly supplying the subak. Intra-subak facilities and relationships are little disturbed. Thus, while some subak property is lost (the physical weir and the right to its management), a great deal of property remains owned and managed by the group--the elaborate distribution facilities and the management of the water rights within the subak command. By terminating its ownership and control at the outlet serving the subak, the state has invested in irrigation in a manner that supports local property rights and fosters continuance of the subak organization.[8]

POLICY IMPLICATIONS

If the property factor is important in mobilizing local action for irrigation activities, what policies might the state follow so as to foster property-based local groups? In the remaining paragraphs, three lines of action are suggested and discussed: (1) rights recognition, (2) indirect investment approaches, and (3) creating share systems.

Rights Recognition

A large portion of contemporary irrigation development involves either the rehabilitation of existing facilities or, as a variation of this, the incorporation of existing small systems into larger public networks. In either case, the fact that the development project is dealing with operating irrigation activities suggests that there also exists some pattern of hydraulic property rights which reflect a pattern of prior investment. Most rehabilitation efforts ignore this reality, often pursuing an officious approach that allows the state to ignore the investment history of the locale. Often this approach results in a new technological apparatus being placed into a muddled property context. Thereafter, the technology is unused or misused and soon inoperative, bypassed, or, if possible, modified to fit the realities of property rights.

There is a straightforward solution to these problems. When doing the reconnaissance and other preproject investigations of the existing situation, effort should be devoted to understanding these established rights, just as attention is directed to investigating the condition of the physical irrigation facilities. That is, a property analysis should be a component of the project planning process.

Indirect Investment Approaches

The conventional investment approach used by irrigation agencies is to plan, design, build, and operate public irrigation facilities using the resources and staff (or contractors) of the state. This I label "direct investment." In other state programs, resources are provided to local irrigation groups for improving those locally owned and managed systems. These resources may be finances, materials, technical advice, or other types of aid. What is avoided is ownership of the facilities by the state itself and hands-on governance of the systems by a state agency. This approach I label "indirect investment."

In view of our discussion of the property factor, the indirect investment strategy is appealing because it provides a means for the state to invest in irrigation development and simultaneously reinforce or create property-based local irrigation groups. Through indirect investment, critical resources are provided to such property-based groups to create, improve, or rebuild irrigation property which they will continue to own and operate. Costs to government can be reduced since indirect investment strategies typically induce some level of investment from the local group to match the state's input and because recurring costs of operation and management (O&M) are with the local group and not with the state.

As will be evident to most readers, the indirect investment approach will have its greatest utility for programs of state investment in small-scale irrigation works (Coward, 1984). Here, the size of the commands to be managed and the technical apparatus of the works is likely to fit the capacities of local, property-based groups.[9]

However, there is another irrigation context that also is susceptible to indirect investment--development of the facilities at the lowest unit in a large public scheme (variously referred to as the chak, watercourse, or tertiary unit). Throughout Asia, the customary practice has been to leave development and elaboration of this portion of the system to the local users themselves. However, more recently, a number of states--sometimes with urging from international donors--have extended their investment activities into the distribution network (Levine and Coward, 1984). Results have been mixed, but frequently the outcomes have been under-utilization and poor maintenance by the user groups. Most of these tertiary development investments have been direct rather than indirect in character. The result is that the programs fail to reinforce or create property-based groups motivated to use and restore the new tertiary facilities. An indirect strategy could ameliorate this situation by providing resources to local groups for their use in improving the tertiary facilities on which they depend. Such an approach would create a property-based group for operating and maintaining the facilities and for mobilizing local resources to be joined with those of the state.

Share-Based Water Rights

As discussed previously, while many community irrigation systems are organized around explicit share systems of water allocation, relatively few public systems have these arrangements. Rather, in many of the agency-managed systems, water allocation is based on more generalized and nonspecified cultivator rights to water. Usually, this nonspecification of water rights creates a setting of negotiated and temporary rights in which either cultivators or agency staff may be placed in dependent and compromising situations.

The alternative is to create social arrangements in which water allocation (and other fundamental irrigation tasks) is based on explicit and specified share concepts--a type of social contract. Those share concepts with which we are familiar achieve specification of the share through some form of volumetric measurement or surrogate measures of volume, such as time, size of proportioning weirs, or size and number of outlets. The creation of a share system can induce property-based group action since typically the shareholders will be able to realize their rights only if co-shareholders act coordinately and only if the agency supplies water to them effectively.

When share systems are in place, they also provide a basis for mobilizing various local resources needed for system operation. Just as the allocation of water is per share, so can labor requirements, material contributions, or cash payments be assigned explicitly and with regard to verification.

Finally, share systems are attractive because they represent an opportunity to introduce concepts of equity into irrigation development. Most irrigation projects, explicitly or implicitly, assume a pattern of water distribution parallel to the existing pattern of land distribution; those with more land are allocated rights to more water. Share systems allow variability on this matter since shares can be allocated to nonlandowners or can be allocated in a pattern inversely correlated with landholding size, or otherwise.[10] In this manner, the new resource created by the state can be distributed to achieve some desirable outcome of social justice and also create a property-based irrigation group.

SUMMARY

Viewing irrigation through a property prism can do much to improve our basic understanding of irrigation behavior and may have promise for development of significant public policies for irrigation development. A fundamental institutional issue for irrigation development in the Asian region involves finding the appropriate mix that will differ from place to place and that will likely change over time. Implicit in that mix are various property rights and relationships that must be made explicit, be understood, and provide incentives consistent with the combination of state-locality actions being planned.

NOTES

1. In this initial sentence, I employ several terms that will be repeated throughout the chapter and which need definition. First, I use the term "irrigation development" to refer to various activities undertaken to build new irrigation schemes, to rehabilitate existing facilities, or to improve the management and performance of irrigation systems. The irrigation systems I am concerned with are those that serve groups of water users and usually are controlled by some group. In this discussion,

I am not concerned with the individually owned well used by a single farm. "State" is used in a generic sense to refer to the legal and administrative apparatus of the national government or its various agents. I use the term "locality" to refer to various forms of local organizations or combinations thereof. I am deliberately avoiding the term "community" because of the sociological assumptions usually associated with it, though, in some cases, the locality which I refer to may be a community. Irrigation facilities created by localities I refer to as "local sector irrigation."

2. We should note that it is not assumed that the basis for social action is any single factor--property or otherwise. Other bases for social action may be kinship, territory, religion, socioeconomic class, gender, etc., and several may be operating concurrently. In part, we label groups differently depending upon the basis for their group formation--families, churches, communities, corporations, and so on.

3. As noted by Hammoudi (forthcoming, 1984), this share can sometimes have an ambiguous form to it.

4. This arrangement may perhaps reflect a common Arabic element in water distribution found in many systems in the greater Arabic world.

5. Much of the writing on subaks has emphasized their independence and autonomy, while minimizing the role of earlier Balinese states in their creation and persistence. However, Geertz' analysis (1980) indicates that the subaks have long been entangled in the administrative apparatus of earlier Balinese states.

6. In the mid-1970s, when the irrigation associations experienced various difficulties, this trust was dissolved.

7. This concept of administrative apparatus is borrowed from Stinchcombe (1983).

8. Unfortunately, more recent investments by the state have begun to modify this earlier pattern--now the state is beginning to invest below the main outlet to the subak. This has taken the form of so-called "tertiary development." One report by a Balinese university research team (Sutawan et al., 1983) suggests this project has very negative consequences for the subak organization.

9. In the Philippines, some imaginative experiments are underway to disaggregate large state systems into smaller segments, each to be under local management (the ownership issue has not been clarified). In this manner, the indirect approach could have even wider utility.

10. Experiments with such novel share systems have been tried in India (Seckler and Joshi, 1981) and proposed for Nepal. These approaches may be most applicable where new systems are being created. If used in the context of existing systems, there will be need to both understand the current property rights and plan for compensation to those whose rights are reduced.

REFERENCES

Barzon, Jacques. "Scholarship Versus Culture," The Atlantic Monthly, 254(5):93-104, 1984.

Bohannan, P. "Land, Tenure, and Land Tenure," African Agrarian Systems, D. Biebuyck, ed. Oxford, England: Oxford University Press, 1963, pp. 101-115.

Coward, E. Walter, Jr. Irrigation and Agricultural Development in Asia: Perspectives from the Social Sciences. Ithaca, New York: Cornell University Press, 1980.

Coward, E. Walter, Jr. "Property in Action: Alternatives for Irrigation Investment," paper presented at Workshop on Water Management and Policy, Khon Kaen, Thailand, 1983.

Coward, E. Walter, Jr. "Improving Policies and Programs for the Development of Small-Scale Irrigation Systems," WMS Report 27, Cornell Studies in Irrigation, Water Management Synthesis Project, Cornell University, Ithaca, New York, 1984.

Demsetz, H. "Toward a Theory of Property Rights," AER Papers and Proceedings, 57:347-352, 1967.

Duewel, John. "Promoting Participatory Approaches to Cultivating Water User Associations: Two Case Studies from Central Java," Technical Report 11, The Determinants of Developing Country Irrigator Project Problems, Cornell University, Ithaca, New York, 1982.

Furubotn, E., and S. Pejovich. "Property Rights and Economic Theory: A Survey of Recent Literature," Journal of Economic Literature, 10:1137-1162, 1972.

Geertz, Clifford. Negara: The Theatre State in Nineteenth Century Bali. Princeton, New Jersey: Princeton University Press, 1980.

Goody, Jack. Death, Property, and the Ancestors. Stanford, California: Stanford University Press, 1962.

Hammoudi, Abdellah. "Substance and Relation: Water Distribution in the Dra Valley," Law and Social Structure in the Middle East, Ann E. Meyer, ed., forthcoming (1984).

Korten, Frances. "Building National Capacity to Develop Water Users' Associations: Experience from the Philippines," World Bank Staff Working Paper No. 528, 1982.

Levine, Gilbert, and E. Walter Coward, Jr. "Irrigation Water Distribution: Options and Opportunities for Production and Equity," unpublished manuscript, Colorado State University, Fort Collins, Colorado, 1984.

Maass, Arthur, and Raymond L. Anderson. ...And the Desert Shall Rejoice: Conflict, Growth, and Justice in Arid Environments. Cambridge, Massachusetts: The MIT Press, 1978.

Malhotra, S. P. "The Warabandi System and Its Infrastructure," Publication No. 157, Central Board of Irrigation and Power, New Delhi, India, 1982.

Martin, Edward, and Robert Yoder. "Water Allocation and Resource Mobilization for Irrigation: A Comparison of Two Systems in Nepal," paper presented at Annual Meeting, Nepal Studies Association, Twelfth Annual Conference on South Asia, University of Wisconsin, Madison, Wisconsin, 1983.

Oad, Ramchand. "Water Management and Relative Water Supply in Irrigation Systems in Indonesia," Ph.D. dissertation, Cornell University, Ithaca, New York, 1982.

Renfro, Raymond, and Edward W. Sparling. "Private Tubewell and Canal Water Trade on Pakistan Punjab Watercourses," paper presented at Water Management and Policy Workshop, Khon Kaen, Thailand, 1983.

Seckler, David, and Deep Joshi. "Sukhomajri: A Rural Development Program in India," mimeographed paper, Colorado State University, Fort Collins, Colorado, 1981.

Stinchombe, Arthur L. Economic Sociology. New York: Academic Press, 1983.

Sutawan, N., et al. Studi Perbandingan Subak Dengan Sistem Pengairan Non-pu Dan Subak Dengan Sistem Pengairan PU. Denpasar, Bali: Universitas Udayana, 1983.

Visaya, Benito P. "The Palsiguan River Multipurpose Project and the Zanjeras," paper presented at Conference on Organization as Strategic Resource in Irrigation Development, Makati, Metro Manila, Philippines, November 1982.

Wang, Sung-Hsing. "Pa Pao Chun: An 18th Century Irrigation System in Central Taiwan," reprinted from the *Bulletin of the Institute of Ethnology, Academia Sinica*, 1972.

17
Water Management: Problems and Potential for Communications in Technology Transfer

Dan Lattimore

Irrigation knowledge developed over the centuries has come to be formidable, sophisticated, and productive. However, observations from around the world suggest that, even with this knowledge, improved water management suffers from disillusionment, misunderstanding, and emphasis on technical detail. Worldwide, we see recurring problems of irrigation system design, construction, operation, maintenance, and rehabilitation (Clyma et al., 1982, p. 1). To break out of the cycle of emphasis on technical details and to increase understanding, we must find innovative but effective ways to transfer improved concepts and processes of water management.

The problem of facilitating technology transfer is an example of one of mankind's oldest habits--borrowing. The story of man is the history of borrowing ideas from others (Bauer and Gergen, 1968). Why, then, should we have trouble transferring technological concepts and ideas from one culture to another? Historically, technology transfer has been one of the most misunderstood concepts. The multidimensional nature of technology in both physical and human contexts does not lend itself to a cookbook definition. There is abundant technology transfer literature, but nothing suggests we have found any universally successful strategies. There is no single model which is empirically tested, standardized, and implemented cross-culturally. Much of the literature is either too abstract or too inconclusive. The process itself is quite complex, with a multitude of variables, many of which are hard to identify and still harder to quantify (Singh, 1983, p. 1).

The purpose of this paper is to synthesize what we know about the communication aspect of technology transfer as it relates to water management and to look at the opportunities available as we seek to improve water management around the world. To begin, though, we must first understand what is meant by technology change and technology transfer and, even more importantly, what we mean by technology itself. Without such definitions, we cannot understand the interrelationships.

TECHNOLOGY

The term "technology" has its origin in the Greek word "techne," which means an art or skill. To use John Joseph Murphy's words (Spencer and Woroniak, 1967):

> The simplest version views a technology as involving only changes in artifacts. A more sophisticated approach adds to the physical objects labor and managerial skills. A third approach views technology as a socio-technological phenomenon; that is, besides involving material and artifact improvements, technology is considered to incorporate a cultural, social, and psychological process as well.

Technology Change

Because technology is difficult to define, there is also considerable misunderstanding about "technological change." Just consider the following definitions (Singh, 1983, p. 6):

- Any kind of shift in the production function. Thus, speedups, improvements in the education of the labor force, and all sorts of things appear as technological changes
- A "residual" measure of the impact on economic development. Residual in this context refers to the share of output not attributable to labor and capital
- An indigenous variable, a device to satisfy wants better than does preexisting knowledge
- Research, economies of scale, improved market, organizational and management ability
- Introduction of an old good in a more appealing way, the introduction of new methods of produc-

tion, the opening of a new market, the conquest of a new source of raw materials, and the carrying out of the new organization of any industry.

On the other hand, technological change can be more narrowly described as "any shift in the production function," the definition commonly used by economists. To the industrial engineer preoccupied with research and development, it implies the process of product development, while the sociologist may confine it simply to change in mechanical devices (Singh, 1983, p. 6)

While technological change is often a positive change, it can also be negative. Evaluation of that change may take years to assess. It is a process involving gradual adaptation of improved techniques or practices in a particular social, economic, and political setting. While the change may be like an atom bomb in its suddenness, it is more often like a glacier that slowly shifts and changes. Technological change, then, is often a slow process influenced by a assortment of variables, such as public opinion, market conditions, management policies, and credit. Technological change is not necessarily restricted to the technological sphere. It often extends to the educational, social, and political realm (Singh, 1983, p. 7)

TECHNOLOGY TRANSFER

In the last few decades, technology transfer has received considerable publicity, but it, too, lacks a clear definition. Popular definitions include these (Singh, 1983, pp. 7-8):

- The process in which an innovation originating in one institution is adopted elsewhere
- A planned and rational movement of information and techniques on how to perform some task, simple or complex
- The simple act of obtaining information from external sources in order to speed up industrial development and, at the same time, conserve technical resources.

Actually, the often-used definition of communication could be added to this list--the art of transmitting ideas and information from one person to another.

Technology transfer is a necessary prerequisite for economic development in the Third World. Without the resources available for basic research, it is necessary for developing countries to use ideas gathered from others to advance as rapidly as is possible. Technology transfer and technological change, in this context, are highly interrelated. Transfer of information often brings about the technological change that increases economic prosperity (Singh, 1983, p. 10) Any technical change sets in motion events that require adjustments. These adjustments "set the stage" for future technological change (Spencer and Woroniak, 1967).

Problems of Technology Transfer

When we consider technology transfer, we must remember that this is a process that has a high degree of interrelated elements. If one element is missing, the other elements will be useless or, at least, their effectiveness will be greatly reduced. To establish modern technology in an underdeveloped world, change is necessary in at least three areas:

- social systems and human attitudes
- knowledge and human skills
- physical implements in which modern technology is embodied.

In each of the above elements, the process of change is slow. Modernization of agricultural technology often will not occur even if fertilizers, irrigation, and equipment are available because social structure and resulting behavior are resistant to change. The needs, though, of developing countries are so severe that they cannot afford to go through the long, arduous process of development that took place over centuries in the West; developing countries must accelerate that process. Formal education has been a major instrument for technological change in the West, but that takes time. No matter how indispensable formal education is, it must be supplemented by training in new ways and processes to make the farmers more efficient in their endeavors. In order to increase the speed of development and meet the pressures of a developing society, a teamwork approach is needed. This requires the collaboration of host countries and scientists from developed countries in an interdisciplinary setting.

Problems in irrigation must be approached with several disciplines complementing each other and filling the gaps that exist within the system in order to accelerate transfer of technology. To accomplish this, several models of technology transfer have been developed.

TECHNOLOGY TRANSFER MODELS

As Lowdermilk et al. (1975) demonstrated in their study in Pakistan's Punjab, besides technological considerations, effectively operating and maintaining a watercourse necessitates attention to social and cultural guidelines. The authors indicate that physical boundaries of a watercourse contain a variety of user subsystems made up of kinship groups which interact with each other along the lines of traditional norms and customs, as well as economic needs. Therefore, the authors conclude, strategies for successful communication of technologies in effective watercourse management should include the study of socioeconomic interactions within the user subsystems of watercourse networks and between the users and society at large. This requires an interdisciplinary focus.

Lowdermilk and his colleagues' user-oriented approach to designing and implementing communication strategies for better operation and maintenance of watercourses has proven to be appropriate. Historically, approaches to dissemination and use of new ideas, technologies, practices and products in agricultural development have experienced three distinct phases: The Diffusion of Innovation Model, The Package Programs Model, and The Induced Innovation Model. Each encompasses the previous model and builds upon it. A comparison of the three models is shown in Table 17.1 and discussed in the following sections.

Diffusion Model

The Diffusion Model, as it is applied in agriculture, focuses on the individual farmer as the unit of adoption and, thus, examines the personality traits which make some farmers more receptive than others. This is a model wherein the vertical communication from the source to user carries the message, and feedback is minimal. Failure of adoption is blamed on the user as a single, isolated unit. No attention is paid to the socioeconomic interactions of

Table 17.1 Kearl's summary of communications considerations involved in successive models of agricultural development (Kearl, 1974, p. 177)

	The Diffusion Model	The Package Programs Model	The Induced Innovation Model
HOW GAINS OCCUR	By wider diffusion of demonstrably improved farming practices	By identification of packages of inputs that dramatically increase output on farms and in regions where they are applicable	By responses within the system that create a steady flow of innovations and needed institutional changes as costs and potential benefits change
PURPOSES OF COMMUNICATION	To motivate farmers to consider the possibility of change; to convey factual data needed in adoption of specific improved practices	To insure that all needed inputs are available at proper times and appropriate locations	To insure that awareness of costs and benefits will generate prompt institutional responses
MAIN DIRECTIONS	Primarily from those charged with identifying better practices to those expected to adopt them	Inter- and intra-agency communication to coordinate availability of all elements in the package	Two-way flow of information about product and factor markets to facilitate prompt responses throughout the system

ROLE OF FEEDBACK	To check adequacy of messages (accuracy, relevance, comprehensibility) as a guide to message revision	To monitor program performance as a basis for modifying its content (but not its goals)	To convey market signals to those allocating scientific and technical resources, those conducting research in agriculture, those who can take initiative in modifying institutions
CRITICAL COMMUNICATION REQUIREMENTS	Mass media; local extension workers and the informational materials to support them	Channels for liaison among agencies Vertical communication from the field to the levels at which agency decisions are made	Modernization of marketing system by creating informational and communication linkages needed for effective functioning tioning of factor and product markets Information channels for effective organization and management of scientific and technical resources Farm organizations to give effective voice to farmers' needs for new technology and for modification of institutions

the user in his environment. The mass media channels information dissemination (Rogers, 1962). The vertical pattern of communication and heavy reliance on mass media make the Diffusion of Innovation Model less attractive for agricultural development in developing countries.

In traditional societies, knowledge moves from the highest authority to the user through a chain of bureaucratic hierarchy (Lerner, 1958). The extension agent (or the change agent) at the village level is the last link of this bureaucratic chain. The agent carries the knowledge to the farmer (or the user) in the form of bureaucratic order, leaving little opportunity for feedback. In this respect, as implemented in traditional settings, the Diffusion of Innovation Model does not lend itself to gathering necessary information about the socioeconomic condition of the user. In this process, the user becomes merely a target for the bureaucratic message delivery system. On the other hand, using the mass media in information dissemination has its shortcomings, also.

Media access is still limited among the rural population of developing countries. Furthermore, the media content, in most instances, is oriented toward urban audiences and their needs. And finally, feedback in mass communication is random and limited.

The Diffusion Model is culture bound and has practical use in developed countries, but its value for communication research in transfer of technologies in developing countries is limited.

Package Programs Model

The Package Programs Model was an answer to frustrations created by the Diffusion Model. This approach is multifaceted and attempts to provide a variety of services necessary for diffusing innovations.

A prototype of the Package Programs was the Puebla Project in Mexico (Biggs, 1972, pp. 11-15). The goal of the project was to encourage small farmers to plant a new high-yield maize seed. The program also provided all the necessary services to implement this innovation. The Puebla Project included financial credits to buy the maize seed and fertilizer, price-support incentives, market analysis and distribution plans, organization of farmers to facilitate the flow of credit and information, and an inter-agency authority to coordinate action and involvement of top-level political leadership to carry out the program.

Promising as it seemed, the Package Programs Model did not achieve continuous success in Mexico. One of the critical shortcomings of the Puebla Project was its inability to gauge long-term variations in applying new technologies and practices.

When the project was initiated, rainfall for the season was perfect for the necessary conditions of planting the new high-yield maize. But in later years, precipitation levels declined and so did the maize returns. The project scientists did not have ready answers to these new conditions. Farmers were heavily in debt to their creditors. Everybody lost interest in the project.

Actually, in its inclusive approach, the Package Programs Model is far more advanced than the Diffusion Model in carrying out newly initiated products and practices. In its multifaceted approach, this program attempts to cover small farmers' needs to initiate an innovation. However, it necessitates a continuous dialogue among governmental agencies, the private sector, political leadership, and small farmers. This program requires an infusion, rather than a mere feedback, for the communication to flow smoothly. Feedback, as generally conceived, is the response to a message received. Infusion, on the other hand, is more of an elicited response to a felt need of the user. It is information-seeking on the part of the diffusion or change agency. Listening to the needs of the user is one of the key elements of a successful change program in agricultural development.

The Package Programs Model is inherently a communication vehicle involving bureaucrats, representatives of the private sector, opinion leaders, agricultural scientists, and small farmers. The small farmers' cooperation is essential to the continuation of the dialogue. In the real world, this is a difficult task and demands painful work and innovation.

In its communication strategies, the Package Programs Model provided information from radio, distributed leaflets, posters, and newsletters, and employed extension agents. However, provisions for input from farmers were limited.

Induced Innovation Model

The Induced Innovation Model emphasizes the importance of the market and market communication with farm organizations but ignores the small farmer. For the small farmer, who has little to buy or sell, marketing and market communications mean little.

This approach is the nonsocialist alternative to a reformist improvement program. The attempt is to develop a framework that meets the needs of high-volume marketing but lets the farmer maintain control of his farm. It stresses the importance of change and looks to "key farmers" to lead the way. Cheap labor, cash crops, and enlightened farmer leadership are assumed. It also necessitates a strong, organization-intensive society to work effectively. In this model, technology transfer occurs through an active extension service working with key farmers (Moris, 1981).

This brief summary of the dominant models of technology transfer in agricultural development indicates that the emphasis needs to be on the user (the small farmer) as the source of initial investigation for successful communication strategies. For continuous success of newly initiated agricultural innovations, "feedforward" (articulation of the changing needs of the small farmer) is indispensable.

In this respect, before we discuss the Problem-Solver Model--which we believe has the necessary flexibility as well as a cyclical process to allow for correction of errors committed--we will present a system which has interesting features.

Training and Visit System. While it is by no means a systematic and analytical model, the Training and Visit System (T and V) was first conceptualized by a World Bank employee, Daniel Benor. It is often still referred to as the Benor System (Moore, 1984, pp. 303-317).

As one of the largest international donor agencies, the World Bank has been quite interested in extension information delivery procedures in developing countries. Conceptualized in the 1970s, the T and V System was implemented in India and Turkey. However, much of the information about the implementation of one system exists within the Indian context (Moore, 1984).

By the time of development and implementation of the T and V System, there was scarcely any agricultural extension in India. It was also almost universally agreed that the village-level workers, who generally had low educational qualifications, had few prospects for promotion. They often had been working in the same posts for 20 or more years, were poorly motivated, and generally exhibited low levels of agricultural knowledge (Moore, 1984).

The basic thrust of the T and V System was to improve farm knowledge through interpersonal communication. That is, to build a system around a publicly employed, village-level agent, whose duty was to teach farmers new practices through personal contact.

The T and V System incorporates six basic principles for structural reorganization of agricultural extension. These six principles were implemented at Chambal, India, in a pilot project (Moore, 1984):

- The creation of a pure extension system concerned solely with the transfer of knowledge between research station and cultivator, and free of all responsibility for the organization of input supply, marketing services, and the dozens of other supplementary tasks normally assigned to extension agencies
- The bringing of all extension staff within the sole control of the Department of Agriculture in India
- A unified extension service in which each agent is responsible for all crops in his area
- A regular and frequent program of training classes through which recommendations of the next phase of the agricultural calendar are transferred in several steps from research station to the farmer via the extension agency
- The limiting of the responsibilities of the field extension agent to meeting about 10 percent of farmers--the so-called "contact" farmers--who are, in turn, responsible for disseminating the information received to other farmers ("follower" farmers)
- A regular and publicly known visit program for the field agent, in which he meets small groups of contact farmers at a fixed time every fortnight and accompanies them to their fields.

These six principles were developed to restructure and enhance the agricultural extension system in India.

The communication at the interpersonal level was designed at two levels: (1) extension agent to contact farmer and (2) contact farmer to follower farmers. In both levels, the feedback was emphasized and was considered important. There was a general emphasis on concentrating messages of wide relevance to many farmers and on the operations to be undertaken in the immediate future (Moore, 1984).

One of the distinguishing features of the T and V System in selection of contact farmers was the de-emphasis of selecting those farmers who were especially good or progressive. Instead, those farmers who were liked and listened to by their neighbors were chosen. In a sense, this is somewhat opposite to the Diffusion Model, wherein innovative farmers take the lead and play a role model to others.

In the 1970s, this T and V system had rapid success. The World Bank poured funds and expertise into the pilot projects in Madhya Pradesh and parts of West Bengal to make them work. These were part of the Command Area Development Authority. These first few years were studied with success stories. In Chambal, for instance, there was an increase in yields despite a decrease in fertilizer use (Moore, 1984).

The interpretations of the T and V System's initial success are varied. One explanation is that the World Bank was intensely involved and did not spare money or effort; therefore, staff of the project were highly motivated, irrigation inputs arrived on time, and adequate incentives were present for change. The combination of effort and funds of the donor agency and the enthusiasm of the staff from the various projects created a "pilot project effect." In fact, recent surveys conducted in the Chambal project have shown that the extension system has "gone to sleep" since the area ceased to be a focus of interest (Moore, 1984).

The fate of the T and V System approach is similar to many of the developmental initiatives launched by donor agencies (see Package Program Model discussed earlier). As long as the donor agency is closely related to the project, things continue to progress, but when the agency pulls out, the project fails.

One of the reasons for the discontinuation of schemes such as the T and V System is that their roots are planted outside. They are promoted by outside sources, not by the host country or host country personnel involved in such projects. It is, perhaps, like human organ transplants that are rejected by the body. For example, the T and V system was criticized by some in India as being preached by the World Bank like a religion (Moore, 1984).

In summary, the T and V System had very reasonable and useful features in its design. For example, emphasizing the interpersonal communication chain between research centers, extension service, contact farmers, and follower farmers attempted to provide the feedback-feed-forward mechanism that would have been quite fruitful if it had been properly applied. However, the system lacked continual monitoring and a built-in flexibility to check and correct errors committed. The next model to be discussed includes this monitoring and redirecting process, which makes it quite flexible.

Problem-Solver Model

One of the recent approaches to transfer of knowledge and practices is called the Problem-Solver Model. This model takes the user as the focus and tries to articulate user needs and characteristics to the source of information so that the user's needs can be effectively met. The Problem-Solver Model consists of stages in a problem-solving cycle inside the user (see Figure 17.1), beginning with a need and ending with satisfaction. These stages include the following steps: (1) realization of the need, (2) diagnosis of the problem, (3) identification and search for resources relevant to the problem, (4) retrieval of potential solutions, (5) fabrication of the solution, and (6) application of the solution to the need (Havelock, 1971).

As Havelock the other initiators of this concept argue, the core assumption of the problem-solver perspective is that self-initiated change has the firmest motivational basis and the best prospects for long-term maintenance. The user must not only accept the innovation, but must internalize it, making it a part of his routine behavior and investing his own energy and enthusiasm in it. The user will be more likely to internalize an innovation that he sees as <u>his own</u>. He is also more likely to accept those innovations that meet his own specific needs or that he has worked on himself to <u>adapt</u> to a specific need.

Two distinct characteristics of this model which make it quite relevant are:

- o It considers the user as part of a social subsystem. In doing this, the model takes into consideration the social and technical variables within that subsystem which may influence the user's decision-making and actions
- o The Problem-Solver Model starts the process of transfer of knowledge and practices from the

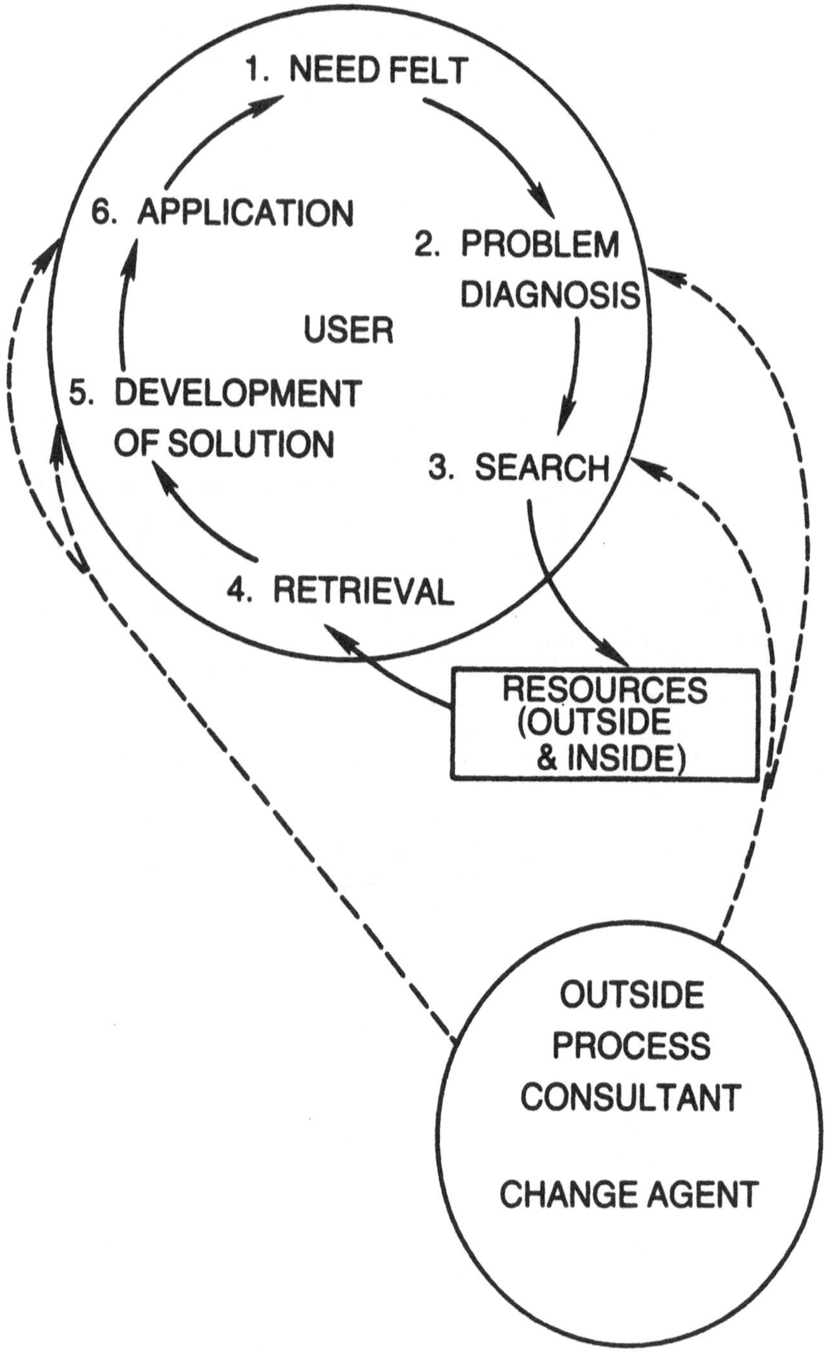

Figure 17.1 Problem-solver model

user's end of the communication continuum and works its way towards the source. This approach, implemented properly, provides the ground for maintaining two-way communication, in that the user can constantly contribute feedforward.

To understand and develop technology transfer strategies in water management, user-oriented communication investigations and strategies should pursue the following goals:

- Assess socioeconomic interactions of the small farmer in his subsystem to discover and describe the factors relevant to his communications behavior with other members of his subsystem and those from outside
- Assess the two-way communication within the organizational structures affecting the transfer of knowledge and practices for effective water management
- Design a monitoring scheme which can be used to continuously observe the patterns of behavior in water management.

The Problem-Solver Model, which is quite similar to the Development Model used by the water management projects that Colorado State University (CSU) has been involved in during the last 15 years, has been used to develop technology transfer activities in water management at CSU (Clyma et al., 1981b).

TECHNOLOGY TRANSFER IN IRRIGATION

Colorado State University has been a leader in developing water management concepts and applying them to development projects around the world. Since the early 1970s, four major water management projects at CSU have been funded through the U.S. Agency for International Development: Pakistan On-Farm Water Management Project, Egypt Water Use and Management Project, Water Management Synthesis I and Water Management Synthesis II. One of the major contract objectives in all four projects was developing and transferring important water management principles and technologies.

The principal means of technology transfer used by these projects has been through direct technical assistance activities, training courses, publications, and audiovisual materials. The beginning point for these activities has been an interdisciplinary audience analysis to understand the "user focus." Attempts have been made (some more formal than others) to determine the target audiences for these technology transfer activities.

Audience Analysis for Technology Transfer

If transfer of water management information is to be effective, identification and evaluation of target audiences must take place prior to beginning that attempted transfer. Audience analysis is basic to the communication process and, thus, to technology transfer. In looking at possible audiences for transfer of technology in water management, one should consider the following five audiences:

- o Water management scientists of host countries. This is peer communication, but is subject to differences in academic backgrounds and orientations, cultural traditions, and governmental constraints
- o Government officials and high-level technocrats of the host country. Communication here tends to be in the area of policy formation and decision-making with regard to water management issues
- o Mid-level government officials, such as irrigation engineers, district directors, and supervisors of irrigation projects
- o Technical assistants in water management, such as extension agents, junior engineers, and community leaders
- o Farmers involved in irrigated agriculture.

Audience analysis should consider why that audience would benefit from the technical change and try to structure the messages to meet those needs, in order to induce the desired change. Understanding psychological principles of persuasion is necessary to be successful.

Technology Transfer Activities

The water management projects at CSU have developed transfer activities aimed at reaching the first two or three audience levels, but seldom have we tried to reach the fourth or fifth level. In order to reach scientists, we have used such techniques as scientific journals, executive summaries, and professional meetings/seminars. To transfer concepts and technologies to the second and third levels of audiences, we have developed planning guides, newsletters, technical reports, slide shows, and videotapes. These materials have tended to emphasize the benefits to the host country for adopting certain concepts or to suggest a sound rationale for using a certain technology. To reach the fourth or fifth levels, the most concrete of all audiences (the technicians and the client-farmers), we have developed technical handbooks, how-to slide shows, and illustrated guides. Language has been a particular problem for this level. We have prepared some material in tne language of the host country, and we have translated some of our videotapes and slides. It is costly, but the results are quite good (Nayman, 1983). Several studies, as part of the Water Management Synthesis Project's Diagnostic Analysis of Farm Irrigation Systems Workshop, have documented major problems with transferring concepts and technologies to the farmer-extension worker levels (Clyma et al., 1981a; Jayaraman et al., 1983)

USER-SOURCE IN TECHNOLOGY TRANSFER OF WATER MANAGEMENT CONCEPTS

Proper handling of information needs of farmers on <u>how</u>, <u>when</u>, and <u>how much</u> to irrigate crops is one of the necessary conditions to proper water management. Jack Keller noted in a recent seminar paper (Keller, 1985):

> I believe a point that is often missed is the need for communication between the users and suppliers so that the physical delivery and application of water can take place in a meaningful way.

Two-way, open, and functional communications between farmers and local and regional irrigation personnel can only help solve the conflicts and problems in irrigation and facilitate better water management. However, results from several diagnostic analysis workshops have highlighted a serious gap in irrigation information between farmers and responsible irrigation personnel.

In their investigation of the Mahi-Kadana Irrigation Project in Gujarat, India, the trainers and participants found that "surprisingly few farmers get information on general farm water use from any source other than a few other farmers" (Jayaraman et al., 1983). The authors indicated that about half the sample farmers on an improved system and about 85 percent of those on unimproved systems received their information from other farmers, particularly in the area of on-farm water management.

Some of the conclusions on poor exchange of information and low source credibility listed by the investigators are:

- Most farmers do not know the local extension worker and receive very little or no assistance from him or the contact farmers
- The major source of information about marketing of farm products and prices is local shopkeepers and market middlemen
- Farmers receive virtually no assistance from anyone except other farmers about <u>how</u>, <u>when</u>, and <u>how much</u> to irrigate.
- Most farmers think agricultural information received is either "sometimes adequate" or "not adequate."
- Farmers consider sources of information about most Command Area Development Authorities (CADA) "fair" to "poor" quality.

Conclusions reached by the authors show that farmers perceived both the quantity and quality of information received from irrigation personnel as poor.

In another study, conducted in the Gambhiri Irrigation Project in India, similar results were obtained with regard to irrigation information (Clyma et al., 1981a). The researchers found that the irrigation system was in near disarray because little information was being passed to the farmers. Their findings also indicated that farmers learned about canal closures and openings from other farmers. Reliable sources of information were virtually nonexistent. Those farmers at the tail of the canals suffered most from this disorganized information pattern. At times, disorganized information dissemination led to chaotic situations at the irrigation system.

In both studies discussed above, the information dissemination on irrigation problems was considered from the perspective of canal users. In other words, these studies looked at how the farmers perceived the ability and availability of irrigation personnel to provide necessary and functional information on irrigation problems. However, if further investigation of how the farmers seek information from irrigation authorities is conducted, a more complete pattern should emerge. We would suggest a reversal of the roles, in that instead of looking at farmers as receivers of information, we should consider them as active seekers of information.

This approach removes the farmer on irrigation systems from becoming a "sink" for some unreliable and, at times, unavailable information and puts him in a position as an active initiator of messages. It opens avenues for two-way, interactive communication between farmers and irrigation authorities.

Looking at farmers' information-seeking behavior can also help us to study what kind of interaction takes place among farmers, irrigation authorities, and village-level workers regarding <u>how</u>, <u>why</u> and <u>how much</u> to irrigate crops.

In order to understand the "user-source" and his information-seeking in regard to water management, we feel that several questions should be posed in audience analysis: (1) who seeks, (2) what kind of irrigation (and other) information, (3) through which channel(s), (4) from whom, and (5) with what results?

1. <u>Who</u> designates the information seeker. In this case, the farmer on the irrigation system is seen as an active seeker, or as the case may be, as a passive, non-seeker of information. With regard to whom, we would suggest the following questions:

- Who among the canal users engages in the <u>most</u> and the <u>fewest</u> information-seeking activities?
- What are the factors which motivate those high information seekers?
- What are the factors that prevent those farmers from seeking information from irrigation authorities and personnel?
- When does the highest frequency of information-seeking occur?
- What are the factors pressuring farmers to seek information from each other instead of proper irrigation sources?
- What are the irrigation system locations of high and low information seekers?

- o What are the seasonal influences on high and low information-seeking?
- o How do the crop varieties influence high and low information-seeking?
- o How do the market conditions influence high and low information-seeking?
- o What are the differences in farmers' information-seeking behavior with regard to irrigation information and other kinds of information on farming matters?
- o Is there any organizational representation in information-seeking by the farmers?
- o How does the organizational mode of information-seeking differ from the individual mode of information-seeking, in terms of obtaining results?
- o How does the proximity of the information source (location of irrigation authorities, canal gatekeeper, village-level worker, etc.) affect the information-seeking behavior of the farmers?
- o Does the close proximity of the source encourage more information-seeking?

2. The Kind of Information. This segment of the investigation is concerned with the problems and conflicts which might prompt the farmer to actively seek information to resolve them.

- o What is the nature of the problems and conflicts which require information-seeking by the farmer?
- o Does the nature of information sought by the farmer change seasonally?
- o Does the information sought by the farmer have any relation to his socioeconomic status and irrigation system location?
- o What kind of information is sought most frequently?
- o How complex is the information sought? How frequently is the information sought conflict oriented?
- o How frequently is the information sought problem-solving oriented?
- o What is the level of specificity of information sought?
- o What percentage of information sought relates to day-to-day operation of irrigation systems?

- o What percentage of information sought is related to long-term policy matters of the irrigation system?

3. Channel. This part relates to channels used or not used by farmers in their information-seeking. It pertains to availability and openness of the channels.

- o What are the strategies used by farmers employing the available channels of information for their benefit?
- o Are there some channels of information used more frequently than others? If so, why?
- o What are the characteristics of these channels which make them more usable than others?
- o How easily available and accessible are the channels of information?
- o How frequently do farmers use alternative channels of information?
- o What are the formal and informal channels of communication for irrigation and other information?
- o What capacities do the channels of information have for handling information?

- o What procedural regulations exist for the use of channels of information?
- o Does the use of some channels require more education and other intellectual attainments than the use of others?

4. Source. This segment concerns source adequacy, availability, and credibiity.

- o How do the farmers agree with the sources on the reliability and accuracy of information provided on irrigation matters?
- o Do the farmers and the source understand the problems and conflicts the same way?
- o What differences exist between farmers' and sources' perceptions of the problems and conflicts with regard to irrigation matters?
- o What differences exist between farmers' and sources' perception of *de jure* and *de facto* practices of irrigation?
- o How aware are the farmers of the capacity of sources providing pertinent and functional information?

- o Where do the conflicts arise between farmers and sources in information-seeking matters?
- o What areas of irrigation practices create more conflict between farmers and information sources?
- o What kinds of procedures exist for conflict resolution between farmers and sources?
- o What bodies exist to negotiate conflicts between farmers and their information sources?

5. Results. The results are an analysis of four previous variables: who, what kind of information, through which channels, and from whom. A pattern of information-seeking behavior of farmers is needed in order to systematically predict what kind of results would emerge from the efforts spent by the farmers in seeking information.

Not much attention has been paid to information dissemination patterns on irrigation systems. This user-source audience needs to be studied systematically to transfer major water management concepts and practices. Monies spent in improving irrigation systems would be more effectively used if the gap of information dissemination was closed between the farmers and lower-level irrigational personnel.

WHITHER TECHNOLOGY TRANSFER?

In this paper, we have discussed the nature and process of technology transfer by keeping our focus on one of the main features of the process: communication.

Historically, the developmental literature has laid a heavy co-emphasis on communication (mass and/or interpersonal) as one of the pivotal factors for the orderly and systematic process of nation building and development. However (as we have discussed in previous parts of this paper), most of the time, failures of models in the realm of development have been due to a variety of reasons, but the single most important among them has been their Western orientation.

From Daniel Lerner's model (which had "empathy" as its major factor) to Rogers and Shoemaker's Diffusion of Innovation theory (which relies heavily on the "innovative personality" for adoption of new practices), practically every single model's theoretical construction and implementation scheme lacks the generic, culture-free, process-oriented, and cyclic approach where Western values and orientations would not assume the outcomes and therefore would not directly affect the results. It has to be noted here, in passim, that it took 26 years for the total adoption of hybrid corn by the farmers in the United States.

In the complicated process of technology transfer between cultures, civilizations, and nation-states, a crucial factor is (at the risk of stating the obvious) genuine acceptance of transferred processes, tools, or technologies by the adopting society (nation-state or peoples of that state). For this to occur, there must be communication among the parties involved in the process of technology transfer. For the sake of clarity, we are defining the communication process (not the channels) as sharing of meaning. For this sharing to occur, two critical communication elements have to be present and achieved: understanding and accuracy.

We briefly define understanding as the overlapping of cognitive perceptions about an object or subject between the communicating parties. In other words, facts and truths about a process of technology must be perceived, to a high degree, similarly between transferring and adopting parties in technology transfer. And this requires, at the initiation of technology transfer, a high degree of communication (interpersonal, preferably) among those involved in the process. Further, to establish understanding, as the Problem-Solver Model suggests, the needs of the adopters must be clearly communicated to the transferring party. As a matter of fact, parties involved in transfer of technology must, at every step of the process, maintain an understanding by exchanging their cognitive perceptions about the process of technology transfer. We believe that, without this two-way communication, the process (no matter how valuable and how important it is for the adopting party or parties) will be doomed to failure.

The second element, the accuracy, is as important as understanding, in that parties involved in technology transfer must perceive each other's cognitive perceptions accurately. In other words, the truths and facts about the need for adoption of a technology must not be a reflection of a select group of people in the adopting society, but of all involved in the process of adoption. This means communication systems designed for the transfer of technology should not only provide feedback on the perceptions of bureaucracies, the elite, or opinion leaders (sometimes called the key informants), but they should also examine the ultimate users of technologies to be transferred in total. By doing this, we will be establishing accuracy in our communication mechanism, whereby parties involved in the process can accurately assess one another's opinions, intentions, wants, and needs.

The concepts of understanding and accuracy described above are the essentials of meaningful, fruitful, and, therefore, functional communications. If transfer of technology in its ultimate objective means solution of problems by sharing knowledge, opinions, and ideas, as well as skills, it has to be based on an equal basis with regard to communication. Technology transfer, no matter in which realm of life it occurs, involves human lives and changes (in degrees) in the lives of those humans adopting new technologies. Therefore, the adopting parties should have a right to communicate and establish with the parties intending to transfer technologies an understanding and accuracy about the nature, process, and consequences of the process. They have a right to share understanding and accuracy at every step of the process. They have a right to communicate.

The emphasis on communication in the process of technology transfer might be appropriate for other reasons as well. Communication is a discipline drawing upon its own research base, as well as being a crossroads of other disciplines. Certainly an interdisciplinary approach is necessary to understand the farmer-client who is an interdisciplinary manager himself. Increasingly important to technology transfer is the application of behavioral science research techniques which can help us to understand the adequacy of understanding and accuracy established between parties involved in technology transfer.

For water management the question may not be, "Whither technology transfer?" but, rather, whether or not we can continue to ignore the necessity to commit financial resources and innovative personnel to research and develop planned communication efforts to genuinely help to establish a system where shared understanding and accuracy could be achieved. Without a focus, our efforts will continue to be haphazard, but with an application of communication research through careful planning, our efforts at transferring technology will be more fruitful, meaningful, and functional.

NOTE

Special acknowledgment should be given to Dr. Oguz B. Nayman, professor of technical journalism and long-time friend and colleague, for the many suggestions for this paper.

REFERENCES

Bauer, Raymond A., and Kenneth J. Gergen, eds. The Study of Policy Formation. New York: The Free Press, 1968.

Biggs, Huntley. "The Puebla Project: Progress and Problems," Water Management Technical Report No. 22, Colorado State University, Fort Collins, Colorado, July 18, 1972.

Clyma, Wayne, S. R. Katariya, L. J. Nelson, S. P. Tomar, J. M. Reddy, S. K. Bakliwal, M. I. Haider, U. R. Mehta, M. K. Lowdermilk, W. R. Laitos, and R. R. Mehta. Diagnostic Analysis of Farm Irrigation Systems on the Gambhiri Irrigation Project, Rajasthan, India, Vols. 1-5. Fort Collins, Colorado: Water Management Synthesis Project, Colorado State University, 1981a.

Clyma, Wayne, Dan Lattimore, and J. Mohan Reddy. "Irrigation Water Management Problems Around the World," paper presented at the Ninth Technical Conference on Irrigation, Drainage, and Flood Control, U.S. Committee on Irrigation, Drainage, and Flood Control, Jackson, Mississippi, 1982.

Clyma, Wayne, Max Lowdermilk, and Dan Lattimore. "On-Farm Water Management for Rural Development," Agricultural Engineering, February 1981b.

Freeman, David. Technology and Society: Issues in Assessment, Conflict, and Choice. Chicago, Ilinois: Rand McNally, 1974.

Havelock, R. G. Planning for Innovation Through Dissemination and Utilization of Knowledge. Ann Arbor, Michigan: Institute for Social Research, University of Michigan, 1971.

Jayaraman, T. K., M. K. Lowdermilk, L. J. Nelson, W. Clyma, J. M. Reddy, and M. I. Haider. Diagnostic Analysis of Farm Irrigation Systems in the Mahi-Kadana Irrigation Project, Gujarat, India. Fort Collins, Colorado: Water Management Synthesis Project, Colorado State University, 1983.

Kearl, B. E. "Communication for Agricultural Development," Communication and Change: The Last Ten Years --and the Next, Schramm and Lerner, eds. Honolulu, Hawaii: An East-West Center Book, University Press of Hawaii, 1974, pp. 175-176.

Keller, Jack. "Irrigation System Management," Chapter 11, Irrigation Management in Developing Countries. Boulder, Colorado: Westview Press, 1985.

Korten, David C. "Community Organizations and Rural Development: A Learning Process Approach," Public Administration Review, Sept.-Oct. 1980.

Lerner, Daniel. The Passing of Traditional Society: Modernizing the Middle East. New York: The Free Press, 1958.

Lowdermilk, Max, Wayne Clyma, and Alan Early. "Physical and Socioeconomic Dynamics of a Watercourse in Pakistan's Punjab: System Constraints and Farmers' Responses," Technical Report No. 42, Pakistan On-Farm Water Management Project, Colorado State University, Fort Collins, Colorado, 1975.

Moore, Mick. "Institutional Development, The World Bank, and India's New Agricultural Extension Programme," The Journal of Development Studies, Vol. 20, No. 4, July 1984.

Moris, Jon. Managing Induced Rural Development. Bloomington, Indiana: International Development Institute, 1981.

Nayman, Oguz B. "Evaluation of Diagnostic Analysis Workshops Audiovisual Training Material," unpublished paper for U.S. Agency for International Development, Water Management Synthesis Project, Colorado State University, Fort Collins, Colorado, 1983.

Rogers, Everitt, and F. F. Shoemaker. The Diffusion of Innovations. New York: The Free Press, 1962.

Schramm, Wilbur, and Daniel Lerner, eds. Communication and Change: The Last Ten Years--and the Next. Honolulu, Hawaii: An East-West Center Book, University Press of Hawaii, 1974.

Singh, Vidya N. Technology Transfer and Economic Development: Models for the Developing Countries. Jersey City, New Jersey: Unz and Company, 1983.

Spencer, Daniel L., and Alexander Woroniak, eds. The Transfer of Technology to Developing Countries. New York: Frederick A. Praeger, 1967.

Contributors

Ian Carruthers, Professor of Agrarian Development, Wye College, University of London.

Wayne Clyma, Professor, Department of Agricultural and Chemical Engineering, Colorado State University.

E. Walter Coward, Jr., Professor, Department of Rural Sociology and Asian Studies, Cornell University.

Warren Fairchild, Sr. Agriculturalist, The World Bank.

Sam Houston Johnson III, Assistant Professor of Agricultural Economics and International Agriculture, University of Illinois.

Jack Keller, Chairman and Professor, Department of Agricultural Engineering, Utah State University.

Dan Lattimore, Associate Professor, Department of Technical Journalism, Colorado State University.

Roberto Lenton, Program Officer, The Ford Foundation.

Max K. Lowdermilk, Project Officer and Irrigation Water Management Specialist, USAID Mission, New Delhi, India.

Donald W. Lybecker, Associate Professor, Department of Agricultural and Natural Resource Economics, Colorado State University.

Kenneth C. Nobe, Professor and Chairman, Department of Agricultural and Natural Resource Economics, Colorado State University.

George Radosevich, Professor, Department of Agricultural and Natural Resource Economics, Colorado State University.

J. Mohan Reddy, Water Management Synthesis Project, Colorado State University.

John A. Replogle, Research Hydraulic Engineer, U.S. Water Conservation Laboratory, ERS, USDA.

Rajan K. Sampath, Professor, Department of Agricultural and Natural Resource Economics, and Managing Director, International School for Agricultural and Resource Development, Colorado State University.

Willard R. Schmehl, Professor, Agronomy Department, Colorado State University.

David W. Seckler, Professor, Department of Agricultural and Natural Resource Economics, and Director, International School for Agricultural and Resource Development, Colorado State University.

Melvin D. Skold, Professor, Department of Agricultural and Natural Resource Economics, Colorado State University.

Dan Yaron, Professor and Chairman, Department of Agricultural Economics and Management Studies, Hebrew University, Israel.

Robert A. Young, Professor, Department of Agricultural and Natural Resource Economics, Colorado State University.

Index

Action research process, 30, 33, 34(fig.), 35(fig.), 41, 54-56. See also Development, model
Adequacy, 101-103
Administrative organizations (AOs), 386-387, 388(fig.)
Administrative systems, 4. See also Organizations, governmental
Afghanistan, 266
Africa, 68
Ali, Syed Hashim, 437
Allocation, 476-478, 487, 503-504
Andersen, J. C., 234, 235
Anderson, Raymond L., 172, 431, 496
Andrews, W., 410(n9)
AOs. See Administrative organizations
Appraisal techniques, 5-6. See also Diagnostic analysis
Argentina, 60, 462, 463, 484
ASTRAN method, 252, 254
Audience analysis, 524-525, 527-530
Australia, 277
Ayers, A. D., 229
Ayers, R. S., 221, 228
Backus, R., 410(n9)
Bagadion, Benjamin U., 445, 449
Bangladesh, 179, 195-214
  Deep Tubewell Irrigation and Credit Program, 203, 204(table)
  Irrigation Management Program, 203, 205, 211
  Kotwali Thana Central Cooperative Association, 199
  Mechanized Cultivation and Power Pump Irrigation Program, 197
  Medium-Term Food Production Plan, 201
  Rural Development Academy, 211-212
  Thana Irrigation Program, 197
  Water Development Board, 196-197, 200, 201
Benefit-cost analysis, 163-169. See also Costs
Benefit estimation, 165-169
Benefit functions, 172-173
Benor, Daniel, 518
Benor System. See Training and Visit System
Bernstein, L., 219, 221

Bias, 297-300, 306-308, 310, 460, 484
Biological research, 81(table), 84
Biswas, A. K., 108
Bohannan, P., 493
Bos, M. G., 140
Bottrall, A., 30, 33, 41
Boulding, Kenneth, 151-152, 155
Bowen, Richard, 169
Bredehoeft, J. D., 172
Bresler, E., 219, 228, 229
Brierly, J.E.C., 461
Bureaucracy, 29, 30, 31(fig.), 438-442, 513-518. See also Institutional framework; Organizations, governmental
Burt, O. R., 169
Busch, L., 410(n9)
Byerlee, D., 80

Canals, 119-120, 140-142
CCC. See Cooperation, coordination, and control
Centralization, 479
CGIAR. See Consultative Group on International Agricultural Research
Chambers, Robert, 5, 52, 53, 63(n5), 98, 435
Cheng, C. P., 410(n9)
Childs, S. W., 219
Chile, 60
China, 432
Client-oriented research, 69-71, 90
Climate. See Geo-climatic conditions
Clyma, Wayne, 6, 23, 27, 67, 91, 318
Coase, R., 159
Cobb-Douglas function, 167
Colbert, Jacques, 425
Colcini, A., 410(n9)
Coleman, Gilroy, 300
Collinson, M. P., 53, 68, 295
Colorado State University, 9, 411(n13), 523, 524
  Water Management Synthesis Project, 6, 91, 525
Command area, 207(table)
Common law, 461. See also Legal systems
Communication, 480, 509-534
Conservation, 132-133, 145, 471
Consultative Group on International Agricultural Research (CGIAR), 49
Cooperation, coordination, and control (CCC), 480, 481(fig.). See also Communication
Corey, Gilbert, 6, 20, 23, 27, 434, 437
Cornell University, 41
Cosen, M. A., 41
Costs, 91, 136, 307-310
  allocation, 161-162
  legal systems and, 482
  recovery, 173-174, 432, 446, 480, 482-483
  See also Pricing
Courts. See Judicial systems; Legal systems
Coward, E. Walter, Jr., 67, 431, 477
Cropping
  intensity, 187(table), 192
  salinity and, 235-237, 250, 253-254
  systems research, 82-83(table), 84
  varieties, 171
Cybernetics, 383-386

DA. See Diagnostic analysis
Data collection, 478-479
  bias, 297-300, 306-308, 310
  error, 297-302, 310
  farm-level, 285-302, 311, 320-321

Data collection (cont'd)
platforms (DCPs), 478
David, R., 461
DCPs. See Data collection, platforms
Delivery, 5, 56-58, 333, 335-336, 337(fig.)
DeLucia, R. J., 172
Demand, 128-131, 163-165, 343-345, 460. See also Hoarding; Market system
Demsetz, H., 493
Dennison, Edward F., 382
Descriptive stage. See Diagnostic analysis
Design stage, 72-80, 104-105, 110(table), 135-136, 399-401, 433, 439-440
Development, 459-461
concepts, 19-46
model, 23-27, 32(fig.), 37(fig.), 41, 43
objectives, 330, 332, 350
strategies, 22-44
Diagnostic analysis, 6-7, 52-54, 67-93, 310-312, 318-319, 449, 525
Dickinson, R. E., 162
Diffusion of Innovation theory, 530
Dikes. See Paddy irrigation systems
Dinar, A., 171
Discipline components, 2
Dispute resolution, 485-486
Donor agencies, 439, 482. See also United States, Agency for International Development; World Bank
Drainage, 250-252, 266-283, 341
economics of, 274, 275, 277-281
in Egypt, 267, 272, 280-281
in Pakistan, 366-367, 374-375
problem sources, 267-274
Drainage (cont'd)
technology, 274-277
Drip irrigation, 253
Drucker, Peter F., 382, 383, 408, 413, 414, 480
Dunn, William N., 464, 467
Duration, 123, 126-128, 131
Dyer, W. M., 280

East Pakistan. See Bangladesh
Economic analysis
farm-level, 290-293, 315-318
Economic efficiency
salinity and, 227-244
See also Efficiency
Economic research, 81(table), 84
Education, 512. See also Training
Efficiency, 1-5, 132-133, 137, 173, 330, 346-347, 460
economic, 227-244
farmer involvement and, 434, 437-438
farm-level, 312-315
in paddy irrigation, 415-417, 418(fig.), 424
pricing and, 346-347
salinity and, 227-244
Egypt, 23, 26, 29, 119, 172-175, 266, 317-318, 462
agricultural costs, 307, 308, 309(table)
data collection in, 306
drainage in, 267, 272, 280-281
reliability in, 445-446
England, 463, 479
Epstein, E., 253
Equity, 56, 57, 111, 174, 213(n3), 336, 338(fig.), 339(fig.), 345, 504
in gravity flow systems, 98, 101, 103

Equity (cont'd)
in paddy systems, 424
as performance measure, 51
Erosion, 104
Euler's Theorum, 167
Evans, R. G., 21
Evenson, Robert E., 443
Evered, R. D., 30
Externalities, 250-252, 286
Eynon, D. G., 80

Fairchild, Warren, 86-88, 90, 410(n9), 459
FAO. See Food and Agriculture Organization
Farmers
communication between, 525-527. See also Communication
involvement, 30, 41, 58-60, 77, 85, 90, 342-343, 427-456, 475-476, 525
technology and, 39(fig.), 40(fig.), 69-70, 75, 77, 315, 317, 521
organizations, 333-334, 342, 398, 430-434, 436, 445-450, 475, 483-385, 494-495, 499-503, 505(n6)
state and, 499. See also Communication
Farming systems, 72-80, 91
budgets, 289, 297, 302-303, 304-305(table), 313, 318-319, 321
defined, 68
fixed costs, 309-310
income, 303, 306-307
information development, 285-289, 302-321
in Pakistan, 368-370
research, 33, 36(fig.), 37(fig.), 38
small, 285-289
variable costs, 307-309, 316
See also Costs
Farming Systems Support Project (FSSP), 84
Farm records, 295, 296-297, 321
Farm surveys, 294-296, 320
Feasibility studies, 399-401
Feddes, F. A., 219
Feedback control, 111-112. See also Performance, monitoring
Feinerman, E., 234, 237, 241, 243
Financial analysis, 316-318. See also Economic analysis
Fixed rotation, 125
Flow rate, 121-122, 125, 129-130
in canals, 141-142
measurement, 138-140, 347
Flumes, 138-140
Food and Agriculture Organization (FAO), 117, 183
Food consumption, 356-357, 409(n1), 409(n2)
France, 105
Freeman, David, 20
Frequency, 122-123, 131. See also Scheduling
FSSP. See Farming Systems Support Project
Furubotn, E., 493

Game theory, 162, 252
Geertz, Clifford, 505(n5)
General approaches, 47-66
Geo-climatic conditions, 459-461
Gibbs, Alexander, 440
Gittinger, J. P., 161
Goody, Jack, 493
Gotsch, C. H., 280
Government
bias, 460, 484
involvement, 413-426, 434-435, 444, 491, 498-503, 535(n8). See also Legal systems; Policy

Government (cont'd)
organizations, 472-473, 479-480, 485-486
See also Administrative systems; Institutional framework
Gravity flow systems, 95-115
Gray, S. L., 165
Groundwater management, 4, 265-283, 468, 470(fig.), 471
investment in, 179-216
in Pakistan, 374-375
See also Drainage; Water

Hanks, R. J., 219, 234-235
Havelock, R. G., 521
Haveman, R. A., 277
Heaney, J. P., 162
Hildebrand, Peter E., 53
Hoarding, 118-119
Hoffman, G. J., 221
Howe, C. H., 252
Hungary, 105
Hutapea, R., 448

IDA. See International Development Agency
IFC. See International Finance Corporation
Implementing agencies, 448-452. See also Institutional framework
Incentives, 444-445
India, 4, 179, 440, 441-442
command area development authorities, 390
communication in, 526
farmer involvement in, 428-429, 430-431, 436, 442, 443-452
Irrigation Water Management Training Project, 406
legal system, 462, 463
ownership in, 496, 506(n10)
paddy farms in, 420
India (cont'd)
relative water supply, 336
reliability in, 437-438
salinity in, 266
scheduling and delivery in, 57
Second Bhakra Main Circle, 50-52
system completions in, 438-439
training in, 443, 518-520
waterlogging in, 268, 272
Indonesia, 59, 448, 463, 474, 475, 483, 500-501, 505(n5), 505(n8)
Information
dissemination patterns, 525-530
farm-level development, 285-289, 302-321
management, 111
Infrastructure, 117
Institutional framework, 10, 457-459, 472-488, 505(n1). See also Organizations
Interdisciplinary approach, 80, 84, 334
International Bank for Reconstruction and Development. See World Bank
International Conference on Global Water Law Systems, 464
International Development Agency (IDA), 354
International donors, 439, 482. See also United States, Agency for International Development; World Bank
International Finance Corporation (IFC), 354
International Irrigation Management Institute, 49
International Research Center, 21, 38

International Rice Research Institute (IRRI), 57
Intervention. See Government, involvement; Legal systems; Policy
Investment
in groundwater management, 179-216
indirect, 502-503
See also Government, involvement
Iran, 266
Iraq, 266, 277
IRRI. See International Rice Research Institute
Irrigation
expansion of, 265, 272, 331(table), 356
field staff, 435
system evolution, 340-342, 348
system structure, 333-334, 335(fig.)
Irrigation development, 504(n1)
Irrigation water management defined, 106
Israel, 219, 225-227, 239, 241, 242(table), 245-251, 253, 254

James, L. D., 161
Japan, 432, 434
Java, 497
Jayaraman, T. K., 439
Johnson, S. H., 266-267
Judicial systems, 473-474. See also Dispute resolution

Keller, Jack, 525
Korea, 432, 434
Korten, Frances F., 21, 29-30, 445, 449

Labor shortfalls, 350
Lattimore, Dan, 20, 67, 268
Leaching, 241, 242(table), 261-263, 270
Leaching requirement (LR), 228
Learning process approach, 30, 31(fig.), 32(fig.), 41
Leathers, K. L., 252
Lee, R. R., 161
Legal systems, 457-458, 461-464, 465(fig.), 466(fig.), 468-472, 482. See also Policy
Lenton, Roberto, 67
Lerner, Daniel, 530
Levine, G., 101
Lieftinck Report, 192
Lituma, G., 410(n9)
LLPs. See Low lift pumps
Loehman, E., 162
Loughlin, J. C., 162
Lowdermilk, Max, 6, 9, 20, 23, 27, 53, 69-70, 80, 106, 513
Low lift pumps (LLPs), 197-199, 200
LR. See Leaching requirement

Maas, E. V., 221
Maass, Arthur, 172, 431, 496
Maierhofer, E. R., 183
Maintenance, 433. See also Farmers, involvement
Malhotra, S. P., 51, 57
Management
defined, 2
levels, 342-345
Management by Objectives, 112, 113(fig.), 382, 383, 480
Management by Results (MBR), 7, 86, 87(fig.), 381-396, 404-406, 408, 409, 480
Management organizations (MOs), 387, 389(fig.), 399, 408
Managers, 114
Marino, M. A., 243

Market system, 152-155, 163-169
Matanga, G. B., 243
MBR. See Management by Results
Methodology, 47-66
Mexico, 223, 244-245, 462, 516-517
Miller, Lawrence, 458
Modernization, 4, 512
Mohamadi, J., 410(n9)
Mohammad, Ghulam, 184, 192
Monitoring. See Flow rate, measurement; Performance, monitoring
Moore, Charles V., 171, 223, 234, 237
Morocco, 105
MOs. See Management organizations
Muligan, P., 410(n9)
Murphy, John Joseph, 510
Murthy, Y. K., 428-429, 443, 451

Nagel, Ernest, 41
Nakayama, M., 252
Needs assessment, 26, 41, 120. See also Diagnostic analysis
Nepal, 496, 506(n10)
Neuman, S. P., 219
NIA. See Philippines, National Irrigation Administration
Nobe, Kenneth C., 7, 86-88, 90, 408, 409, 410(n9), 480
No-share systems, 497-498

ODI. See Overseas Development Institute
Olian, A., 239
Operating deficits, 206, 208
Organizations, 458, 472-476, 505(n2)
  change, 41
  cost recovery and, 482-483
Organizations (cont'd)
  governmental, 472-473, 479-480, 485-486
  judicial, 473-474
  non-governmental, 474-475. See also Farmers, organizations
  planning and, 478-479
  user, 333-334, 342, 398, 430-434, 436, 445-450, 475, 483-485, 494-495, 499-503, 505(n6)
Over-irrigation, 132-133, 417, 420-424. See also Drainage; Waterlogging
Overseas Development Institute (ODI), 21
Oyarzabal-Tamargo, F., 171, 223

Paddy irrigation systems, 413-426
Pakistan, 6, 23, 26, 27, 29, 266, 355-356, 359-361, 367-368, 432, 445-446, 496
  Agricultural Development Bank, 370
  Command Water Management Project, 7, 8, 358-359, 362-366, 370-381, 386, 387, 390-411
  Department of Irrigation and Power of Sind, 279
  drainage project, 274-275, 276, 278-279
  farmer participation in, 59, 432, 433, 440, 442, 450
  groundwater management in, 179-216
  implementation in, 99, 513
  Indus Irrigation System, 361, 365
  institutional structure, 361-362, 390, 391, 392(fig.), 393-395, 403-405, 409, 410-411(n12)

Pakistan (cont'd)
irrigation intensity, 272, 273(fig.)
Irrigation Systems Rehabilitation Project, 371
legal system, 462, 463
local level organization, 201, 203
On-Farm Water Management Project, 20, 211, 407
Revised Action Program, 355
Salinity Control and Reclamation Project (SCARP), 183-195, 203, 208, 209-210
subsidies in, 206, 208
training in, 211-212, 378-379, 398, 405-407, 411(n13)
Water and Power Development Authority (WAPDA), 183, 201, 210, 279, 397, 400-401
Waterlogging in, 266-267, 270, 272
Water Users Associations, 398
Pant, Niranjan, 430
Parkinson, J. K., 234, 235
Pejovich, S., 493
Performance, 106-107, 329
of gravity flow systems, 99
influences on, 110(table)
monitoring, 50-52, 88, 90, 107-112, 449
See also Diagnostic analysis
Personnel. See Training
Peterson, Dean, 329
Philippines, 41
cost recovery, 446
farmer organizations in, 483
farmer participation in, 59, 432-433, 434, 443, 448, 450
governmental involvement in, 505(n9)
government organizations, 473
Philippines (cont'd)
IRRI/NIA project, 57
legal system, 462, 463
National Irrigation Administration (NIA), 57, 59, 473, 476, 480
ownership in, 496
paddy systems in, 422, 423(table)
water laws, 471, 473
Phillip, P. F., 21, 33, 38, 45
Pipelines, 129
Planning, 103-104, 110(table), 350-351, 478-480
farmer involvement in, 433, 448
frameworks, 27-29
organizations and, 458
with saline irrigation, 239-244
Podmore, C. A., 80
Policy, 132-133, 464, 467. See also Government, involvement; Legal systems
Polovin, A., 243
Pricing, 155-162, 173-175
efficiency and, 346-347
shadow, 163-169
See also Costs
Productivity, 30, 171, 265, 356-358
farmer involvement and, 434
as performance measure, 51
pricing and, 160
See also Yields
Property, 491-508
creation, 492
rights, 492-494, 501-502
See also Water, ownership

Qamar, U., 410(n9)

Raheja, S. K., 51
Ram, Rati, 443
Rao, K. L., 2-3
Rationing, 415, 424. See

Rationing (cont'd)
also Water, control
Ratner, A., 250
Relative water supply (RWS), 336, 338(fig.), 340, 345
Reliability, 101, 340, 433, 437, 445-447, 483
Renfro, Raymond, 496
Replogle, J. A., 140
Research, 33-40
biological, 81(table), 84
economic, 81(table), 84
experimental, 293-294, 320
See also Data collection
Resource allocation, 152-155
Return-flow development, 341
Return flows. See Drainage
Revelle, Roger, 184
Riley, Jay P., 162
Robinson, F. E., 222
Rogers, Everitt, 530
Rosenzweig, Mark R., 443
Rotation, 125-128, 422-423. See also Scheduling
Russia, 266
RWS. See Relative water supply

Salinity, 27, 217-263, 266-267
efficiency and, 227-244
income losses from, 222-227, 234-239, 241, 254-255
increase in, 271(fig.)
scheduling and, 229-232, 233(fig.), 239, 241, 243
See also Soil, salinity
Saudi Arabia, 463
SCARP. See Pakistan, Salinity Control and Reclamation Project
Scheduling, 56-58, 119-135, 143-145
flexible, 128-131
policy, 132-133
salinity and, 229-232,
Scheduling (cont'd)
233(fig.), 239, 241, 243
terminology, 123-131
Schmehl, W. R., 20, 21, 33, 38
Scientific method, 40-41, 42(fig.), 293-294, 320
Seckler, David, 7, 51, 408, 409, 480
Shadow prices, 163-169
Shalhevet, J., 221
Shaner, W. W., 21, 33, 38, 71, 80, 85, 91
"Share" systems, 495-497, 503-504, 506(n10)
Shoemaker, F. F., 530
Simulation, 243
Skogerboe, G. V., 21, 27-29
Smith, Adam, 153
Soil
moisture content, 171
salinity, 3, 219, 220(table), 221, 229-230, 239, 240 (fig.), 241, 274
texture, 136
Spain, 431, 462, 463, 474, 475, 483, 496
Sparling, Edward W., 80, 496
Spillage, 130
Sprinklers, 129, 253
Sri Lanka, 26, 41, 59, 443, 448, 450, 473
Mahawili Development Board, 144
Sritharan, Subramaniaiyer, 43
Subsidies, 206, 208, 346-347, 480
Sundar, A., 114
Surface systems, 136-137
Susman, G. I., 30
Suzuki, M., 252
System
defined, 68
Systems approach, 5-7

Tail enders, 281, 420-421, 422. See also Drainage; Equity
Taiwan, 432, 434, 500
Taxes, 173, 174-175, 433
Technology, 253-254
appropriate, 39(fig.), 40(fig.), 212-213, 441-442
change, 510-511
defined, 510
Diffusion Model, 513-516
drainage, 274-277
farmer acceptance of, 69-70, 75, 77, 315, 317, 521
Induced Innovation Model, 514(table), 517-518
Package Programs Model, 514(table), 516-517
Problem-Solver Model, 521-523
transfer, 67, 68-70, 75, 77, 78(table), 84, 86, 509-534
Testing stage, 80-86
Thailand, 59, 473, 474, 482, 484
Training, 7-10, 348, 349(fig.), 442-443, 445
of extension personnel, 407
for Pakistan's project, 211-212, 378-379, 398, 405-407, 411(n13)
of water users, 211-212
Training and Visit System, 518-521
Tubewells, 184, 340
in Bangladesh, 197-199, 200, 201, 203, 205, 211, 212, 214(n7)
in Pakistan, 205, 208-209, 212, 275, 276, 367
in United States, 276-277
See also Pakistan, Salinity Control and Reclamation Project
Tukase, Kunio, 436

United Nations, 117
Food and Agriculture Organization (FAO), 117, 183
World Bank and, 354
United States, 133, 134-135, 136-137, 342, 496
Agency for International Development (USAID), 4, 379-380, 397, 400, 401-403, 409, 410(n11), 443
compensation claim against, 244-245
cost recovery in, 480, 482
farmers organizations in, 431, 475, 484
government organizations, 473
legal system, 462, 463
management in, 382-383
pricing in, 160
relative water supply, 336
salinity in, 222, 223
subsidies in, 346-347
tubewells, 276-277
water courts in, 474
Uphoff, N., 21, 29
USAID. See United States, Agency for International Development
Users. See Farmers

Valera, A., 413, 417, 420
Value estimation, 163-169. See also Pricing
Varied rotation, 125-128
Verma, R. K., 430
Voet, H., 239
Volume. See Flow rate

Wade, R., 98
Wadleigh, C. H., 229
Walker, W. R., 21
WAPDA. See Pakistan, Water and Power Development Authority

Water
available supplies. See Groundwater management
conservation, 132-133, 145, 471
consumption, 154
control, 99-106, 107, 137-138, 141-142. See also Rationing; Scheduling
laws. See Legal systems
markets, 153-155
ownership, 476-478, 495-506
quality, 27, 171, 244, 468, 471
quantity substitution, 245-250
reuse, 424
tables, 265-283. See also Waterlogging
user associations. See Farmers, organizations
value, 155, 165, 169, 170-171. See also Pricing
volume. See Flow rate
See also Delivery; Flow rate; Pricing; Salinity; Scheduling
Waterlogging, 3, 27, 266-267
causes of, 267-274
increase in, 270-274
Water Management Synthesis Project, 29
Weirs, 138, 140
Welch, Finis, 443
Westcot, D. W., 221, 228
Wickham, T. H., 413, 417, 420
Wiener, Aaron, 7, 8, 451
Wild flooding, 104-105
World Bank, 4
operations, 353-355
Pakistan project and, 381, 397, 401-403
procedures, 355
Training and Visit System, 518, 520
training programs, 443

Yaron, D., 171, 219, 225, 229, 232, 234, 239, 243, 245, 247
Yields, 19, 170, 171-172, 213(n4), 272, 368-370
estimates, 303, 306-307
optimum, 312-313, 314(fig.)
in paddy systems, 415-417, 418(fig.), 419(table)
salinity and, 217-222, 229, 254, 261-263
See also Productivity
Young, H. P., 162, 252
Young, R. A., 165, 169, 171, 172, 223, 252, 277
Young, T. J., 252

Zandstra, H. G., 80